S0-ANX-909

Fritz Oberhettinger · Larry Badii

Tables of Laplace Transforms

Springer-Verlag
New York Heidelberg Berlin 1973

Fritz Oberhettinger
Professor of Mathematics, Oregon State University, Corvallis, Oregon, U.S.A.

Larry Badii
Associate Professor of Mathematics, Eastern Michigan University,
Ypsilanti, Michigan, U.S.A.

AMS Subject Classifications (1970): 44 A 10

ISBN 0-387-06350-1 Springer-Verlag New York Heidelberg Berlin
ISBN 3-540-06350-1 Springer-Verlag Berlin Heidelberg New York

This work is subject to copyright. All rights are reserved, whether the whole or part of the material is concerned, specifically those of translation, reprinting, re-use of illustrations, broadcasting, reproduction by photocopying machine or similar means, and storage in data banks. Under § 54 of the German Copyright Law where copies are made for other than private use, a fee is payable to the publisher, the amount of the fee to be determined by agreement with the publisher. © by Springer-Verlag Berlin · Heidelberg 1973. Printed in Germany. Library of Congress Catalog Card Number 73-81328.
Offset printing: fotokop wilhelm weihert kg, Darmstadt · Binding: Konrad Triltsch, Graphischer Betrieb, Würzburg.

PHYSICS
QA
432
.O12

1419535

Preface

This material represents a collection of integrals of the Laplace- and inverse Laplace Transform type. The usefulness of this kind of information as a tool in various branches of Mathematics is firmly established. Previous publications include the contributions by A. Erdélyi and Roberts and Kaufmann (see References). Special consideration is given to results involving higher functions as integrand and it is believed that a substantial amount of them is presented here for the first time. Greek letters denote complex parameters within the given range of validity. Latin letters denote (unless otherwise stated) real positive parameters and a possible extension to complex values by analytic continuation will often pose no serious problem. The authors are indebted to Mrs. Jolan Eröss for her tireless effort and patience while typing this manuscript.

Oregon State University
Corvallis, Oregon

Eastern Michigan University
Ypsilanti, Michigan

The Authors

Contents

Part II. Inverse Laplace Transforms

Part I. Laplace Transforms

Introduction

The function $g(p)$ of the complex variable p defined by the integral

$$g(p) = \int_0^\infty f(t)e^{-pt}dt \tag{1}$$

is called the one sided Laplace transform of $f(t)$ where $f(t)$ is a function of the real variable $t, (0 < t < \infty)$ which is integrable in every finite interval. If the integral converges at a point $p = p_0$, then it converges for every p such that $\text{Re } p > \text{Re } p_0$. The behavior of the integral (1) in the p-plane may be one of the following:

(a) Divergent everywhere

(b) Convergent everywhere

(c) There exists a number β such that (1) converges, when $\text{Re } p > \beta$ and diverges when $\text{Re } p < \beta$. The number β which is the greatest lower bound of $\text{Re } p$ for the set of all p's in the p-plane at which (1) converges is called the abscissa of convergence. The line $\text{Re } p = \beta$ is called the axis of convergence. Similarly one defines the abscissa α of absolute convergence and the axis of absolute convergence. If the integral (1) is absolutely convergent at the point p with $\text{Re } p = \alpha$ then it is absolutely and uniformly convergent in the half plane $\text{Re } p \geqq \alpha$. Obviously $\alpha \geqq \beta$. The Laplace transform $g(p)$ often denoted as $L\{f(t)\}$, is an analytic function of the variable p in the half-plane $\text{Re } p > \beta$ and it uniquely defines the function $f(t)$ almost everywhere. The integral (1) may also be defined as a Fourier integral whose integrand is $e^{-\sigma t}f(t)$, when $p = \sigma + i\tau$. It is because of this similarity that it is avoided to give further explanations about the integral (1) here. The inversion theorem of (1) reads: Let $f(t)$ be of bounded variation in the vicinity of a point $t \geqq 0$ then the following inversion formula holds.

$$\text{(2)} \quad \text{P.V.} \frac{1}{2\pi i} \int_{c-i\infty}^{c+i\infty} g(p) e^{pt} dp = \begin{cases} \frac{1}{2}[f(t+0)+f(t-0)] & t > 0 \\ \frac{1}{2}f(0) & t = 0 \\ 0 & t < 0 \end{cases}$$

Here c is any value larger than the abscissa α of absolute convergence. Integrals of the form (1) and (2) are listed in Part I and Part II respectively.

REFERENCES

Churchill, R. V., 1958. Operational Mathematics. McGraw-Hill.

Doetsch, G., 1950-1956. Handbuch der Laplace Transformation 3 vols. Verlag Birkhauser, Basel.

Doetsch, G., 1961. Guide to the Application of Laplace Transforms. Von Nostrand.

Doetsch, G., 1947. Tabellen zur Laplace Transformation. Springer Verlag.

Erdélyi, A. et.al. 1952. Tables of Integral Transforms, Vol. 1. McGraw-Hill, 1954.

McLachlan, N. W. and Humbert, P. 1950. Formulaires pour le calcul symbolique. Gauthier-Villars.

Nixon, F. E. 1960. Handbook of Laplace Transformation. Prentice Hall.

Roberts, G. E. and H. Kaufmann, 1966. Tables of Laplace Transforms. W. B. Saunders Co.

Sneddon, I. N. 1972. The Use of Integral Transforms. McGraw-Hill.

Widder, D. V., 1971. An Introduction to Transform Theory. Academic Press.

Widder, D. V., 1941. The Laplace Transform. Princeton University Press.

1.1 General Formulas

	$f(t)$	$g(p) = \int_0^\infty f(t)e^{-pt}dt$
1.1	$f(t)$	$g(p) = \int_0^\infty e^{-pt}f(t)dt$
1.2	$(2\pi i)^{-1} \int_{c-i\infty}^{c+i\infty} e^{ut}g(u)du$	$g(p)$
1.3	$f(at)$	$a^{-1}g(p/a)$ $\quad a > 0$
1.4	$f(t+a)$	$e^{ap}[g(p) - \int_0^a e^{-pu}f(u)du]$ $\quad a \geqq 0$
1.5	$f(t+a) = f(t)$	$(1-e^{-ap})^{-1} \int_0^a e^{-pt}f(t)dt$
1.6	$f(t+a) = -f(t)$	$(1+e^{-ap})^{-1} \int_0^a e^{-pt}f(t)dt$
1.7	$0 \quad t < b/a$ $f(at-b) \quad t > b/a$	$a^{-1}\exp(-bp/a)g(p/a)$ $\quad a,b > 0$
1.8	$e^{-at}f(t)$	$g(p+a)$
1.9	$t^n f(t)$	$(-1)^n \frac{d^n}{dp^n}[g(p)]$

	$f(t)$	$g(p) = \int_0^\infty f(t) e^{-pt} dt$
1.10	$t^{-n} f(t)$	$\int_p^\infty \cdots \int_p^\infty g(p)(dp)^n$ n-th repeated integral
1.11	$t^{\mu} f(t)$	$p^{\nu+1}[\Gamma(\nu+1)]^{-1}$ $\cdot \int_1^\infty (x-1)^{\nu} [\int_0^\infty t^{\mu-\nu-1} f(t) e^{-pxt} dt] dx$ $\operatorname{Re} \nu > -1$
1.12	$f_1(t) f_2(t)$	$(2\pi i)^{-1} \int_{c-i\infty}^{c+i\infty} g_1(u) g_2(p-u) du$
1.13	$t^{\nu-1} f(t^{-1})$ $\operatorname{Re} \nu > -1$	$p^{-\frac{1}{2}\nu} \int_0^\infty u^{\frac{1}{2}\nu} J_{\nu}[2(up)^{\frac{1}{2}}] g(u) du$
1.14	$f(t^2)$	$\pi^{-\frac{1}{2}} \int_0^\infty \exp(-\frac{1}{4}p^2/u^2) g(u^2) du$
1.15	$t^{\nu} f(t^2)$	$(2\pi)^{-\frac{1}{2}} \int_0^\infty u^{\nu-2} \exp(-\frac{1}{4}p^2 u^2) D_{\nu}(pu) g(\frac{1}{2}u^{-2}) du$
1.16	$t^{\nu-1} f(t^{-1})$ $\operatorname{Re} \nu > -1$	$p^{-\frac{1}{2}\nu} \int_0^\infty u^{\frac{1}{2}\nu} J_{\nu}[2(up)^{\frac{1}{2}}] g(u) du$

	$f(t)$	$g(p) = \int_0^\infty f(t)e^{-pt}dt$
1.17	$f(ae^t - a)$	$[a\Gamma(1+p)]^{-1} \int_0^\infty e^{-u}u^p g(u/a)\,du$ $\quad a > 0$
1.18	$f(a \sinh t)$	$\int_0^\infty J_p(au)g(u)\,du$ $\quad a > 0$
1.19	$f'(t)$	$pg(p) - f(0)$
1.20	$f^{(n)}(t)$ $n=1,2,3,\cdots$	$p^n g(p) - p^{n-1}f(0) - p^{n-2}f'(0) - \cdots - f^{(n-1)}(0)$
1.21	$(t \frac{d}{dt})^n f(t)$	$-(p \frac{d}{dp})^n g(p)$
1.22	$(\frac{d}{dt} t)^n f(t)$	$(-p \frac{d}{dp})^n g(p)$
1.23	$(t^{-1} \frac{d}{dt})^n f(t)$ if $(t^{-1} \frac{d}{dt})^k f(t)=0$ t or t=0, $k=0,1,2,\cdots n-1$	$\int_p^\infty p \int_p^\infty \cdots p \int_p^\infty pg(p)(dp)^n$

	$f(t)$	$g(p) = \int_0^\infty f(t)e^{-pt}dt$
1.24	$t^m f^{(n)}(t) \quad m \geqq n$ $m,n = 0,1,2,\cdots$	$(-\frac{d}{dp})^m [p^n g(p)]$
1.25	$t^m f^{(n)}(t)$ $m < n$ $m,n = 0,1,2,\cdots$	$(-\frac{d}{dp})^m [p^n g(p)] + (-1)^{m-1}$ $\cdot [\frac{(n-1)!}{(n-m-1)!} p^{n-m-1} f(0)$ $+ \frac{(n-2)!}{(n-m-2)!} p^{n-m-2} f'(0)$ $+ \cdots + m!\, f^{(m-n-1)}(0)]$
1.26	$\frac{d^n}{dt^n} [t^m f(t)]$ $m \geqq n, m,n=0,1,2,\cdots$	$(-1)^m p^n g^{(m)}(p)$
1.27	$\frac{d^n}{dt^n} [t^m f(t)]$ $m < n$ $m,n=0,1,2,\cdots$	$(-1)^m p^n g^{(m)}(p) - m! p^{n-m-1} f(0)$ $- \frac{(m+1)!}{1!} p^{n-m-2} f'(0) - \cdots$ $- \frac{(n-1)!}{(n-m-1)!} f^{(n-m-1)}(0)$
1.28	$(e^t \frac{d}{dt})^n f(t)$ $n=1,2,3,\cdots$	$(p-1)(p-2)\cdots(p-n)g(p-n)$ if $f^{(k)}(0)=0$ for $k=0,1,\cdots,n-1$

	$f(t)$	$g(p) = \int_0^\infty f(t)e^{-pt}dt$
1.29	$\frac{\partial}{\partial a} f(t,a)$	$\frac{\partial}{\partial a} g(p,a)$
1.30	$\int_{a_o}^{a} f(t,u)du$	$\int_{a_o}^{a} g(p,v)dv$
1.31	$\int_o^t f_1(u)f_2(t-u)du$	$g_1(p)g_2(p)$
1.32	$\int_o^t \cdots \int_o^t f(u)(du)^n$	$p^{-n}g(p)$
1.33	$\int_a^t f(u)du$	$p^{-1}g(p)-p^{-1}\int_a^o f(u)du$
1.34	$\int_o^t u^{-1}f(u)du$	$p^{-1}\int_p^\infty g(v)dv$
1.35	$\int_t^\infty u^{-1}f(u)du$	$p^{-1}\int_o^p g(v)dv$
1.36	$\int_o^\infty \frac{t^{u-1}}{\Gamma(u)} f(u)du$	$g(\log p)$

	$f(t)$	$g(p) = \int_0^\infty f(t)e^{-pt}dt$
1.37	$\int_0^\infty \frac{t^{au-1}}{\Gamma(au)} f(u)du$	$g(\log p^a)$
1.38	$\int_0^\infty \frac{t^u}{\Gamma(1+u)} f(u)du$	$p^{-1}g(\log p)$
1.39	$t^{-\frac{1}{2}} \int_0^\infty \exp(-\frac{1}{4}u^2/t)f(u)du$	$(\pi/p)^{\frac{1}{2}}g(p^{\frac{1}{2}})$
1.40	$t^{-3/2} \int_0^\infty u \exp(-\frac{1}{4}u^2/t)f(u)du$	$2\pi^{\frac{1}{2}}g(p^{\frac{1}{2}})$
1.41	$\int_0^t (t-u)^{-\frac{1}{2}} \exp[-\frac{1}{4}u^2(t-u)^{-1}]f(u)du$	$(\pi/p)^{\frac{1}{2}}g(p+p^{\frac{1}{2}})$
1.42	$\int_0^\infty u^{-\frac{1}{2}}\sin[2(tu)^{\frac{1}{2}}]f(u)du$	$\pi^{\frac{1}{2}}p^{-3/2}g(p^{-1})$
1.43	$t^{-\frac{1}{2}} \int_0^\infty \cos[2(tu)^{\frac{1}{2}}]f(u)du$	$(\pi/p)^{\frac{1}{2}}g(p^{-1})$
1.44	$\int_0^\infty u^{-\frac{1}{2}}\sinh[2(tu)^{\frac{1}{2}}]f(u)du$	$\pi^{\frac{1}{2}}p^{-3/2}g(-p^{-1})$

	$f(t)$	$g(p)=\int_0^\infty f(t)e^{-pt}dt$
1.45	$t^{-\frac{1}{2}}\int_0^\infty \cosh[2(tu)^{\frac{1}{2}}]f(u)du$	$(\pi/p)^{\frac{1}{2}}g(-p^{-1})$
1.46	$\int_0^t J_0[a(t^2-u^2)^{\frac{1}{2}}]f(u)du$	$(p^2+a^2)^{-\frac{1}{2}}g[(p^2+a^2)^{\frac{1}{2}}]$
1.47	$\int_0^t J_0[2(tu-u^2)^{\frac{1}{2}}]f(u)du$	$p^{-1}g(p+p^{-1})$
1.48	$\int_0^\infty J_0[2(tu)^{\frac{1}{2}}]f(u)du$	$p^{-1}g(p^{-1})$
1.49	$g(t)-a\int_0^t f[(t^2-u^2)^{\frac{1}{2}}]J_1(au)du$	$g[(p^2+a^2)^{\frac{1}{2}}]$
1.50	$f(t)-at\int_0^t (t^2-u^2)^{-\frac{1}{2}}$ $\cdot J_1[a(t^2-u^2)^{\frac{1}{2}}]f(n)du$	$p(p^2+a^2)^{-\frac{1}{2}}g[(p^2+a^2)^{\frac{1}{2}}]$
1.51	$t^{-\frac{1}{2}}\int_0^\infty e^{-bu}(t+2u)^{-\frac{1}{2}}f(u)$ $\cdot J_1[a(t^2+2tu)^{\frac{1}{2}}]udu$	$a^{-1}g(b)-a^{-1}g[(p^2+a^2)^{\frac{1}{2}}-p+b]$

	$f(t)$	$g(p) = \int_0^\infty f(t)e^{-pt}dt$
1.52	$t^{\nu} \int_0^\infty u^{-\nu} J_{2\nu}[2(tu)^{\frac{1}{2}}] f(u) du$ Re $\nu > -\frac{1}{2}$	$p^{-2\nu-1} g(p^{-1})$
1.53	$\int_0^t u^{-\nu}(t-u)^{\nu} J_{2\nu}[2(atu-au^2)^{\frac{1}{2}}]$ $\cdot f(u)du$	$a^{\nu} p^{-2\nu-1} g(p+ap^{-1})$
1.54	$\int_0^t (\frac{t-u}{t+u})^{\nu} J_{2\nu}[a(t^2-u^2)^{\frac{1}{2}}]$ $\cdot f(u)du$, Re $\nu > -\frac{1}{2}$	$a^{2\nu}(p^2+a^2)^{-\frac{1}{2}}[(p^2+a^2)^{\frac{1}{2}}+p]^{-2\nu}$ $\cdot g[(p^2+a^2)^{\frac{1}{2}}]$
1.55	$t^{\nu} \int_0^\infty (t+2u)^{-\nu} J_{2\nu}[at^{\frac{1}{2}}(t+2u)^{\frac{1}{2}}]$ $\cdot f(u)du$ Re $\nu > -\frac{1}{2}$	$a^{-2\nu}(p^2+a^2)^{-\frac{1}{2}}[(p^2+a^2)^{\frac{1}{2}}-p]^{2\nu}$ $\cdot g[(p^2+a^2)^{\frac{1}{2}}-p]$
1.56	$\int_0^t (\frac{t-u}{t+u})^{\nu} I_{2\nu}[a(t^2-u^2)^{\frac{1}{2}}]$ $\cdot f(u)du$, Re $\nu > -\frac{1}{2}$	$a^{2\nu}(p^2-a^2)^{-\frac{1}{2}}[p+(p^2-a^2)^{\frac{1}{2}}]^{-2\nu}$ $\cdot g[(p^2-a^2)^{\frac{1}{2}}]$
1.57	$f(t)+a \int_0^t f[(t^2-u^2)^{\frac{1}{2}}] I_1(u) du$	$g[(p^2-a^2)^{\frac{1}{2}}]$

	$f(t)$	$g(p) = \int_0^\infty f(t)e^{-pt}dt$
1.58	$f(t)+at \int_0^t (t^2-u^2)^{-\frac{1}{2}}$ $\cdot I_1[a(t^2-u^2)^{\frac{1}{2}}]f(u)du$	$p(p^2-a^2)^{-\frac{1}{2}}g[(p^2-a^2)^{\frac{1}{2}}]$
1.59	$t^{-\frac{1}{2}} \int_0^\infty e^{-bu}(t+2u)^{-\frac{1}{2}}$ $\cdot I_1[at^{\frac{1}{2}}(t+2u)^{\frac{1}{2}}]uf(u)du$	$a^{-1}g[b+(p^2-a^2)^{\frac{1}{2}}-p]-a^{-1}g(b)$
1.60	$t^{\nu} \int_0^\infty (t+2u)^{-\nu}$ $\cdot I_{2\nu}[a(t^2+2tu)^{\frac{1}{2}}]f(u)du$ $\mathrm{Re}\,\nu > -\frac{1}{2}$	$a^{-2\nu}(p^2-a^2)^{-\frac{1}{2}}[p-(p^2-a^2)^{\frac{1}{2}}]^{2\nu}$ $\cdot g[(p^2-a^2)^{\frac{1}{2}}-p]$
1.61	$t^{-\frac{1}{2}-\frac{1}{2}n} \int_0^\infty \exp(-\frac{1}{4}u^2/t)$ $\cdot He_n[(2t)^{-\frac{1}{2}}u]f(u)du$	$\pi^{\frac{1}{2}}2^{\frac{1}{2}n}p^{-\frac{1}{2}+\frac{1}{2}n}g(p^{\frac{1}{2}})$
1.62	$t^{-\nu-1} \int_0^\infty \exp(-\frac{1}{8}u^2/t)$ $\cdot D_{2\nu+1}[(2t)^{-\frac{1}{2}}u]f(u)du$	$\pi^{\frac{1}{2}}2^{\nu+\frac{1}{2}}p^{\nu}g(p^{\frac{1}{2}})$

1.2 Algebraic Functions

	$f(t)$	$g(p) = \int_0^\infty f(t)e^{-pt}dt$
2.1	$0 \quad t < a$ $1 \quad a<t<b$ $0 \quad t > b$	$p^{-1}(e^{-ap}-e^{-bp})$
2.2	1	p^{-1} Re $p > 0$
2.3	$0 \quad t < b$ $t^n \quad t > b$ $n = 1,2,3,\cdots$	$p^{-n-1}n!\ e^{-bp} \sum_{k=0}^{n} (bp)^k/k!$ Re $p > 0$
2.4	$t^n \quad t < b$ $0 \quad t > b$ $n = 1,2,3,\cdots$	$n!p^{-n-1} - p^{-n-1}n!\ e^{-bp} \sum_{k=0}^{n} (bp)^k/k!$ Re $p > 0$
2.5	t^n $n = 1,2,3,\cdots$	$n!\ p^{-n-1}$ Re $p > 0$
2.6	$0 \quad t < b$ $(t+a)^{-1} \quad b<t<c$ $0 \quad t > c$ $-a$ not between b and c	$e^{ap}[Ei(-ap-cp)$ $- Ei(-ap-bp)]$

	$f(t)$	$g(p) = \int_0^\infty f(t)e^{-pt}dt$
2.7	$0 \quad t < b$ $(t+a)^{-1} \quad t > b$	$-e^{ap}Ei(-ap-bp)$ $a < b \qquad \mathrm{Re}\, p > 0$
2.8	$(t+a)^{-1} \quad t < b$ $0 \quad t > b$	$e^{ap}[Ei(-ap-bp)-Ei(-ap)]$
2.9	$(t+a)^{-1}$	$-e^{ap}Ei(-ap) \qquad \mathrm{Re}\, p > 0$
2.10	$0 \quad t < b$ $(t-a)^{-1} \quad t > b$	$-e^{-ap}Ei(ap-bp) \qquad \mathrm{Re}\, p > 0$ $a < b$
2.11	$(a-t)^{-1} \quad t < b$ $0 \quad t > b$	$e^{-ap}[\overline{Ei}(ap) - \overline{Ei}(ap-bp)]$
2.12	$(t-a)^{-1}$ (Cauchy principal value)	$-e^{-ap}\overline{Ei}(ap) \qquad \mathrm{Re}\, p > 0$
2.13	$0 \quad t < a$ $t^{-1}(t-a)^{1/2} \quad t > a$	$(\pi/p)^{1/2}e^{-ap} - \pi a^{1/2}\mathrm{Erfc}[(ap)^{1/2}]$ $\mathrm{Re}\, p > 0$

	$f(t)$	$g(p) = \int_0^\infty f(t)e^{-pt}dt$
2.14	$0 \quad t < b$ $(t+a)^{-n} \quad t > b$ $n = 2,3,4,\cdots$	$[(n-1)!]^{-1}(-p)^{n-1}[\sum_{k=1}^{n-1}(k-1)!(-ap-bp)^{-k}$ $- e^{ap}\,\mathrm{Ei}(-ap-bp)]$ $\mathrm{Re}\, p \geqq 0$
2.15	$t^n(t+a)^{-1}$ $n = 1,2,3,\cdots$	$(-1)^{n-1}a^n e^{ap}\,\mathrm{Ei}(-ap)$ $+ (-a)^n \sum_{k=1}^{n}(k-1)!(-ap)^{-k}$ $\mathrm{Re}\, p > 0$
2.16	$(t^2+a^2)^{-1}$	$a^{-1}[\mathrm{Ci}(ap)\sin(ap)-\mathrm{si}(ap)\cos(ap)]$ $\mathrm{Re}\, p > 0$
2.17	$t(t^2+a^2)^{-1}$	$- \mathrm{Ci}(ap)\cos(ap)-\mathrm{si}(ap)\sin(ap)$ $\mathrm{Re}\, p > 0$
2.18	$t^{-\frac{1}{2}}(a^2+t^2)^{-1}$ $\lvert\arg a\rvert<\frac{1}{2}\pi$	$2^{\frac{1}{2}}\pi a^{-3/2}\{\cos(ap)[\frac{1}{2}-S(ap)]$ $- \sin(ap)[\frac{1}{2}-C(ap)]\}$ $\quad \mathrm{Re}\, p \geqq 0$
2.19	$(at+b)(t^2+c^2)^{-1}$	$-[a\cos(cp)-\frac{b}{c}\sin(cp)]\mathrm{Ci}(cp)$ $- [a\sin(cp) + \frac{b}{c}\cos(cp)]\mathrm{si}(cp)$ $\mathrm{Re}\, p > 0$

	$f(t)$	$g(p) = \int_0^\infty f(t)e^{-pt}dt$
2.20	$0 \quad a<t<b$ $(t^2-a^2)^{-1} \quad t>b$	$-(2a)^{-1}[e^{-ap}Ei(ap-bp)$ $-e^{ap}Ei(-ap-bp)] \quad \mathrm{Re}\, p \geqq 0$
2.21	$(t^2-a^2)^{-1}$ (Cauchy principal value)	$-\frac{1}{2}a^{-1}[e^{-ap}\,\overline{Ei}(ap)-e^{ap}\,Ei(-ap)]$ $\mathrm{Re}\, p \geqq 0$
2.22	$(at+b)(t^2-c^2)^{-1}$ $\lvert\arg\pm c\rvert < \pi$	$-\frac{1}{2}(a-b/c)e^{cp}Ei(-cp)$ $-\frac{1}{2}(a+b/c)e^{-cp}\overline{Ei}(cp) \quad \mathrm{Re}\, p > 0$
2.23	$(at+b)(t^2-c^2)^{-1}$ (Cauchy principal value)	$-\frac{1}{2}(a-b/c)e^{cp}Ei(-cp)$ $-\frac{1}{2}(a+b/c)e^{-cp}\overline{Ei}(cp)$ $\mathrm{Re}\, p > 0$
2.24	$f(t)=(a^2-t^2)^{-1} \quad t<b$ $0 \quad t>b$ $a>b$	$(2a)^{-1}\{e^{-ap}[\overline{Ei}(ap)-\overline{Ei}(ap-bp)]$ $+e^{ap}[Ei(-ap-bp)-Ei(-ap)]\}$ $\mathrm{Re}\, p > 0$
2.25	$t^{\frac{1}{2}}$	$2\pi^{\frac{1}{2}}p^{-\frac{3}{2}} \quad \mathrm{Re}\, p > 0$
2.26	$0 \quad t<b$ $t^{-\frac{1}{2}} \quad t>b$	$(\pi/p)^{\frac{1}{2}}\mathrm{Erfc}[(bp)^{\frac{1}{2}}]$ $\mathrm{Re}\, p > 0$

	$f(t)$	$g(p) = \int_0^\infty f(t)e^{-pt}dt$
2.27	$t^{-\frac{1}{2}}$	$(\pi/p)^{\frac{1}{2}}$ Re $p > 0$
2.28	$t^{-\frac{1}{2}}$ $t < b$ 0 $t > b$	$(\pi/p)^{\frac{1}{2}}\,\mathrm{Erf}[(bp)^{\frac{1}{2}}]$
2.29	0 $t < b$ $t^{-3/2}$ $t > b$	$2b^{-\frac{1}{2}}e^{-bp} - 2(\pi p)^{\frac{1}{2}}\mathrm{Erfc}[(bp)^{\frac{1}{2}}]$ Re $p \geqq 0$
2.30	$(t+a)^{-\frac{1}{2}}$ $\lvert\arg a\rvert < \pi$	$(\pi/p)^{\frac{1}{2}}\, e^{ap}\, \mathrm{Erfc}[(ap)^{\frac{1}{2}}]$ Re $p > 0$
2.31	$(t+a)^{-3/2}$ $\lvert\arg a\rvert < \pi$	$2a^{-\frac{1}{2}} - 2(\pi p)^{\frac{1}{2}}e^{ap}\mathrm{Erfc}[(ap)^{\frac{1}{2}}]$ Re $p \geqq 0$
2.32	$t^{\frac{1}{2}}(a+t)^{\frac{1}{2}}$ $\lvert\arg a\rvert < \pi$	$\frac{1}{2}a/p\; e^{\frac{1}{2}a/p}\, K_1(\frac{1}{2}ap)$ Re $p > 0$
2.33	$t^{\frac{1}{2}}(a+t)^{-1}$ $\lvert\arg a\rvert < \pi$	$(\pi/p)^{\frac{1}{2}} - \pi a^{\frac{1}{2}}e^{ap}\mathrm{Erfc}[(ap)^{\frac{1}{2}}]$
2.34	$t^{-\frac{1}{2}}(a+t)^{-1}$ $\lvert\arg a\rvert < \pi$	$\pi a^{-\frac{1}{2}}e^{ap}\, \mathrm{Erfc}[(ap)^{\frac{1}{2}}]$ Re $p \geqq 0$

	$f(t)$	$g(p) = \int_0^\infty f(t)e^{-pt}dt$
2.35	$0 \qquad t < a$ $t^{-1}(t-a)^{-\frac{1}{2}} \qquad t > a$	$\pi a^{-\frac{1}{2}} \mathrm{Erfc}[(ap)^{\frac{1}{2}}] \quad \mathrm{Re}\, p \geq 0$
2.36	$t^{-\frac{1}{2}}(1+2at)$	$\pi^{\frac{1}{2}} p^{-3/2}(p+a) \quad \mathrm{Re}\, p > 0$
2.37	$(t^2+a^2)^{\frac{1}{2}}$ $\lvert \arg a \rvert < \frac{1}{2}\pi$	$\frac{1}{2}\pi[\mathbf{H}_1(ap) - Y_1(ap)]$ $\mathrm{Re}\, p > 0$
2.38	$(t^2+a^2)^{-\frac{1}{2}}$ $\lvert \arg a \rvert < \frac{1}{2}\pi$	$S_{0,0}(ap) = \frac{1}{2}\pi[\mathbf{H}_0(ap) - Y_0(ap)]$ $\mathrm{Re}\, p > 0$
2.39	$t^{-\frac{1}{2}}(a^2+t^2)^{-\frac{1}{2}}$ $\lvert \arg a \rvert < \frac{1}{2}\pi$	$\frac{1}{4}\pi^{3/2} p^{\frac{1}{2}}\{[J_{\frac{1}{4}}(\frac{1}{2}ap)]^2 + [Y_{\frac{1}{4}}(\frac{1}{2}ap)]^2\}$ $\mathrm{Re}\, p \geq 0$
2.40	$0 \qquad t < 1$ $(a+t)^{-1}(t^2-1)^{-\frac{1}{2}}$ $t > 1$	$e^{ap} \int_p^\infty e^{-ax} K_0(x)dx$ $\mathrm{Re}\, p \geq 0$
2.41	$0 \qquad t < a$ $t(t^2-a^2)^{-\frac{1}{2}} \qquad t > a$	$a K_1(ap) \quad \mathrm{Re}\, p > 0$

	$f(t)$	$g(p) = \int_0^\infty f(t)e^{-pt}dt$
2.42	$0 \quad t < a$ $t^{-\frac{1}{2}}(t^2-a^2)^{-\frac{1}{2}} \quad t > a$	$(\frac{1}{2}p/\pi)^{\frac{1}{2}}[K_{\frac{1}{4}}(\frac{1}{2}ap)]^2$ $\mathrm{Re}\, p \geqq 0$
2.43	$0 \quad t < a$ $t^{\frac{1}{2}}(t^2-a^2)^{-\frac{1}{2}} \quad t > a$	$a(\frac{1}{2}p/\pi)^{\frac{1}{2}}K_{\frac{1}{4}}(\frac{1}{2}ap)K_{3/4}(\frac{1}{2}ap)$ $\mathrm{Re}\, p > 0$
2.44	$0 \quad t < a$ $t^{-3/2}(t^2-a^2)^{-\frac{1}{2}}$ $t > a$	$p(\frac{1}{2}p/\pi)^{\frac{1}{2}}\{[K_{3/4}(\frac{1}{2}ap)]^2-[K_{\frac{1}{4}}(\frac{1}{2}ap)]^2\}$ $\mathrm{Re}\, p \geqq 0$
2.45	$t(a^2-t^2)^{-\frac{1}{2}} \quad t < a$ $0 \quad t > a$	$\frac{1}{2}\pi a[L_1(ap) - I_1(ap)]+a$
2.46	$t^{\frac{1}{2}}(a^2-t^2)^{-\frac{1}{2}} \quad t < a$ $0 \quad t > a$	$\frac{1}{2}\pi a(\frac{1}{2}\pi p)^{\frac{1}{2}}[I_{\frac{1}{4}}(\frac{1}{2}ap)I_{-3/4}(\frac{1}{2}ap)$ $- I_{-\frac{1}{4}}(\frac{1}{2}ap)I_{3/4}(\frac{1}{2}ap)]$
2.47	$t^{-\frac{1}{2}}(a^2-t^2)^{-\frac{1}{2}} \quad t < a$ $0 \quad t > a$	$(\pi p)^{\frac{1}{2}}K_{\frac{1}{4}}(\frac{1}{2}ap)$ $\cdot\ [I_{\frac{1}{4}}(\frac{1}{2}ap) + I_{-\frac{1}{4}}(\frac{1}{2}ap)]$
2.48	$t^{-\frac{1}{2}}(a+t)^{-\frac{1}{2}}$	$e^{ap}K_o(ap) \quad \mathrm{Re}\, p > 0$

	$f(t)$	$g(p) = \int_0^\infty f(t)e^{-pt}dt$
2.49	$(t+a)(t^2+2at)^{-\frac{1}{2}}$ $\lvert\arg a\rvert < \pi$	$ae^{ap} K_1(ap)$ Re $p > 0$
2.50	$(a-t)(2at-t^2)^{-\frac{1}{2}}$ $t < 2a$ 0 $t > 2a$	$\pi ae^{-ap} I_1(ap)$ Re $p > 0$
2.51	$[t+(t^2+a^2)^{\frac{1}{2}}]^{-1}$	$\frac{1}{2}\pi(ap)^{-1}[\mathbf{H}_1(ap)-Y_1(ap)] - (ap)^{-2}$ Re $p > 0$
2.52	$[at+(1+a^2t^2)^{\frac{1}{2}}]^n$ $+[at-(1+a^2t^2)^{\frac{1}{2}}]^n$ $n = 0,1,2,\cdots$	$2a^{-1} O_n(p/a)$ Re $p > 0$
2.53	$(1+a^2t^2)^{-\frac{1}{2}}[at+(1+a^2t^2)^{\frac{1}{2}}]^n$ $n = 0,1,2,\cdots$	$(2a)^{-1}[S_n(p/a)-\pi\mathbf{E}_n(p/a)$ $-\pi Y_n(p/a)]$ Re $p > 0$
2.54	$(1+a^2t^2)^{-\frac{1}{2}}[at-(1+a^2t^2)^{\frac{1}{2}}]^n$ $n = 0,1,2,\cdots$	$-(2a)^{-1}[S_n(p/a)+\pi\mathbf{E}_n(p/a)$ $+ \pi Y_n(p/a)]$ Re $p > 0$

	$f(t)$	$g(p) = \int_0^\infty f(t)e^{-pt}dt$
2.55	$t^{-\frac{1}{2}}(t+a)^{-\frac{1}{2}}[a+(t+a)^{\frac{1}{2}}]^{\frac{1}{2}}$ Re $a > 0$	$2^{3/4}\pi^{\frac{1}{2}}p^{-\frac{1}{4}}e^{\frac{1}{2}ap}D_{-\frac{1}{2}}[(2ap)^{\frac{1}{2}}]$ $= 2^{\frac{1}{2}}a^{\frac{1}{4}}\ e^{\frac{1}{2}ap}\ K_{\frac{1}{4}}(\frac{1}{2}ap)$ Re $p > 0$
2.56	$t^{-\frac{1}{2}}(t^2+a^2)^{-\frac{1}{2}}[t+(t^2+a^2)^{\frac{1}{2}}]^{\frac{1}{2}}$ Re $a > 0$	$2^{-\frac{1}{2}}\pi[\sin(\frac{1}{2}ap)J_0(\frac{1}{2}ap)$ $-\cos(\frac{1}{2}ap)Y_0(\frac{1}{2}ap)]$ Re $p > 0$
2.57	$t^{-\frac{1}{2}}(t^2+a^2)^{-\frac{1}{2}}[(t^2+a^2)^{\frac{1}{2}}-t]^{\frac{1}{2}}$ Re $a > 0$	$2^{-\frac{1}{2}}\pi[\cos(\frac{1}{2}ap)J_0(\frac{1}{2}ap)$ $+\sin(\frac{1}{2}ap)Y_0(\frac{1}{2}ap)]$ Re $p > 0$
2.58	$t^{-\frac{1}{2}}(t^2+a^2)^{-\frac{1}{2}}[(t^2+a^2)^{\frac{1}{2}}+t]^{3/2}$ Re $a > 0$	$2^{-\frac{1}{2}}\pi a[\sin(\frac{1}{2}ap)J_1(\frac{1}{2}ap)$ $-\cos(\frac{1}{2}ap)Y_1(\frac{1}{2}ap)]$ Re $p > 0$
2.59	$t^{-\frac{1}{2}}(t^2+a^2)^{-\frac{1}{2}}[(t^2+a^2)^{\frac{1}{2}}-t]^{3/2}$ Re $a > 0$	$-2^{-\frac{1}{2}}\pi a[\cos(\frac{1}{2}ap)J_1(\frac{1}{2}ap)$ $+\sin(\frac{1}{2}ap)Y_1(\frac{1}{2}ap)]$ Re $p > 0$

1.3 Powers of Arbitrary Order

	$f(t)$	$g(p) = \int_0^\infty f(t) e^{-pt} dt$
3.1	$0 \quad t < a$ $t^\nu \quad t > a$	$p^{-\nu-1}\Gamma(\nu+1,ap) \qquad \mathrm{Re}\, p > 0$
3.2	$t^\nu \quad t < a$ $0 \quad t > a$ $\mathrm{Re}\,\nu > -1$	$p^{-\nu-1}\gamma(\nu+1,ap)$
3.3	$t^\nu \qquad \mathrm{Re}\,\nu > -1$	$p^{-\nu-1}\Gamma(\nu+1) \qquad \mathrm{Re}\, p > 0$
3.4	$0 \quad t < a$ $(t-a)^\nu \quad t > a$ $\mathrm{Re}\,\nu > -1$	$p^{-\nu-1}e^{-ap}\Gamma(\nu+1) \qquad \mathrm{Re}\, p > 0$
3.5	$(a-t)^\nu \quad t < a$ $0 \quad t > a$ $\mathrm{Re}\,\nu > -1$	$p^{-\nu-1}e^{-ap}\gamma(\nu+1,-ap)$
3.6	$(t+a)^\nu$ $\lvert\arg a\rvert < \pi$	$p^{-\nu-1}e^{ap}\Gamma(\nu+1,ap) \qquad \mathrm{Re}\, p > 0$
3.7	$0 \quad t < a$ $t^{-1}(t-a)^\nu \quad t > a$ $\mathrm{Re}\,\nu > -1$	$a^\nu\Gamma(\nu+1)\Gamma(-\nu,ap)$

	$f(t)$	$g(p) = \int_0^\infty f(t)e^{-pt}dt$
3.8	$t^{\nu}(t+a)^{-1}$ $\mathrm{Re}\,\nu > -1,\ \lvert\arg a\rvert < \pi$	$a^{\nu}e^{ap}\Gamma(\nu+1)\Gamma(-\nu,ap)$ $\mathrm{Re}\,p > 0$
3.9	$t^{\nu-1}(t+a)^{-\nu-\frac{1}{2}}$ $\mathrm{Re}\,\nu > 0,\ \lvert\arg a\rvert < \pi$	$2^{\nu}a^{-\frac{1}{2}}e^{\frac{1}{2}ap}\Gamma(\nu)$ $\cdot D_{-2\nu}[(2ap)^{\frac{1}{2}}]$ $\mathrm{Re}\,p \geqq 0$
3.10	$t^{\nu-1}(t+a)^{\frac{1}{2}-\nu}$ $\mathrm{Re}\,\nu > 0,\ \lvert\arg a\rvert < \pi$	$2^{\nu-\frac{1}{2}}p^{-\frac{1}{2}}e^{\frac{1}{2}ap}\Gamma(\nu)$ $\cdot D_{1-2\nu}[(2ap)^{\frac{1}{2}}]$ $\mathrm{Re}\,p > 0$
3.11	$t^{\nu-1}(t^2+a^2)^{-1}$ $\mathrm{Re}\,\nu > 0$ $\mathrm{Re}\,a > 0$	$a^{\nu-3/2}p^{\frac{1}{2}}\Gamma(\nu)S_{\frac{1}{2}-\nu,\frac{1}{2}}(ap)$ $\mathrm{Re}\,p > 0$
3.12	$(t^2+a^2)^{\nu-\frac{1}{2}}$ $\mathrm{Re}\,a > 0$	$\frac{1}{2}\pi^{\frac{1}{2}}(2a/p)^{\nu}\Gamma(\nu+\frac{1}{2})$ $\cdot[\mathbf{H}_{\nu}(ap)-Y_{\nu}(ap)]$ $\mathrm{Re}\,p > 0$
3.13	0 $t < a$ $(t^2-a^2)^{\nu-\frac{1}{2}}$ $t > a$ $\mathrm{Re}\,\nu > -\frac{1}{2}$	$\pi^{-\frac{1}{2}}(2a/p)^{\nu}\Gamma(\frac{1}{2}+\nu)K_{\nu}(ap)$ $\mathrm{Re}\,p > 0$
3.14	$(a^2-t^2)^{\nu-\frac{1}{2}}$ $t < a$ 0 $t > a$ $\mathrm{Re}\,\nu > -\frac{1}{2}$	$\frac{1}{2}\pi^{\frac{1}{2}}(2a/p)^{\nu}\Gamma(\frac{1}{2}+\nu)$ $\cdot[I_{\nu}(ap) - \mathbf{L}_{\nu}(ap)]$

	$f(t)$	$g(p) = \int_0^\infty f(t)e^{-pt}dt$
3.15	$(2at-t^2)^{\nu-\frac{1}{2}}$ $t < 2a$ 0 $t > 2a$ $\mathrm{Re}\,\nu > -\frac{1}{2}$	$\pi^{\frac{1}{2}}(2a/p)^{\nu}e^{-ap}$ $\cdot\Gamma(\frac{1}{2}+\nu)I_{\nu}(ap)$
3.16	$(2at+t^2)^{\nu-\frac{1}{2}}$ $\mathrm{Re}\,\nu > -\frac{1}{2}, \lvert\arg a\rvert < \pi$	$\pi^{-\frac{1}{2}}(2a/p)^{\nu}e^{ap}$ $\cdot\Gamma(\frac{1}{2}+\nu)K_{\nu}(ap)$ $\mathrm{Re}\,p > 0$
3.17	$(2iat+t^2)^{\nu-\frac{1}{2}}$ $\mathrm{Re}\,\nu > -\frac{1}{2}$	$-\frac{1}{2}i\pi^{\frac{1}{2}}(2a/p)^{\nu}e^{iap}$ $\cdot\Gamma(\frac{1}{2}+\nu)H_{\nu}^{(2)}(ap)$ $\mathrm{Re}\,p > 0$
3.18	$(-2iat+t^2)^{\nu-\frac{1}{2}}$ $\mathrm{Re}\,\nu > -\frac{1}{2}$	$\frac{1}{2}i\pi^{\frac{1}{2}}(2a/p)^{\nu}e^{-iap}$ $\cdot\Gamma(\frac{1}{2}+\nu)H_{\nu}^{(1)}(ap)$ $\mathrm{Re}\,p > 0$
3.19	0 $t < a$ $(t-a)^{\nu-1}(t+a)^{\frac{1}{2}-\nu}$ $t > a$ $\mathrm{Re}\,\nu > 0$	$2^{\nu-\frac{1}{2}}p^{-\frac{1}{2}}\Gamma(\nu)D_{1-2\nu}[2(ap)^{\frac{1}{2}}]$ $\mathrm{Re}\,p > 0$
3.20	0 $t < 2b$ $(t+2a)^{\nu}(t-2b)^{\nu}$ $t > 2b$ $\mathrm{Re}\,\nu < -1$	$2^{\frac{1}{2}+\nu}\pi^{-\frac{1}{2}}(a+b)^{\nu+\frac{1}{2}}\Gamma(1+\nu)p^{-\nu-\frac{1}{2}}$ $\cdot e^{p(a-b)}K_{\nu+\frac{1}{2}}[p(a+b)]$ $\mathrm{Re}\,p > 0$

	$f(t)$	$g(p) = \int_0^\infty f(t)e^{-pt}dt$
3.21	$0 \qquad t < a$ $(t-a)^{\nu-1}(t+a)^{-\nu-\frac{1}{2}}$ $t > a$ $\mathrm{Re}\,\nu > 0$	$2^{\nu-\frac{1}{2}}a^{-\frac{1}{2}}\Gamma(\nu)$ $\cdot D_{-2\nu}[2(ap)^{\frac{1}{2}}]$ $\qquad \mathrm{Re}\,p > 0$
3.22	$t^{\mu}(t+a)^{\nu}$ $\mathrm{Re}\,\mu > -1, \ \lvert\arg a\rvert<\pi$	$a^{\frac{1}{2}\mu+\frac{1}{2}\nu}p^{-1-\frac{1}{2}\mu-\frac{1}{2}\nu}e^{\frac{1}{2}ap}$ $\cdot\Gamma(\mu+1)W_{\frac{1}{2}\nu-\frac{1}{2}\mu,\frac{1}{2}+\frac{1}{2}\nu+\frac{1}{2}\mu}(ap)$ $\mathrm{Re}\,p > 0$
3.23	$0 \qquad t < a$ $(t-a)^{2\mu-1}(b-t)^{2\nu-1}$ $a<t<b$ $0 \qquad t > b$ $\mathrm{Re}\,\mu > 0, \ \mathrm{Re}\,\nu > 0$	$(b-a)^{\mu+\nu-1}p^{-\mu-\nu}B(2\mu,2\nu)$ $\cdot\exp[-\frac{1}{2}(a+b)p]$ $\cdot M_{\mu-\nu,\mu+\nu-\frac{1}{2}}(bp-ap)$
3.24	$0 \qquad t < b$ $(t+a)^{2\mu-1}(t-b)^{2\nu-1}$ $t > b$ $\mathrm{Re}\,\nu > 0, \ \lvert\arg a\rvert<\pi$	$(a+b)^{\mu+\nu-1}p^{-\mu-\nu}\exp[\frac{1}{2}(a-b)p]$ $\cdot\Gamma(2\nu)W_{\mu-\nu,\mu+\nu-\frac{1}{2}}(ap+bp)$ $\mathrm{Re}\,p > 0$

	$f(t)$	$g(p) = \int_0^\infty f(t)e^{-pt}dt$
3.25	$t^{\nu}(t+a)^{-\frac{1}{2}}[(t+a)^{\frac{1}{2}}+a^{\frac{1}{2}}]^{-\nu}$ $\mathrm{Re}\ \nu > -1$	$2(2p)^{-\frac{1}{2}\nu}\ \Gamma(1+\nu)$ $\cdot\exp(\frac{1}{2}ap)\ D_{-\nu-1}[(2ap)^{\frac{1}{2}}]$ $\mathrm{Re}\ p > 0$
3.26	$[(t+2a)^{\frac{1}{2}}+t^{\frac{1}{2}}]^{2\nu}$ $-[(t+2a)^{\frac{1}{2}}-t^{\frac{1}{2}}]^{2\nu}$	$2(2a)^{\nu}\nu p^{-1}e^{ap}K_{\nu}(ap)$ $\mathrm{Re}\ p > 0$
3.27	$0 \qquad t < a$ $[(t+a)^{\frac{1}{2}}+(t-a)^{\frac{1}{2}}]^{2\nu}$ $-[(t+a)^{\frac{1}{2}}-(t-a)^{\frac{1}{2}}]^{2\nu}$ $t > a$	$2(2a)^{\nu}\nu p^{-1}K_{\nu}(ap)$ $\mathrm{Re}\ p > 0$
3.28	$[(t^2+1)^{\frac{1}{2}}+t]^{\nu}$	$p^{-1}+\pi\nu(p\ \sin\ \pi\nu)^{-1}$ $\cdot[J_{-\nu}(p) - \mathbf{J}_{-\nu}(p)]$ $= p^{-1}s_{1,\nu}(p) + \nu p^{-1}s_{0,\nu}(p)$ $\mathrm{Re}\ p > 0$
3.29	$(t^2+a^2)^{-\frac{1}{2}}[(t^2+a^2)^{\frac{1}{2}}+t]^{\nu}$	$\pi a^{\nu}\csc(\pi\nu)$ $\cdot[J_{-\nu}(ap) - \mathbf{J}_{-\nu}(ap)]$ $\mathrm{Re}\ p > 0$

	$f(t)$	$g(p) = \int_0^\infty f(t)e^{-pt}dt$
3.30	$t^{-\frac{1}{2}}(t^2+a^2)^{-\frac{1}{2}}$ $\cdot[(t^2+a^2)^{\frac{1}{2}}+t]^{2\nu}$	$(\frac{1}{2}\pi)^{3/2}a^{2\nu}p^{\frac{1}{2}}$ $\cdot[J_{\frac{1}{4}-\nu}(\frac{1}{2}ap)Y_{-\frac{1}{4}-\nu}(\frac{1}{2}ap)$ $- J_{-\frac{1}{4}-\nu}(\frac{1}{2}ap)Y_{\frac{1}{4}-\nu}(\frac{1}{2}ap)]$ Re $p > 0$
3.31	$t^{-\nu-1}(a^2+t^2)^{-\frac{1}{2}}$ $\cdot[(a^2+t^2)^{\frac{1}{2}}+a]^{\nu+\frac{1}{2}}$	$(\frac{1}{2}a)^{-\frac{1}{2}}\Gamma(-\nu)$ $\cdot D_\nu[(2api)^{\frac{1}{2}}]D_\nu[(-2api)^{\frac{1}{2}}]$ Re $p > 0$
3.32	$0 \quad t < a$ $[t+(t^2-a^2)^{\frac{1}{2}}]^\nu$ $-[t-(t^2-a^2)^{\frac{1}{2}}]^\nu \quad t > a$	$2\nu a^\nu p^{-1}K_\nu(ap)$ Re $p > 0$
3.33	$[(t^2+a^2)^{\frac{1}{2}}+t]^\nu$ $+ [(t^2+a^2)^{\frac{1}{2}}-t]^\nu$	$2p^{-1}a^\nu S_{1,\nu}(ap)$ Re $p > 0$
3.34	$[(t^2+a^2)^{\frac{1}{2}}+t]^\nu$ $- [(t^2+a^2)^{\frac{1}{2}}-t]^\nu$	$2\nu(\nu^2-1)^{-1}[a^\nu S_{2,\nu}(ap) - a^{\nu+1}]$ Re $p > 0$

	$f(t)$	$g(p) = \int_0^\infty f(t)e^{-pt}dt$
3.35	$0 \qquad t < a$ $(t^2-a^2)^{-\frac{1}{2}}\{[t+(t^2-a^2)^{\frac{1}{2}}]^\nu$ $+ [t-(t^2-a^2)^{\frac{1}{2}}]^\nu\}$ $t > a$	$2a^\nu K_\nu(ap)$ Re $p > 0$
3.36	$(t^2+a^2)^{-\frac{1}{2}}\{[(t^2+a^2)^{\frac{1}{2}}+t]^\nu$ $+[(t^2+a^2)^{\frac{1}{2}}-t]^\nu\}$	$2a^\nu S_{0,\nu}(ap)$ Re $p > 0$
3.37	$(t^2+a^2)^{-\frac{1}{2}}\{[(t^2+a^2)^{\frac{1}{2}}+t]^\nu$ $-[(t^2+a^2)^{\frac{1}{2}}-t]^\nu\}$	$2\nu a^\nu S_{-1,\nu}(ap)$ Re $p > 0$
3.38	$0 \qquad t < a$ $t^{-\frac{1}{2}}(t^2-a^2)^{-\frac{1}{2}}$ $\cdot\{[t+(t^2-a^2)^{\frac{1}{2}}]^\nu$ $+[t-(t^2-a^2)^{\frac{1}{2}}]^\nu\}$ $t > a$	$a^\nu(2p/\pi)^{\frac{1}{2}}$ $\cdot K_{\frac{1}{4}+\frac{1}{2}\nu}(\frac{1}{2}ap)K_{\frac{1}{4}-\frac{1}{2}\nu}(\frac{1}{2}ap)$ Re $p > 0$

1.4 Sectionally Rational - and Rows of Delta Functions

	f(t)		$g(p) = \int_0^\infty f(t)e^{-pt}dt$
4.1	0	$2na<t<(2n+1)a$	$p^{-1}(e^{ap}+1)^{-1}$
	1	$(2n-1)a<t<2na$	Re $p > 0$
4.2	0	$(4n-1)a<t<(4n+1)a$	$p^{-1}\mathrm{sech}(ap)$
	2	$(4n+1)a<t<(4n+3)a$	Re $p > 0$
4.3	$\frac{1}{2}$	$2na<t<(2n+1)a$	$\frac{1}{2}p^{-1}\tanh(\frac{1}{2}ap)$
	$-\frac{1}{2}$	$(2n-1)a<t<2na$	Re $p > 0$
4.4	n	$na<t<(n+1)a$	$p^{-1}(e^{ap}-1)^{-1}$ Re $p > 0$
4.5	$n+1$	$na<t<(n+1)a$	$p^{-1}(1-e^{-ap})^{-1}$ Re $p > 0$
4.6	$2n+1$	$2na<t<2(n+1)a$	$p^{-1}\coth(ap)$ Re $p > 0$
4.7	0	$0<t<a$	$p^{-1}\mathrm{csch}(ap)$
	$2n$	$(2n-1)a<t<(2n+1)a$	Re $p > 0$
4.8	n	$\log(2n-1)<t<\log(2n+1)$	$p^{-1}(1-2^{2-p})\zeta(p-1)$
		$n = 1,2,3,\cdots$	Re $p > 0$
4.9	n	$\log n<t<\log(n+1)$	$p^{-1}\zeta(p)$ Re $p > 0$

	$f(t)$	$g(p) = \int_0^\infty f(t)e^{-pt}dt$
4.10	$\sum_{0\leq \log n \leq t} (t-\log n)^{a-1}$ Re $a > 0$	$\Gamma(a)p^{-a}\zeta(p)$ Re $p > 0$
4.11	$(1-a)^{-1}(1-a^n)$ $nb<t<(n+1b$	$p^{-1}(e^{bp}-a)^{-1}$ Re $p>0$, b Re $p>$Re$(\log a)$
4.12	$\binom{n}{m}$ $\quad na<t<(n+1)a$	$p^{-1}e^{-ap}(e^{ap}-1)^{-m}$ Re $p > 0$
4.13	n^m $\quad na<t<(n+1)a$	$p^{-1}(-a)^m(1-e^{-ap})$ $\cdot \frac{d^m}{dp^m}(1-e^{-ap})^{-1}$ Re $p > 0$
4.14	$t \quad 0<t<1$ $1 \quad t>1$	$p^{-2}(1-e^{-p})$ Re $p > 0$
4.15	$t \quad 0<t<1$ $2-t \quad 1<t<2$ $0 \quad t>2$	$p^{-2}(1-e^{-p})^2$ Re $p > 0$

	$f(t)$	$g(p) = \int_0^\infty f(t)e^{-pt}dt$
4.16	$(2n+1)t-2an(n+1)$ $2na<t<2(n+1)a$	$p^{-2}\coth(ap)$ $\operatorname{Re} p > 0$
4.17	$a-(-1)^n(2na+a-t)$ $2na<t<2(n+1)a$	$p^{-2}\tanh(ap)$ $\operatorname{Re} p > 0$
4.18	$a(t-nb)$ $nb<t<(n+1)b$	$ap^{-2}-\frac{1}{2}ab\,p^{-1}$ $\cdot\,[\coth(\frac{1}{2}bp)-1]$ $\operatorname{Re} p > 0$
4.19	$0 \quad 0<t<a$ $t-(-1)^n(t-2na)$ $(2n-1)a<t<(2n+1)a$ $n\geq 1$	$p^{-2}\operatorname{sech}(ap)$ $\operatorname{Re} p > 0$
4.20	$0 \quad 0<t<a$ $2n(t-na)$ $(2n-1)a<t<(2n+1)a$ $n\geq 1$	$p^{-2}\operatorname{csch}(ap)$ $\operatorname{Re} p > 0$

	$f(t)$	$g(p) = \int_0^\infty f(t)e^{-pt}dt$
4.21	$(t-na)^2 \quad na<t<(n+1)a$	$2p^{-3}-p^{-2}(a^2+2ap) \cdot (e^{ap}-1)^{-1}$ Re $p > 0$
4.22	$H(t-c) \quad c>0$	$p^{-1}e^{-cp}$ Re $p > 0$
4.23	$H(t+c) \quad c>0$	p^{-1} Re $p > 0$
4.24	$\sum_{n=0}^{\infty} H(t-b-na)$ $a>0,\ b\geqq 0$	$(2p)^{-1}\operatorname{csch}(\frac{1}{2}ap) \cdot \exp[-p(b-\frac{1}{2}a)]$ Re $p > 0$
4.25	$\sum_{n=0}^{\infty} (-1)^n H(t-b-na)$ $a>0,\ b\geqq 0$	$(2p)^{-1}\operatorname{sech}(\frac{1}{2}ap) \cdot \exp[-p(b-\frac{1}{2}a)]$ Re $p > 0$
4.26	$\sum_{n=0}^{\infty} H(t+b-na)$ $a>0,\ (m-1)a<b<ma$ $m = 1,2,3,\cdots$	$mp^{-1}+(2p)^{-1}e^{\frac{1}{2}ap}\operatorname{csch}(\frac{1}{2}ap) \cdot \exp[-p(ma-b)]$ Re $p > 0$

	$f(t)$	$g(p) = \int_0^\infty f(t)e^{-pt}dt$
4.27	$\sum_{n=0}^{\infty} (-1)^n H(t+b-na)$ $a>0,\ (m-1)a<b<ma$ $m = 2,4,6,\cdots$	$(-1)^m e^{\frac{1}{2}ap}\operatorname{sech}(\tfrac{1}{2}ap)(2p)^{-1}$ $\cdot \exp[-p(ma-b)]$ Re $p>0$ (If $m = 1,3,5,\cdots$ add the term p^{-1}.)
4.28	$\delta(t-c)$, $c>0$	e^{-cp} Re $p > 0$
4.29	$\sum_{n=0}^{\infty} \delta(t-b-na)$ $a>0,\ b\geq 0$	$\tfrac{1}{2}\operatorname{csch}(\tfrac{1}{2}ap)\exp[-p(b-\tfrac{1}{2}a)]$ Re $p > 0$
4.30	$\sum_{n=0}^{\infty} (-1)^n\delta(t-b-na)$ $a>0,\ b\geq 0$	$\tfrac{1}{2}\operatorname{sech}(\tfrac{1}{2}ap)\exp[-p(b-\tfrac{1}{2}a)]$ Re $p > 0$
4.31	$\sum_{n=0}^{\infty} \delta(t+b-na)$ $a>0,\ (m-1)a<b<ma$ $m = 1,2,3,\cdots$	$\tfrac{1}{2}e^{\frac{1}{2}ap}\operatorname{csch}(\tfrac{1}{2}ap)$ $\cdot \exp[-p(ma-b)]$ Re $p > 0$

	$f(t)$	$g(p) = \int_0^\infty f(t)e^{-pt}dt$
4.32	$\sum_{n=0}^{\infty} (-1)^n \delta(t+b-na)$ $a>0,\ (m-1)a<b<ma$ $m = 0,1,2,\cdots$	$\frac{1}{2}(-1)^m \operatorname{sech}(\frac{1}{2}ap)e^{\frac{1}{2}ap}$ $\cdot \exp[-p(ma-b)]$ $\operatorname{Re} p > 0$
4.33	$n^2 \qquad n \leqq t < n+1$	$p^{-1}(1+e^p)(e^p-1)^2$ $\operatorname{Re} p > 0$
4.34	$n^3 \qquad n \leqq t < n+1$	$p^{-1}(1+e^{2p}+4e^p)(e^p-1)^{-3}$ $\operatorname{Re} p > 0$
4.35	$n^4 \qquad n \leqq t < n+1$	$p^{-1}(1+e^{3p}+11e^{2p}+11e^p)$ $\cdot(e^p-1)^{-4} \qquad \operatorname{Re} p > 0$
4.36	$n^m \qquad n \leqq t < n+1$ $m = 0,1,2,\cdots$	$(-1)^m p^{-1}(1-e^{-p})$ $\cdot \frac{d^m}{dp^m} [(1-e^{-t})^{-1}]$ $\operatorname{Re} p > 0$
4.37	$a^n \qquad n \leqq t < n+1$	$p^{-1}(e^p-1)(e^p-a)^{-1}$ $\operatorname{Re} p > 0$

	$f(t)$	$g(p) = \int_0^\infty f(t)e^{-pt}dt$
4.33	$0 \quad t < \pi^2/4$ $(2n+2) \quad \pi^2(n+\frac{1}{2})^2 < t < \pi^2(n+\frac{3}{2})^2$ $n = 0,1,2,\cdots$	$p^{-1}\theta_2(0\|p)$
4.39	$(2n+1)$ $n^2\pi^2 < t < (n+1)^2\pi^2$ $n = 0,1,2,\cdots$	$p^{-1}\theta_3(0\|p)$
4.40	$1 \quad (2n)^2\pi^2 < t < (2n+1)^2\pi^2$ $-1 \quad (2n+1)^2\pi^2 < t < (2n+2)^2\pi^2$	$p^{-1}\theta_4(0\|p)$
4.41	$\sum_{n=0}^{k} \cos[(2n+1)a]$ $k = [-\frac{1}{2}+t^{\frac{1}{2}}/\pi]$ $f(t) = 0 \quad \text{if} \quad t < \pi^2/4$	$\frac{1}{2}p^{-1}\theta_2(a\|p)$
4.42	$\sum_{n=0}^{k} (-1)^n \sin[(2n+1)a]$ k as in 41 $f(t) = 0 \quad \text{if} \quad t < \pi^2/4$	$\frac{1}{2}p^{-1}\theta_1(a\|p)$

	$f(t)$	$g(p) = \int_0^\infty f(t)e^{-pt}dt$
4.43	$\sum_{n=0}^{k} \varepsilon_n \cos(2\pi na)$ $k = [t^{1/2}/\pi]$	$p^{-1}\theta_3(a\|p)$
4.44	$\sum_{n=0}^{k} (-1)^n \varepsilon_n \cos(2\pi na)$ k as in 43	$p^{-1}\theta_4(a\|p)$
4.45	$\sum_{n=0}^{k} (-1)^n \sin[(2n+1)\pi a]$ $\cdot[t-(n+\frac{1}{2})^2\pi^2]^{\nu-1}$ k as in (41) $f(t) = 0$ if $t < \pi^2/4$	$\frac{1}{2}\Gamma(\nu)p^{-\nu}\theta_1(a\|p)$ $\mathrm{Re}\,\nu > 0$
4.46	$\sum_{n=0}^{k} \cos[(2n+1)\pi a]$ $\cdot[t-(n+\frac{1}{2})^2\pi^2]^{\nu-1}$ k as in (41), $f(t)=0$ if $t<\pi^2/4$	$\frac{1}{2}\Gamma(\nu)p^{-\nu}\theta_2(a\|p)$ $\mathrm{Re}\,\nu > 0$

	$f(t)$	$g(p) = \int_0^\infty f(t)e^{-pt}dt$
4.47	$\sum_{n=0}^{k} \varepsilon_n \cos(2\pi na)$ $\cdot(t-n^2\pi^2)^{\nu-1}$ k as in (43)	$\Gamma(\nu)\theta_3(a\|p)$ $\mathrm{Re}\ \nu > 0$
4.48	$\sum_{n=0}^{k} (-1)^n \varepsilon_n \cos(2\pi na)$ $\cdot(t-n^2\pi^2)^{\nu-1}$ k as in (43)	$\Gamma(\nu)\theta_4(a\|p)$ $\mathrm{Re}\ \nu > 0$

.5 Exponential Functions

	$f(t)$	$g(p) = \int_0^\infty f(t) e^{-pt} dt$
.1	e^{-at}	$(p+a)^{-1}$ $\qquad \operatorname{Re}(p+a) > 0$
.2	$t\, e^{-at}$	$(p+a)^{-2}$ $\qquad \operatorname{Re}(p+a) > 0$
.3	$t^{\nu-1}\, e^{-at}$ $\qquad \operatorname{Re} \nu > 0$	$\Gamma(\nu)(p+a)^{-\nu}$ $\qquad \operatorname{Re}(p+a) > 0$
.4	$t^{-1}(e^{-at} - e^{-bt})$	$\log\left(\frac{p+b}{p+a}\right)$ $\operatorname{Re} p > \operatorname{Max}(-\operatorname{Re} a, -\operatorname{Re} b)$
.5	$t^{-3/2}(e^{-at} - e^{-bt})$	$2\pi^{1/2}[(p+b)^{1/2} - (p+a)^{1/2}]$ $\operatorname{Re} p > \operatorname{Max}(-\operatorname{Re} a, -\operatorname{Re} b)$
.6	$t^{-2}(1-e^{-at})^2$	$(p+2a)\log(p+2a) + p \log p$ $- 2(p+a)\log(p+a)$ $\operatorname{Re} p \geqq 0, \operatorname{Re}(p+2a) \geqq 0$
.7	$(1-e^{-t/a})^n$ $n = 0,1,2,\cdots$	$\frac{n!}{p(1+ap)_n}$ $\operatorname{Re} p > 0, \operatorname{Re}(p + \frac{n}{a}) > 0$

	$f(t)$	$g(p) = \int_0^\infty f(t)e^{-pt}dt$
5.8	$(1-e^{-at})^{\nu}$ Re $\nu>-1$, Re $a>0$	$a^{-1}B(\nu+1,\ p/a)$ Re $p > 0$
5.9	$t(1-e^{-t})^{\nu}$ Re $\nu>-2$	$B(p,\nu+1)[\psi(p+\nu+1)-\psi(p)]$ Re $p > 0$
5.10	$t(1-e^{-t})^{-1}$	$\psi'(p) = \zeta(2,p)$ Re $p > 0$
5.11	$t^n(1-e^{-at})^{-1}$ Re $a>0$	$(-a)^{-n-1}\ \psi^{(n)}(p/a)$ Re $p > 0$
5.12	$t^{\nu-1}(e^{at}-1)^{-1}$ Re $\nu>1$, Re $a>0$	$a^{-\nu}\Gamma(\nu)\zeta(\nu,1+p/a)$ Re$(p+a)>0$
5.13	$t^{\nu-1}(e^{t}-z)^{-1}$ Re $\nu>0, \lvert\arg z\rvert<\pi$	$\Gamma(\nu)\ \sum_{k=0}^{\infty}\ z^k(k+1+p)^{-\nu}$ Re $p > -1$
5.14	$t^{-1}(1-e^{-t})^{-1}-t^{-2}-\tfrac{1}{2}t^{-1}$	$p(1-\log p)+\tfrac{1}{2}\log(\tfrac{1}{2}\pi p)$ $+ \log\Gamma(p)$ Re $p > 0$
5.15	$(1-e^{-t})^{-1}(1-e^{-at})$	$\psi(p+a)-\psi(p)$ Re $p>\mathrm{Max}(0,-\mathrm{Re}\ a)$

	$f(t)$	$g(p) = \int_0^\infty f(t)e^{-pt}dt$
5.16	$(1+e^{-t})^{-1}$	$\frac{1}{2}[\psi(\frac{1}{2}+\frac{1}{2}p)-\psi(\frac{1}{2}p)]$ Re $p > 0$
5.17	$t^{\nu-1}(1+e^{-t})^{-1}$	$\Gamma(\nu)(2^{-\nu}+p^{-\nu})$ $\cdot [\zeta(\nu,\frac{1}{2}+\frac{1}{2}p)-\Gamma(\nu,\frac{1}{2}p)]$ Re $p > -1$
5.18	$t^{-1}(1+e^{-t})^{-1}(1-e^{-at})$	$\log[\Gamma(\frac{1}{2}p)\Gamma(\frac{1}{2}+\frac{1}{2}a+\frac{1}{2}p)]$ $- \log[\Gamma(\frac{1}{2}+\frac{1}{2}p)\Gamma(\frac{1}{2}a+\frac{1}{2}p)]$ Re $p>$Max$(0,-$Re $a)$
5.19	$(e^{-ct}-e^{-bt})(1-e^{-at})^{-1}$	$\frac{1}{a}[\psi(\frac{p+b}{a}) - \psi(\frac{p+c}{a})]$ Re $a > 0$ Re $p>$Max$(-$Re $b,-$Re $c)$
5.20	$t^{-1}(1+e^{-\frac{1}{2}t})^{-1}(e^{-at}-e^{-bt})$	$\log[\Gamma(p+a)\Gamma(\frac{1}{2}+p+b)]$ $- \log[\Gamma(p+b)\Gamma(\frac{1}{2}+p+a)]$ Re $p>$Max$(-$Re $a,-$Re $b)$
5.21	$t^{-1}(1+e^{-ct})^{-1}(e^{-at}-e^{-bt})$	$\log\{\Gamma[\frac{1}{2c}(a+p)]\Gamma[\frac{1}{2c}(b+c+p)]\}$ $- \log\{\Gamma[\frac{1}{2c}(b+p)]\Gamma[\frac{1}{2c}(a+c+p)]\}$ Re$>p$ Max$[-$Re $a,-$Re $b, -$Re$(a+c), -$Re$(b+c)]$

	$f(t)$	$g(p) = \int_0^\infty f(t) e^{-pt} dt$
5.22	$(1-e^{-t})^{\nu-1}(1-ze^{-t})^{-\mu}$ $\operatorname{Re}\nu>0, \ \lvert\arg(1-z)\rvert<\pi$	$B(p,\nu)$ $\cdot {}_2F_1(\mu,p;p+\nu;z) \qquad \operatorname{Re} p > 0$
5.23	$e^t(e^t-1)^{\nu-1}(1-z+ze^{-t})^{-\mu}$	$\Gamma(\nu)\Gamma(p-\nu)[\Gamma(p)]^{-1}$ $\cdot {}_2F_1(\mu,\nu;p;z) \qquad \operatorname{Re} p > \operatorname{Re}\nu$
5.24	$(1-e^{-t})^{-1}(1-e^{-at})(1-e^{-bt})$	$\psi(p+a)+\psi(p+b)-\psi(p)$ $-\psi(p+a+b)$ $\operatorname{Re} p>\operatorname{Max}[0,-\operatorname{Re} a,-\operatorname{Re} b,$ $-\operatorname{Re}(a+b)]$
5.25	$t^{-1}(1-e^{-t})^{-1}(1-e^{-at})(1-e^{-bt})$	$\log\Gamma(p)+\log\Gamma(p+a+b)$ $-\log\Gamma(p+a)-\log\Gamma(p+b)$ $\operatorname{Re} p>\operatorname{Max}[0,-\operatorname{Re} a,-\operatorname{Re} b,$ $-\operatorname{Re}(a+b)]$
5.26	$t^{-1}(1-e^{-t})^{-1}(1-e^{-at})(1-e^{-bt})$ $\cdot(1-e^{-ct})$	$\log\Gamma(p)+\log\Gamma(p+b+c)$ $+\log\Gamma(p+a+c)+\log\Gamma(p+a+b)$ $-\log\Gamma(p+a)-\log\Gamma(p+b)$ $-\log\Gamma(p+c)-\log\Gamma(p+a+b+c)$ $\operatorname{Re} p>\operatorname{Max}[0,-\operatorname{Re} a,-\operatorname{Re} b,-\operatorname{Re} c,$ $-\operatorname{Re}(a+b)-\operatorname{Re}(a+c)-\operatorname{Re}(b+c)]$

	$f(t)$		$g(p) = \int_0^\infty f(t)e^{-pt}dt$	
5.27	$e^{-a/t}$	Re $a \geqq 0$	$2(a/p)^{\frac{1}{2}}K_1[2(ap)^{\frac{1}{2}}]$	Re $p > 0$
5.28	$t^{-\frac{1}{2}}e^{-a/t}$	Re $a \geqq 0$	$(\pi/p)^{\frac{1}{2}}\exp[-2(ap)^{\frac{1}{2}}]$	Re $p > 0$
5.29	$t^{-1}e^{-a/t}$	$a \neq 0$, Re $a \geqq 0$	$2K_0[2(ap)^{\frac{1}{2}}]$	Re $p > 0$
5.30	$t^{-3/2}e^{-a/t}$	Re $a > 0$	$(\pi/a)^{\frac{1}{2}}\exp[-2(ap)^{\frac{1}{2}}]$	Re $p > 0$
5.31	$t^{\frac{1}{2}}e^{-a/t}$	Re $a \geqq 0$	$\frac{1}{2}\pi^{\frac{1}{2}}p^{-3/2}\exp[-2(ap)^{\frac{1}{2}}] \cdot [1+2(ap)^{\frac{1}{2}}]$	Re $p > 0$
5.32	$t^{-\frac{1}{2}}(e^{-a/t}-1)$	Re $a \geqq 0$	$(\pi/p)^{\frac{1}{2}}\{\exp[-2(ap)^{\frac{1}{2}}]-1\}$	Re $p \geqq 0$
5.33	$t^{-\frac{1}{2}} \frac{d^n}{da^n} e^{-a/t}$	Re $a > 0$	$(-1)^n\pi^{\frac{1}{2}}p^{\frac{1}{2}n-\frac{1}{2}}\exp[-2(ap)^{\frac{1}{2}}]$	Re $p > 0$
5.34	$t^{\nu-1}e^{-a/t}$	Re $a > 0$	$2(a/p)^{\frac{1}{2}\nu}K_\nu[2(ap)^{\frac{1}{2}}]$	Re $p > 0$

	$f(t)$	$g(p) = \int_0^\infty f(t)e^{-pt}dt$
5.35	$t^{-\frac{1}{2}}(at+2bc)^{-1}$ $\cdot\exp[-t^{-1}(b+c)^2]$	$\pi(2abc)^{-\frac{1}{2}}$ $\cdot\exp[\frac{1}{2}a(b+c)^2/(bc)]$ $\cdot\exp(2bcp/a)$ $\cdot\mathrm{Erfc}[(\frac{1}{2}ac/b)^{\frac{1}{2}}+(\frac{1}{2}ab/c)^{\frac{1}{2}}+(2bcp/a)^{\frac{1}{2}}]$ Re $p > 0$
5.36	$e^{-at^{\frac{1}{2}}}$ Re $a > 0$	$p^{-1}-\frac{1}{2}\pi^{\frac{1}{2}}ap^{-3/2}\exp(\frac{1}{4}a^2/p)$ $\cdot\mathrm{Erfc}(\frac{1}{2}ap^{-\frac{1}{2}})$ Re $p > 0$
5.37	$t^{\frac{1}{2}}e^{-at^{\frac{1}{2}}}$ Re $a > 0$	$-\frac{1}{2}ap^{-2}+\frac{1}{4}\pi^{\frac{1}{2}}p^{-5/2}(a^2+2p)$ $\cdot\exp(\frac{1}{4}a^2/p)\mathrm{Erfc}(\frac{1}{2}ap^{-\frac{1}{2}})$ Re $p > 0$
5.38	$t^{-\frac{1}{2}}e^{-at^{\frac{1}{2}}}$ Re $a > 0$	$(\pi/p)^{\frac{1}{2}}\exp(\frac{1}{4}a^2/p)\mathrm{Erfc}(\frac{1}{2}ap^{-\frac{1}{2}})$ Re $p > 0$
5.39	$t^{-3/4}e^{-at^{\frac{1}{2}}}$ Re $a > 0$	$2^{-3/4}ap^{-\frac{1}{2}}\exp(1/8\,a^2/p)$ $\cdot K_{\frac{1}{4}}(1/8\,a^2/p)$ Re $p > 0$
5.40	$t^{\nu-1}e^{-at^{\frac{1}{2}}}$ Re $a>0$, Re $\nu>0$	$2(2p)^{-\nu}\Gamma(2\nu)\exp(1/8\,a^2/p)$ $\cdot D_{-2\nu}[(\frac{1}{2}a^2/p)^{\frac{1}{2}}]$ Re $p > 0$

	$f(t)$	$g(p) = \int_0^\infty f(t)e^{-pt}dt$
5.41	e^{-at^2} Re $a > 0$	$\frac{1}{2}(\pi/a)^{\frac{1}{2}}\exp(\frac{1}{4}p^2/a)$ $\cdot\mathrm{Erfc}(\frac{1}{2}pa^{-\frac{1}{2}})$
5.42	0 $t < b$ e^{-at^2} $t > b$ Re $a > 0$	$\frac{1}{2}(\pi/a)^{\frac{1}{2}}\exp(\frac{1}{4}p^2/a)$ $\cdot\mathrm{Erfc}(\frac{1}{2}pa^{-\frac{1}{2}}+ba^{\frac{1}{2}})$
5.43	$t\ e^{-at^2}$ Re $a > 0$	$\frac{1}{2}a^{-1}[1-\frac{1}{2}(\pi/a)^{\frac{1}{2}}p$ $\cdot\exp(\frac{1}{4}p^2/a)\mathrm{Erfc}(\frac{1}{2}pa^{-\frac{1}{2}})$
5.44	$t^{-\frac{1}{2}}e^{-at^2}$ Re $a > 0$	$\frac{1}{2}(p/a)^{\frac{1}{2}}\exp(\frac{1}{8}p^2/a)$ $\cdot K_{\frac{1}{4}}(\frac{1}{8}p^2/a)$
5.45	$t^{\frac{1}{2}}e^{-at^2}$ Re $a > 0$	$\frac{1}{8}(p/a)^{\frac{3}{2}}\ \exp(\frac{1}{8}p^2/a)$ $\cdot[K_{\frac{3}{4}}(\frac{1}{8}p^2(a)-K_{\frac{1}{4}}(\frac{1}{8}p^2/a)]$
5.46	$t^{\nu-1}e^{-at^2}$ Re $\nu > 0$, Re $a > 0$	$(2a)^{-\frac{1}{2}\nu}\Gamma(\nu)\exp(\frac{1}{8}p^2/a)$ $\cdot D_{-\nu}[(\frac{1}{2}p^2/a)^{\frac{1}{2}}]$
5.47	$e^{-a^3t^3}$	$2p^{-1}(\frac{p}{3a})^{\frac{3}{2}}S_{o,\frac{1}{3}}[2(\frac{p}{3a})^{\frac{3}{2}}]$

	$f(t)$	$g(p) = \int_0^\infty f(t)e^{-pt}dt$
5.48	$t^{-\frac{1}{2}}(a^2+t^2)^{-\frac{1}{2}}$ $\cdot\exp[-b(a^2+t^2)^{\frac{1}{2}}]$ Re $b>0$	$(p/\pi)^{\frac{1}{2}}K_{\frac{1}{4}}(\frac{1}{2}z_1)K_{\frac{1}{4}}(\frac{1}{2}z_2)$ $z_{\frac{1}{2}}=a[b\pm(b^2-p^2)^{\frac{1}{2}}]$ Re $(p+b) > 0$
5.49	$(a^2+t^2)^{-\frac{1}{2}}[(a^2+t^2)^{\frac{1}{2}}+a]^{\frac{1}{2}}$ $\cdot\exp[-b(a^2+t^2)^{\frac{1}{2}}]$ Re $b>0$	$(\frac{1}{2}\pi)^{\frac{1}{2}}p(b^2-p^2)^{-\frac{1}{2}}$ $\cdot\{z_2^{-\frac{1}{2}}\exp[-a(b^2-p^2)^{\frac{1}{2}}]$ $\cdot\mathrm{Erfc}[(z_2)^{\frac{1}{2}}]$ $-z_1^{-\frac{1}{2}}\exp[a(b^2-p^2)^{\frac{1}{2}}]$ $\cdot\mathrm{Erfc}[(z_1)^{\frac{1}{2}}]\}$ $z_{\frac{1}{2}}=a[b\pm(b^2-p^2)^{\frac{1}{2}}]$ Re $(p+b) > 0$
5.50	$(a^2+t^2)^{-\frac{1}{2}}[(a^2+t^2)^{\frac{1}{2}}+a]^{-\frac{1}{2}}$ $\cdot\exp[-b(a^2+t^2)^{\frac{1}{2}}]$ Re $b>0$	$\pi(2a)^{-\frac{1}{2}}e^{ab}$ $\cdot\mathrm{Erfc}(z_1^{\frac{1}{2}})\mathrm{Erfc}(z_2^{\frac{1}{2}})$ $z_{\frac{1}{2}}=a[b\pm(b^2-p^2)^{\frac{1}{2}}]$ Re $(p+b) > 0$
5.51	$t^{-2k-\frac{1}{2}}(a^2+t^2)^{-\frac{1}{2}}[(a^2+t^2)^{\frac{1}{2}}+a]^{2k}$ $\cdot\exp[-b(a^2+t^2)^{\frac{1}{2}}]$ Re $b>0$, Re $k<\frac{1}{4}$	$(\frac{1}{2}a)^{-\frac{1}{2}}\Gamma(\frac{1}{2}-2k)$ $D_{2k-\frac{1}{2}}[(2z_1)^{\frac{1}{2}}]D_{2k-\frac{1}{2}}[(2z_2)^{\frac{1}{2}}]$ $z_{\frac{1}{2}}=a[b\pm(b^2-p^2)^{\frac{1}{2}}]$ Re $(p+b) > 0$

	$f(t)$	$g(p) = \int_0^\infty f(t)e^{-pt}dt$
5.52	$0 \quad t < a$ $(t-a)^{-\frac{1}{2}}\exp[-b(t^2-a^2)^{\frac{1}{2}}]$ $t > a$ $\operatorname{Re} b > 0$	$(\frac{1}{2}\pi)^{\frac{1}{2}}b(p^2-b^2)^{-\frac{1}{2}}$ $\cdot\{y_2^{-\frac{1}{2}}\exp[-a(p^2-b^2)^{\frac{1}{2}}]$ $\cdot\operatorname{Erfc}(z_2^{\frac{1}{2}})$ $-y_1^{-\frac{1}{2}}\exp[a(p^2-b^2)^{\frac{1}{2}}]$ $\cdot\operatorname{Erfc}(y_1^{\frac{1}{2}})\}$ $y_{\frac{1}{2}} = a[p\pm(p^2-b^2)^{\frac{1}{2}}]$ $\operatorname{Re}(p+b) > 0$
5.53	$0 \quad t < a$ $(t^2-a^2)^{-3/4}\exp[-b(t^2-a^2)^{\frac{1}{2}}]$ $t > a$ $\operatorname{Re} b > 0$	$(b/\pi)^{\frac{1}{2}}K_{\frac{1}{4}}(\frac{1}{2}y_1)K_{\frac{1}{4}}(\frac{1}{2}y_2)$ $y_{\frac{1}{2}} = a[p\pm(p^2-b^2)^{\frac{1}{2}}]$ $\operatorname{Re}(p+b) > 0$
5.54	$0 \quad t < a$ $(t-a)^{-\frac{1}{2}}(t+a)^{-1}$ $\exp[-b(t^2-a^2)^{\frac{1}{2}}]$ $t > a$ $\operatorname{Re} b > 0$	$\pi(2a)^{-\frac{1}{2}}e^{ap}$ $\cdot\operatorname{Erfc}(y_1^{\frac{1}{2}})\operatorname{Erfc}(y_2^{\frac{1}{2}})$ $y_{\frac{1}{2}} = a[p\pm(p^2-b^2)^{\frac{1}{2}}]$ $\operatorname{Re}(p+b) > 0$

	$f(t)$	$g(p) = \int_0^\infty f(t)e^{-pt}dt$
5.55	$0 \qquad t < a$ $(t^2-a^2)^{-k-3/4}(t+a)^{2k}$ $\cdot\exp[-b(t^2-a^2)^{\frac{1}{2}}] \qquad t > a$ $\operatorname{Re} k<1/4,\ \operatorname{Re} b>0$	$(\frac{1}{2}a)^{-\frac{1}{2}}\Gamma(\frac{1}{2}-2k)$ $\cdot D_{2k-\frac{1}{2}}[(2y_1)^{\frac{1}{2}}]D_{2k-\frac{1}{2}}[(2y_2)^{\frac{1}{2}}]$ $y_{\frac{1}{2}} = a[p\pm(p^2-b^2)^{\frac{1}{2}}]$
5.56	$\exp(-ae^{-t})$ $\exp(-ae^{t}) \qquad \operatorname{Re} a>0$	$a^{p}\gamma(p,a) \qquad \operatorname{Re} p > 0$ $a^{-p}\Gamma(-p,a)$
5.57	$(1-e^{-t})^{\nu-1}\exp(ae^{-t})$ $\operatorname{Re} \nu>0$	$B(\nu,p)a^{-\frac{1}{2}\nu-\frac{1}{2}p}e^{\frac{1}{2}a}$ $\cdot M_{\frac{1}{2}\nu-\frac{1}{2}p,\frac{1}{2}\nu+\frac{1}{2}p}(a)$ $\operatorname{Re} p > 0$
5.58	$(1-e^{-t})^{\nu-1}\exp(-ae^{t})$ $\operatorname{Re} \nu>0,\ \operatorname{Re} a>0$	$\Gamma(\nu)a^{\frac{1}{2}p-\frac{1}{2}}e^{-\frac{1}{2}a}$ $\cdot W_{\frac{1}{2}-\frac{1}{2}p-\nu,-\frac{1}{2}p}(a)$ $\operatorname{Re} p>\operatorname{Re} \nu$
5.59	$\{[z+(1-e^{-t})^{\frac{1}{2}}]^{\nu}+[z-(1-e^{-t})^{\frac{1}{2}}]^{\nu}\}$ $\cdot(1-e^{-t})^{-\frac{1}{2}}$	$2^{p+1}e^{(p+\nu)\pi i}\Gamma(p)[\Gamma(-\nu)]^{-1}$ $\cdot(z^2-1)^{\frac{1}{2}p+\frac{1}{2}\nu}q_{p-1}^{-\nu-p}(z)$ $= 2^{\frac{1}{2}+p}\pi^{\frac{1}{2}}\Gamma(p)(z^2-1)^{\frac{1}{2}(p+\nu-\frac{1}{2})}$ $\cdot P_{\nu+p-\frac{1}{2}}^{\frac{1}{2}-p}[z(z^2-1)^{-\frac{1}{2}}]$ $\operatorname{Re} p > 0$

	$f(t)$	$g(p) = \int_0^\infty f(t)e^{-pt}dt$
5.60	$0 \qquad t < a$ $(e^{-a}-e^{-t})^{\nu}(1-e^{-2t})^{\frac{1}{2}\nu-\frac{1}{2}}$ $t > a$ $\mathrm{Re}\,\nu > -1$	$2^{\frac{1}{2}\nu+\frac{1}{2}p}\Gamma(1+\frac{1}{2}\nu+\frac{1}{2}p)B(\nu+1,p)$ $\cdot e^{-\frac{1}{2}a(\nu+p)}P_{\frac{1}{2}-\frac{1}{2}p}^{-\frac{1}{2}p-\frac{1}{2}\nu}[(1-e^{-2a})^{\frac{1}{2}}]$ $\mathrm{Re}\,p > 0$
5.61	$(e^{t}-1)^{\nu-1}\exp[-a(e^{t}-1)^{-1}]$ $\mathrm{Re}\,a > 0$	$\Gamma(1+p-\nu)a^{\frac{1}{2}\nu-\frac{1}{2}}e^{\frac{1}{2}a}$ $\cdot W_{\frac{1}{2}\nu-\frac{1}{2}-p,\frac{1}{2}\nu}(a)$ $\mathrm{Re}(p-\nu) > -1$
5.62	$0 \qquad t < a$ $z^{-1}[(t+z)^{\nu}e^{bz}$ $+(t-z)^{\nu}e^{-bz}] \qquad t > a$ $z = (t^2-a^2)^{\frac{1}{2}}$	$2a^{\nu}(p+b)^{\frac{1}{2}\nu}(p-b)^{-\frac{1}{2}\nu}$ $\cdot K_{\nu}[a(p^2-b^2)^{\frac{1}{2}}]$
5.63	$\exp(-a \sinh t)$	$\pi \csc(\pi p)[\mathbf{J}_p(a)-J_p(a)]$

1.6 Logarithmic Functions

	$f(t)$	$g(p) = \int_0^\infty f(t)e^{-pt}dt$
6.1	$\log t$	$-p^{-1}(\gamma+\log p)$ Re $p > 0$
6.2	$(\log t)^2$	$p^{-1}[\frac{\pi^2}{6} + (\gamma+\log p)^2]$ Re $p > 0$
6.3	0 $t < a$ $\log(t/a)$ $t > a$	$-p^{-1}\,\mathrm{Ei}(-ap)$ Re $p > 0$
6.4	0 $t < a$ $\log t$ $t > a$	$p^{-1}[e^{-ap}\log a-\mathrm{Ei}(-ap)]$ Re $p > 0$
6.5	$\log(b+t)$ $t < a$ 0 $t > a$	$p^{-1}[\log b-e^{-ap}\log(a+b)$ $-e^{bp}\mathrm{Ei}(-bp)+e^{bp}\mathrm{Ei}(-ap-bp)]$ Re $p > 0$
6.6	$\log(a+t)$ $\lvert\arg a\rvert < \pi$	$p^{-1}[\log a-e^{ap}\mathrm{Ei}(-ap)]$ Re $p > 0$
6.7	$\log(a+bt)$ $\lvert\arg(b/a)\rvert < \pi$	$p^{-1}[\log a-e^{ap/b}\mathrm{Ei}(-ap/b)]$ Re $p > 0$
6.8	$\log(b-t)$ $t < a$ 0 $t > a$ $b > a$	$p^{-1}\{\log b-e^{-ap}\log(b-a)$ $+ e^{-bp}[\overline{\mathrm{Ei}}(bp-ap)-\overline{\mathrm{Ei}}(bp)]\}$

	$f(t)$	$g(p) = \int_0^\infty f(t)e^{-pt}dt$
6.9	$t(1-\log t)$	$p^{-2}(\gamma+\log p)$ Re $p > 0$
6.10	$\log\|a-t\|$ $a > 0$	$p^{-1}[\log a-e^{-ap}\overline{Ei}(ap)]$ Re $p > 0$
6.11	$\log(a^2+t^2)$ Re $a > 0$	$2p^{-1}[\log a-Ci(ap)\cos(ap)$ $-\ si(ap)\sin(ap)]$ Re $p > 0$
6.12	$\log(t^2-a^2)$ $\|\text{Im } a\| > 0$	$p^{-1}[2\ \log a-e^{ap}Ei(-ap)$ $-\ e^{-ap}\overline{Ei}(ap)]$ Re $p > 0$
6.13	$\log\|t^2-a^2\|$ $a > 0$	$p^{-1}[2\log a-e^{ap}Ei(-ap)$ $-\ e^{ap}\overline{Ei}(ap)]$ Re $p > 0$
6.14	$\log(b^2-t^2)$ $t < a$ 0 $t > a$ $b > a$	$p^{-1}\{2\log b-e^{-ap}\log(b^2-a^2)$ $-\ e^{-bp}[\overline{Ei}(bp)-\overline{Ei}(bp-ap)]$ $-\ e^{-bp}[Ei(-bp)-Ei(-bp-ap)]\}$ Re $p > 0$
6.15	$t^{-1}\log(1+a^2t^2)$	$[Ci(p/a)]^2 + [si(p/a)]^2$ Re $p > 0$

	$f(t)$	$g(p) = \int_0^\infty f(t)e^{-pt}dt$
6.16	$t^{-1}\log\|1-a^2t^2\|$	$\overline{Ei}(p/a)Ei(-p/a)$ Re $p > 0$
6.17	0 $t < a$ $(t+a)^{-1}\log(t/a)$ $t > a$	$\tfrac{1}{2}e^{ap}[Ei(-ap)]^2$ Re $p > 0$
6.18	0 $t < a$ $(t+a)^{-1}\log t$ $t > a$	$e^{ap}\{\tfrac{1}{2}[Ei(-ap)]^2-\log a\ Ei(-2ap)\}$ Re $p > 0$
6.19	0 $t < a+b$ $t^{-1}\log[(t-a)(t-b)/ab]$ $t > a+b$	$Ei(-ap)Ei(-bp)$ Re $p > 0$
6.20	0 $t < a$ $\log\{(2a)^{-\frac{1}{2}}[(t+a)^{\frac{1}{2}}+(t-a)^{\frac{1}{2}}]\}$ $t > a$	$(2p)^{-1}K_0(ap)$ Re $p > 0$
6.21	0 $t < a$ $(t^2-a^2)^{-\frac{1}{2}}\log(t^2-a^2)$ $t > a$	$-[\gamma+\log(2p/a)]K_0(ap)$ Re $p > 0$

	$f(t)$	$g(p) = \int_0^\infty f(t)e^{-pt}dt$
.22	$0 \quad t < a$ $\log[t+(t^2-a^2)^{1/2}] \quad t > a$	$p^{-1}[K_0(ap)+e^{-ap}\log a]$ $\mathrm{Re}\, p > 0$
.23	$\log[t+(t^2+a^2)^{1/2}]$	$p^{-1}\log a + \frac{\pi}{2}p^{-1}[\mathbf{H}_0(ap)-Y_0(ap)]$ $\mathrm{Re}\, p > 0$
.24	$t^{-1}\log[t+(t^2+1)^{1/2}]$	$\frac{1}{2}\pi \int_p^\infty x^{-1}[\mathbf{H}_0(x)-Y_0(x)]dx$ $\mathrm{Re}\, p > 0$
.25	$(t^2+1)^{-1/2}\log[t+(t^2+1)^{1/2}]$	$S_{-1,0}(p) \quad \mathrm{Re}\, p > 0$
.26	$(t^2+a^2)^{-1/2}\log[t+(t^2+a^2)^{1/2}]$	$\frac{1}{2}\pi\log a[\mathbf{H}_0(ap)-Y_0(ap)]$ $- S_{-1,0}(ap) \quad \mathrm{Re}\, p > 0$
.27	$\log\{a^{-1/2}[t+(t^2+a^2)^{1/2}]^{1/2}\}$	$\frac{1}{4}\pi p^{-1}[\mathbf{H}_0(ap)-Y_0(ap)]$ $\mathrm{Re}\, p > 0$
.28	$(t+a+b)^{-1}\log[(t+a)(t+b)]$	$e^{(a+b)p}[\mathrm{Ei}(-ap)\mathrm{Ei}(-bp)$ $- \log(ab)\mathrm{Ei}(-ap-bp)]$ $\mathrm{Re}\, p > 0$

	$f(t)$	$g(p) = \int_0^\infty f(t)e^{-pt}dt$
6.29	$(t+a+b)^{-1}\log[(t+a)(t+b)/ab]$	$e^{(a+b)p}\mathrm{Ei}(-ap)\mathrm{Ei}(-bp)$ $\mathrm{Re}\, p > 0$
6.30	$\log\{a^{-\frac{1}{2}}[t^{\frac{1}{2}}+(t+a)^{\frac{1}{2}}]\}$	$\frac{1}{2}p^{-1}e^{\frac{1}{2}ap}K_0(\frac{1}{2}ap)$ $\mathrm{Re}\, p > 0$
6.31	$t^{-1}e^{-a/t}\log t$ $\mathrm{Re}\, a > 0$	$\log(a/p)K_0[2(ap)^{\frac{1}{2}}]$ $\mathrm{Re}\, p > 0$
6.32	$t^{-\frac{1}{2}}e^{-a/t}\log t$ $\mathrm{Re}\, a > 0$	$(\pi/p)^{\frac{1}{2}}\{\frac{1}{2}\exp[-2(ap)^{\frac{1}{2}}]\log(a/p)$ $- \exp[2(ap)^{\frac{1}{2}}]\mathrm{Ei}[-4(ap)^{\frac{1}{2}}]\}$ $\mathrm{Re}\, p > 0$
6.33	$t^{-3/2}e^{-a/t}\log t$ $\mathrm{Re}\, a > 0$	$(\pi/a)^{\frac{1}{2}}\{\frac{1}{2}\exp[-2(ap)^{\frac{1}{2}}]\log(a/p)$ $+ \exp[2(ap)^{\frac{1}{2}}]\mathrm{Ei}[-4(ap)^{\frac{1}{2}}]\}$ $\mathrm{Re}\, p > 0$
6.34	$t^{-\frac{1}{2}}\log t$	$-(\pi/p)^{\frac{1}{2}}[\gamma+\log(4p)]$ $\mathrm{Re}\, p > 0$
6.35	$t^n \log t$ $n = 1,2,3,\cdots$	$n!p^{-n-1}[\psi(n+1)-\log p]$ $= n!p^{-n-1}(1+\frac{1}{2}+\cdots+\frac{1}{n} -\gamma - \log$ $\mathrm{Re}\, p > 0$

	$f(t)$	$g(p) = \int_0^\infty f(t)e^{-pt}dt$
6.36	$t^{n-\frac{1}{2}} \log t$ $n = 1,2,3,\cdots$	$\Gamma(n+\frac{1}{2})p^{-n-\frac{1}{2}}[\psi(n+\frac{1}{2})-\log p]$ $= \pi^{\frac{1}{2}}2^{-2n}p^{-n-\frac{1}{2}}(2n)!/n!$ $\cdot\ [2(1+\frac{1}{3}+\cdots+\frac{1}{2n-1})-\gamma-\log(4p)]$ $\mathrm{Re}\ p > 0$
6.37	$t^{\nu-1} \log t$ $\mathrm{Re}\ \nu > 0$	$\Gamma(\nu)p^{-\nu}[\psi(\nu)-\log p]$ $\quad \mathrm{Re}\ p > 0$
6.38	$(\log t)^2$	$p^{-1}[\frac{\pi^2}{6} + (\gamma+\log p)^2]$ $\quad \mathrm{Re}\ p > 0$
6.39	$t^{\nu-1} (\log t)^2$ $\mathrm{Re}\ \nu > 0$	$\Gamma(\nu)p^{-\nu}\{[\psi(\nu)-\log p]^2 + \psi'(\nu)\}$ $\mathrm{Re}\ p > 0$
6.40	$\log(1+e^{-at})$	$p^{-1}[\log 2-\frac{1}{2}\psi(1+\frac{1}{2}p/a)$ $+ \frac{1}{2}\psi(\frac{1}{2}+\frac{1}{2}p/a)]$ $\quad \mathrm{Re}\ p > 0$
.41	$\log(1-e^{-at})$	$p^{-1}[\gamma+\psi(1+p/a)]$ $\quad \mathrm{Re}\ p > 0$
.42	$\log(e^{at}-1)$	$-\ p^{-1}[\gamma+\psi(p/a)]$ $\quad \mathrm{Re}\ p > 0$
.43	$\log[at(e^{at}-1)^{-1}]$	$p^{-1}[\psi(p/a)-\log(p/a)]$ $\quad \mathrm{Re}\ p > 0$

1.7 Trigonometric Functions

	$f(t)$	$g(p) = \int_0^\infty f(t)e^{-pt}dt$
7.1	$\sin(at)$	$a(p^2+a^2)^{-1}$ Re $p > \lvert \mathrm{Im}\, a \rvert$
7.2	$\cos(at)$	$p(p^2+a^2)^{-1}$ Re $p > \lvert \mathrm{Im}\, a \rvert$
7.3	$\lvert \sin(at) \rvert$	$a(p^2+a^2)^{-1}\coth(\tfrac{1}{2}\pi p/a)$ Re $p > \lvert \mathrm{Im}\, a \rvert$
7.4	$\lvert \cos(at) \rvert$	$(p^2+a^2)^{-1}[p+a\,\mathrm{csch}(\tfrac{1}{2}\pi p/a)]$ Re $p > \lvert \mathrm{Im}\, a \rvert$
7.5	$t^{-1}\sin(at)$	$\arctan(a/p)$ Re $p > \lvert \mathrm{Im}\, a \rvert$
7.6	$t^{-1}(1-\cos at)$	$\tfrac{1}{2}\log(1+a^2/p^2)$ Re $p > \lvert \mathrm{Im}\, a \rvert$
7.7	$t^{-1}(\cos at - \cos bt)$	$\tfrac{1}{2}\log[(p^2+b^2)(p^2+a^2)^{-1}]$ Re $p > \mathrm{Max}[\mathrm{Im}\, a, \mathrm{Im}\, b]$
7.8	$t^{-2}(t-\sin t)$	$\tfrac{1}{2}\log(1+p^{-2}) + p\,\mathrm{arccot}\, p - 1$ Re $p > \lvert \mathrm{Im}\, a \rvert$
7.9	$t^{-2}(1-\cos at)$	$a\arctan(a/p) - \tfrac{1}{2}p\log(1+a^2/p^2)$ Re $p > \lvert \mathrm{Im}\, a \rvert$

	$f(t)$	$g(p) = \int_0^\infty f(t)e^{-pt}dt$
7.10	$t^{-2}(\cos at - \cos bt)$	$\frac{1}{2}p \log[(p^2+a^2)(p^2+b^2)^{-1}]$ $+ b\arctan(b/p) - a\arctan(a/p)$ $\operatorname{Re} p > \lvert\operatorname{Im} a\rvert$
7.11	$t^{-3}(t - \sin t)$	$\frac{\pi}{4} - \frac{1}{2}\arctan p + \frac{1}{2}p$ $- \frac{1}{2}p \log(1+p^{-2}) - \frac{1}{2}p^2 \operatorname{arccot} p$ $\operatorname{Re} p > \lvert\operatorname{Im} a\rvert$
7.12	$t^{-\frac{1}{2}}\sin(at)$	$(\frac{1}{2}\pi)^{\frac{1}{2}}(p^2+a^2)^{-\frac{1}{2}}[(p^2+a^2)^{\frac{1}{2}}-p]^{\frac{1}{2}}$ $\operatorname{Re} p > \lvert\operatorname{Im} a\rvert$
7.13	$t^{-\frac{1}{2}}\cos(at)$	$(\frac{1}{2}\pi)^{\frac{1}{2}}(p^2+a^2)^{-\frac{1}{2}}[(p^2+a^2)^{\frac{1}{2}}+p]^{\frac{1}{2}}$ $\operatorname{Re} p > \lvert\operatorname{Im} a\rvert$
7.14	$t^{-3/2}\sin(at)$	$(2\pi)^{\frac{1}{2}}[(p^2+a^2)^{\frac{1}{2}}-p]^{\frac{1}{2}}$ $\operatorname{Re} p > \lvert\operatorname{Im} a\rvert$
7.15	$t^{\nu-1}\sin(at)$ $\operatorname{Re} \nu > -1$	$\Gamma(\nu)(p^2+a^2)^{-\frac{1}{2}\nu}$ $\cdot \sin[\nu \arctan(a/p)]$ $\operatorname{Re} p > \lvert\operatorname{Im} a\rvert$
7.16	$t^{\nu-1}\cos(at)$ $\operatorname{Re} \nu > 0$	$\Gamma(\nu)(p^2+a^2)^{-\frac{1}{2}\nu}$ $\cdot \cos[\nu \arctan(a/p)]$ $\operatorname{Re} p > \lvert\operatorname{Im} a\rvert$

	$f(t)$	$g(p) = \int_0^\infty f(t)e^{-pt}dt$
7.17	$t^{-1}(e^{at}-1)\sin(bt)$	$\operatorname{arccot}[(p^2-ap+b)/(ab)]$ $\operatorname{Re} p > \operatorname{Max}[\lvert\operatorname{Im} b\rvert, \lvert\operatorname{Im} b\rvert+\operatorname{Re} a]$
7.18	$t^{-1}\cos(ct)(e^{-bt}-e^{-at})$	$\tfrac{1}{2}\log\{[(p+a)^2+c^2][(p+b)^2+c^2]^{-1}\}$ $\operatorname{Re} p > \operatorname{Im} c - \operatorname{Min}(\operatorname{Re} a, \operatorname{Re} b)$
7.19	$(e^t-1)^{-1}\sin(at)$	$\tfrac{1}{2}i[\psi(p+1-ia)-\psi(p+1+ia)]$ $\operatorname{Re} s > -1 + \lvert\operatorname{Im} a\rvert$
7.20	$(1-e^{-t})^{-1}\sin(at)$	$\tfrac{1}{2}i[\psi(p-ia) - \psi(p+ia)]$ $\operatorname{Re} p > \lvert\operatorname{Im} a\rvert$
7.21	$(1-e^{-bt})^{\nu-1}\sin(at)$ $\operatorname{Re} \nu > -1,\ b > 0$	$\tfrac{1}{2}ib^{-1}[B(\nu, \frac{p+ia}{b}) - B(\nu, \frac{p-ia}{b})]$ $\operatorname{Re} p > \lvert\operatorname{Im} a\rvert$
7.22	$(1-a^{-bt})^{\nu-1}\cos(at)$ $\operatorname{Re} \nu > 0,\ b > 0$	$\tfrac{1}{2}b^{-1}[B/\nu, \frac{p+ia}{b}) + B(\nu, \frac{p-ia}{b})]$ $\operatorname{Re} p > \lvert\operatorname{Im} a\rvert$
7.23	$t^{-\frac{1}{2}}e^{-a^2/t}\sin(bt)$	$\pi^{\frac{1}{2}}(p^2+b^2)^{-\frac{1}{2}}e^{-2au}$ $\cdot\ [u\sin(2av) + v\cos(2av)]$ $\begin{matrix}u\\v\end{matrix} = 2^{-\frac{1}{2}}[(p^2+b^2)^{\frac{1}{2}} \pm p]^{\frac{1}{2}}$ $\operatorname{Re} p > 0$

	$f(t)$	$g(p) = \int_0^\infty f(t)e^{-pt}dt$
7.24	$t^{-\frac{1}{2}}e^{-a^2/t}\cos(bt)$	$\pi^{\frac{1}{2}}(p^2+b^2)^{-\frac{1}{2}}e^{-2au}$ $\cdot\ [u\cos(2av)-v\sin(2av)]$ Re $p > 0$, (u,v as before)
7.25	$t^{-3/2}e^{-a^2/t}\sin(bt)$	$\pi^{\frac{1}{2}}a^{-1}e^{-2au}\sin(2av)$ Re $p > 0$, (u,v as before)
7.26	$t^{-3/2}e^{-a^2/t}\cos(bt)$	$\pi^{\frac{1}{2}}a^{-1}e^{-2au}\cos(2av)$ Re $p > 0$, (u,v as before)
7.27	$t^{\nu-1}e^{-a^2/t}\sin(bt)$	$i\ a^{\nu}\{(p+ib)^{-\frac{1}{2}\nu}K_{\nu}[2a(p+ib)^{\frac{1}{2}}]$ $-\ (p-ib)^{-\frac{1}{2}\nu}K_{\nu}[2a(p-ia)^{\frac{1}{2}}]$
7.28	$t^{\nu-1}e^{-a^2/t}\cos(bt)$	$a^{\nu}\{(p+ib)^{-\frac{1}{2}\nu}K_{\nu}[2a(p+ib)^{\frac{1}{2}}]$ $+\ (p-ib)^{-\frac{1}{2}\nu}K_{\nu}[2a(p-ib)^{\frac{1}{2}}]\}$
7.29	$e^{-at^2}\sin(bt)$	$-\frac{1}{4}i(\pi/a)^{\frac{1}{2}}\{\exp[(p-ib)^2/(4a)]$ $\cdot\ \mathrm{Erfc}[\frac{1}{2}a^{-\frac{1}{2}}(p-ib)]$ $-\ \exp[(p+ib)^2/4a]$ $\cdot\ \mathrm{Erfc}[\frac{1}{2}a^{-\frac{1}{2}}(p+ib)]\}$

	$f(t)$	$g(p) = \int_0^\infty f(t)e^{-pt}dt$
7.30	$e^{-at^2}\cos(bt)$	$\frac{1}{4}(\pi/a)^{\frac{1}{2}}\{\exp[(p-ib)^2/(4a)]$ $\cdot \mathrm{Erfc}[\frac{1}{2}a^{-\frac{1}{2}}(p-ib)]$ $+ \exp[(p+ib)^2/(4a)]$ $\cdot \mathrm{Erfc}[\frac{1}{2}a^{-\frac{1}{2}}(p+ib)]\}$
7.31	$t^{\nu-1}e^{-at^2}\sin(bt)$ $\mathrm{Re}\,\nu > -1$	$\frac{1}{2}i\Gamma(\nu)(2a)^{-\frac{1}{2}\nu}\exp[(p^2-b^2)/8a]$ $\cdot \{\exp(ibp/4a)D_{-\nu}[(2a)^{-\frac{1}{2}}(p+ib)]$ $- \exp(-ibp/4a)D_{-\nu}[(2a)^{-\frac{1}{2}}(p-ib)]\}$
7.32	$t^{\nu-1}e^{-at^2}\cos(bt)$ $\mathrm{Re}\,\nu > 0$	$\frac{1}{2}\Gamma(\nu)(2a)^{-\frac{1}{2}\nu}\exp[(p^2-b^2)/8a]$ $\cdot \{\exp(ibp/4a)D_{-\nu}[(2a)^{-\frac{1}{2}}(p+ib)]$ $+ \exp(-ibp/4a)D_{-\nu}[(2a)^{-\frac{1}{2}}(p-ib]\}$
7.33	$\log t \sin(at)$	$(p^2+a^2)^{-1}[p \arctan(a/p)-a\gamma$ $-\frac{1}{2}a \log(a^2+p^2)]$ $\mathrm{Re}\,p > \lvert \mathrm{Im} a\rvert$
7.34	$\log t \cos(at)$	$-(p^2+a^2)^{-1}[a \arctan(a/p) + p\gamma$ $+ \frac{1}{2}p \log(p^2+a^2)]$ $\mathrm{Re}\,p > \lvert \mathrm{Im} a\rvert$

	$f(t)$	$g(p) = \int_0^\infty f(t)e^{-pt}dt$
7.35	$t^{-1}\log t \sin(at)$	$-[\gamma+\frac{1}{2}\log(p^2+a^2)]\arctan(a/p)$ $\operatorname{Re} p > \lvert\operatorname{Im} a\rvert$
7.36	$t^{\nu-1}\log t \sin(at)$ $\operatorname{Re}\nu > -1$	$\Gamma(\nu)(p^2+a^2)^{-\frac{1}{2}\nu}\sin[\nu\arctan(a/p)]$ $\cdot\{\psi(\nu) - \frac{1}{2}\log(p^2+a^2)$ $+ \arctan(a/p)\cot[\nu\arctan(a/p)]\}$ $\operatorname{Re} p > \lvert\operatorname{Im} a\rvert$
7.37	$t^{\nu-1}\log t \cos(at)$ $\operatorname{Re}\nu > 0$	$\Gamma(\nu)(p^2+a^2)^{-\frac{1}{2}\nu}\cos[\nu\arctan(a/p)]$ $\cdot\{\psi(\nu) - \frac{1}{2}\log(p^2+a^2)$ $- \arctan(a/p)\tan[\nu\arctan(a/p)]\}$ $\operatorname{Re} p > \lvert\operatorname{Im} a\rvert$
7.38	$\sin^2(at)$	$2a^2p^{-1}(p^2+4a^2)^{-1}$ $\quad \operatorname{Re} p > 2\lvert\operatorname{Im} a\rvert$
7.39	$\cos^2(at)$	$p^{-1}(p^2+2a^2)(p^2+4a^2)^{-1}$ $\quad \operatorname{Re} p > 2\lvert\operatorname{Im} a\rvert$
7.40	$\sin^{2n}(at)$ $n = 1,2,3,\cdots$	$(2n)!a^{2n}p^{-1}(p^2+4a^2)^{-1}(p^2+16a^2)^{-1}\cdots(p^2+4n^2a^2)^{-1}$ $\operatorname{Re} p > 2n\lvert\operatorname{Im} a\rvert$

	$f(t)$	$g(p) = \int_0^\infty f(t)e^{-pt}dt$
7.41	$\cos^{2n}(at)$ $n = 1,2,3,\cdots$	$(2n)!a^{2n}p^{-1}(p^2+4a^2)^{-1}\cdots(p^2+4n^2a^2)^{-1}$ $\cdot \{1+ \frac{p^2}{2!a^2} + \frac{p^2(4a^2+p^2)}{4!a^4} +\cdots$ $+ \frac{p^2(p^2+4a^2)\cdots[p^2+4(n-1)^2a^2]}{(2n)!\ a^{2n}}\}$ $\mathrm{Re}\ p > 2n\lvert\mathrm{Im}a\rvert$
7.42	$\sin^{2n+1}(at)$ $n = 0,1,2,\cdots$	$(2n+1)!a^{2n+1}$ $\cdot (p^2+a^2)^{-1}(p^2+9a^2)^{-1}\cdots[p^2+(2n+1)^2a^2]^{-1}$ $\mathrm{Re}\ p > (2n+1)\lvert\mathrm{Im}a\rvert$
7.43	$\cos^{2n+1}(at)$ $n=0,1,2,\cdots$	$(2n+1)!a^{2n}p(p^2+a^2)^{-1}\cdots[p^2+(2n+1)^2a^2]^{-1}$ $\cdot\{1+\frac{p^2+a^2}{3!a^2}+\cdots+\frac{(p^2+a^2)(p^2+9a^2)\cdots[p^2+(2n+1)^2a^2]}{(2n+1)!a^{2n}}\}$ $\mathrm{Re}\ p > (2n+1\lvert\mathrm{Im}a\rvert$
7.44	$\sin(at)\sin(bt)$	$2abp[p^2+(a+b)^2]^{-1}[p^2+(a-b)^2]^{-1}$ $\mathrm{Re}\ p > \lvert\mathrm{Im}(\pm a\pm b)\rvert$
7.45	$\cos(at)\cos(bt)$	$p(p^2+a^2+b^2)$ $\cdot [p^2+(a+b)^2]^{-1}[p^2+(a-b)^2]^{-1}$ $\mathrm{Re}\ p > \lvert\mathrm{Im}(\pm a\pm b)\rvert$

	$f(t)$	$g(p) = \int_0^\infty f(t)e^{-pt}dt$
7.46	$\cos(at)\sin(bt)$	$b(p^2-a^2+b^2) \cdot [p^2+(a+b)^2]^{-1}[p^2+(a-b)^2]^{-1}$
7.47	$t^{-1}\sin(at)\sin(bt)$	$\frac{1}{4}\log\{[p^2+(a+b)^2][p^2+(a-b)^2]^{-1}\}$ $\mathrm{Re}\, p > \lvert \mathrm{Im}(\pm a \pm b)\rvert$
7.48	$\csc t \sin[(2n+1)t]$ $n=1,2,3,\cdots$	$\frac{1}{p} + \sum_{m=1}^{n} 2p(p^2+4m^2)$ $\mathrm{Re}\, p > 0$
7.49	$t^{-1}\sin^2(at)$	$\frac{1}{4}\log(1+4a^2/p^2)$ $\mathrm{Re}\, p > 2\lvert \mathrm{Im} a\rvert$
7.50	$t^{-1}\sin^3(at)$	$\frac{3}{4}\arctan(a/p) - \frac{1}{4}\arctan(3a/p)$ $\mathrm{Re}\, p > 3\lvert \mathrm{Im} a\rvert$
7.51	$t^{-1}\sin^4(at)$	$\frac{1}{8}\log[p^{-3}(p^2+4a^2)^2] - \frac{1}{16}\log(p^2+16a^2)$
7.52	$t^{-1}\sin(at)\cos(bt)$	$\frac{1}{2}\arctan[2ap(p^2-a^2+b^2)^{-1}]$ $\mathrm{Re}\, p > \mathrm{Max}[\lvert \mathrm{Im}(a+b)\rvert, \lvert \mathrm{Im}(a-b)\rvert]$

	$f(t)$	$g(p) = \int_0^\infty f(t)e^{-pt}dt$
7.53	$t^{-2}\sin(at)\sin(bt)$	$\frac{1}{2}a\arctan[2bp(p^2+a^2-b^2)^{-1}]$ $+\frac{1}{2}b\arctan[2ap(p^2-a^2+b^2)^{-1}]$ $+\frac{1}{4}p\log\{[p^2+(a-b)^2][p^2+(a+b)^2]^{-1}\}$ $\operatorname{Re} p \geqq \lvert\operatorname{Im}(\pm a\pm b)\rvert$
7.54	$\lvert\sin(ax)\rvert^{\nu}$ $\operatorname{Re}\nu > -1$	$\pi 2^{-\nu-1}a^{-1}\Gamma(1+\nu)\operatorname{csch}(\frac{1}{2}\pi p/a)$ $\cdot[\Gamma(1+\frac{1}{2}\nu+\frac{1}{2}ip/a)\Gamma(1+\frac{1}{2}\nu-\frac{1}{2}ip/a)]^{-1}$ $\operatorname{Re} p > \operatorname{Re}\nu\lvert\operatorname{Im} a\rvert$
7.55	$\sin(at)$ $\quad t < b$ 0 $\quad t > b$	$(p^2+a^2)^{-1}[a-ae^{-bp}\cos(ab)$ $-pe^{-bp}\sin(ab)]$
7.56	0 $\quad t < a$ $\sin(ct)$ $\quad a < t < b$ 0 $\quad t > b$	$(p^2+c^2)^{-1}$ $\cdot\{e^{-ap}[c\cos(ac)+p\sin(ac)]$ $-e^{-bp}[c\cos(bc)+p\sin(bc)]\}$
7.57	0 $\quad t < a$ $\cos(ct)$ $\quad a < t < b$ 0 $\quad t > b$	$(p^2+c^2)^{-1}$ $\cdot\{e^{-ap}[p\cos(ac)-c\sin(ac)]$ $-e^{-bp}[p\cos(bc)-c\sin(bc)]\}$

	$f(t)$	$g(p) = \int_0^\infty f(t)e^{-pt}dt$
7.58	$0 \quad t < a$ $\sin^2(ct) \quad a<t<b$ $0 \quad t < b$	$(p^2+4c^2)^{-1}\{e^{-ap}[2c^2p^{-1}+p\sin^2(ac)$ $+ c\sin(2ac)]-e^{-bp}[2c^2p^{-1}$ $+ p\sin^2(bc)+c\sin(2bc)]\}$
7.59	$0 \quad t < a$ $\cos^2(ct) \quad a<t<b$ $0 \quad t > b$	$(p^2+4c^2)^{-1}\{e^{-ap}[2c^2p^{-1}+p\cos^2(ac)$ $- c\sin(2ac)] - e^{-bp}[2c^2p^{-1}$ $+ p\cos^2(bc) - c\sin(2bc)]\}$
7.60	$\sin^\nu t \quad t < \pi$ $0 \quad t > \pi$ $\mathrm{Re}\,\nu > -1$	$2^{-\nu}\pi\Gamma(1+\nu)e^{-\frac{1}{2}\pi s}$ $\cdot\ [\Gamma(1+\frac{1}{2}\nu+\frac{1}{2}ip)\Gamma(1+\frac{1}{2}\nu-\frac{1}{2}ip)]^{-1}$
7.61	$\sin^{2n}t \quad t < \frac{1}{2}\pi$ $0 \quad t > \frac{1}{2}\pi$ $n=1,2,3,\cdots$	$(2n)!p^{-1}(p^2+4)^{-1}(p^2+16)^{-1}\cdots(p^2+4n^2)^{-1}$ $\{1-e^{-\frac{1}{2}\pi p}[1+\frac{p^2}{2!}+\cdots+\frac{p^2(p^2+4)\cdots[p^2+4(n-1)^2]}{(2n)!}]\}$
7.62	$\cos^{2n}t \quad t < \frac{1}{2}\pi$ $0 \quad t > \frac{1}{2}\pi$ $n=1,2,3,\cdots$	$(2n)!p^{-1}(p^2+4)^{-1}(p^2+16)^{-1}\cdots(p^2+4n^2)^{-1}$ $\cdot\{-e^{-\frac{1}{2}\pi p}+1+\frac{p^2}{2!}+\cdots+\frac{p^2(p^2+4)\cdots[p^2+4(n-1)^2]}{(2n)!}\}$

	$f(t)$	$g(p) = \int_0^\infty f(t)e^{-pt}dt$
7.63	$\sin^{2n}t \quad t < m\pi$ $0 \quad t > m\pi$ $n,m=1,2,3,\cdots$	$p^{-1}(p^2+4)^{-1}(p^2+16)^{-1}\cdots(p^2+4n^2)^{-1}$ $\cdot\ (2n)!(1-e^{-m\pi p})$
7.64	$0 \quad t < \frac{1}{2}\pi$ $\cos^{2n}t \quad \frac{\pi}{2}<t<(m+\frac{1}{2})\pi$ $0 \quad t>(m+\frac{1}{2})\pi$ $n,m=1,2,3,\cdots$	$p^{-1}(p^2+4)^{-1}(p^2+16)^{-1}\cdots(p^2+4n^2)^{-1}$ $\cdot\ (2n)!e^{-\frac{1}{2}\pi p}(1-e^{-m\pi p})$
7.65	$\sin^{2n+1}t \quad t < \frac{1}{2}\pi$ $0 \quad t > \frac{1}{2}\pi$ $n=1,2,3,\cdots$	$(2n+1)!(p^2+1)^{-1}(p^2+9)^{-1}\cdots[p^2+(2n+1)^2]^{-1}$ $\cdot\ \{1-pe^{-\frac{1}{2}\pi p}[1+\frac{p^2+1}{3!}+\cdots$ $+\frac{(p^2+1)(p^2+9)\cdots[p^2+(2n-1)^2]}{(2n+1)!}]\}$
7.66	$\cos^{2n+1}t \quad t < \frac{1}{2}\pi$ $0 \quad t > \frac{1}{2}\pi$ $n=0,1,2,\cdots$	$(2n+1)!(p^2+1)^{-1}(p^2+9)^{-1}\cdots[p^2+(2n+1)^2]^{-1}$ $\cdot\ \{e^{-\frac{1}{2}\pi p}+p[1+\frac{p^2+1}{3!}+\cdots$ $+\frac{(p^2+1)(p^2+9)\cdots[p^2+(2n-1)^2]}{(2n+1)!}]\}$
7.67	$\sin^{2n+1}t \quad t < m\pi$ $0 \quad t > m\pi$ $n=0,1,2,\cdots;$ $m=1,2,3,\cdots$	$(2n+1)!(p^2+1)^{-1}(p^2+9)^{-1}\cdots[p^2+(2n+1)^2]^{-1}$ $\cdot\ [1-(-1)^m e^{-m\pi p}]$

	$f(t)$	$g(p) = \int_0^\infty f(t)e^{-pt}dt$
7.68	$0 \quad t<\frac{1}{2}\pi$ $t\sin t \quad t>\frac{1}{2}\pi$	$(1+p^2)^{-2}e^{-\frac{1}{2}\pi p}[p^2-1+\frac{1}{2}\pi p(p^2+1)]$ $\mathrm{Re}\, p > 0$
7.69	$0 \quad t < \frac{1}{2}\pi$ $t\cos t \quad t > \frac{1}{2}\pi$	$-(1+p^2)^{-2}e^{-\frac{1}{2}\pi p}[2p+\frac{1}{2}\pi(p^2+1)]$ $\mathrm{Re}\, p > 0$
7.70	$t\sin t \quad t < \frac{1}{2}\pi$ $0 \quad t > \frac{1}{2}\pi$	$(1+p^2)^{-2}\{2p-e^{-\frac{1}{2}\pi p}[\frac{1}{2}\pi p(1+p^2)$ $+ p^2-1]\}$ $\quad \mathrm{Re}\, p > 0$
7.71	$t\cos t \quad t < \frac{1}{2}\pi$ $0 \quad t > \frac{1}{2}\pi$	$(1+p^2)^{-2}\{p^2-1+e^{-\frac{1}{2}\pi p}[\frac{1}{2}\pi(1+p^2)+2p]\}$ $\mathrm{Re}\, p > 0$
7.72	$(\cosh a+\cos t)^{-1}$	$p\,\mathrm{csch}\,a \sum_{n=0}^{\infty} (-1)^n \varepsilon_n (p^2+n^2)^{-1}e^{-na}$ $\mathrm{Re}\, p > 0$
7.73	$(\cosh a-\cos t)^{-1}$	$p\,\mathrm{csch}\,a \sum_{n=0}^{\infty} \varepsilon_n (p^2+n^2)^{-1}e^{-na}$ $\mathrm{Re}\, p > 0$
7.74	$\sin(at^{\frac{1}{2}})$	$\frac{1}{2}a\pi^{\frac{1}{2}}p^{-\frac{3}{2}}\exp(-\frac{1}{4}a^2/p)$ $\mathrm{Re}\, p > 0$

	$f(t)$	$g(p) = \int_0^\infty f(t)e^{-pt}dt$
7.75	$\cos(at^{\frac{1}{2}})$	$p^{-1}+\frac{1}{2}ia\ p^{-3/2}\pi^{\frac{1}{2}}\exp(-\frac{1}{4}a^2/p)$ $\cdot\ \mathrm{Erf}(\frac{1}{2}ia\ p^{-\frac{1}{2}})$ Re $p > 0$
7.76	$t^{-1}\sin(at^{\frac{1}{2}})$	$\pi\mathrm{Erf}(\frac{1}{2}ap^{-\frac{1}{2}})$ Re $p > 0$
7.77	$t^{\frac{1}{2}}\sin(at^{\frac{1}{2}})$	$\frac{1}{2}a\ p^{-2}-i\pi^{\frac{1}{2}}p^{-5/2}(\frac{1}{2}p-\frac{1}{4}a^2)$ $\cdot\ \exp(-\frac{1}{4}a^2/p)\ \mathrm{Erf}(\frac{1}{2}ia\ p^{-\frac{1}{2}})$ Re $p > 0$
7.78	$t^{\frac{1}{2}}\cos(at^{\frac{1}{2}})$	$\pi^{\frac{1}{2}}p^{-5/2}(\frac{1}{2}p-\frac{1}{4}a^2)\exp(-\frac{1}{4}a^2/p)$ Re $p > 0$
7.79	$t^{-\frac{1}{2}}\sin(at^{\frac{1}{2}})$	$-i(\pi/p)^{\frac{1}{2}}\exp(-\frac{1}{4}a^2/p)\mathrm{Erf}(\frac{1}{2}ia\ p^{-\frac{1}{2}})$ Re $p > 0$
7.80	$t^{-\frac{1}{2}}\cos(at^{\frac{1}{2}})$	$(\pi/p)^{\frac{1}{2}}\exp(-\frac{1}{4}a^2/p)$ Re $p > 0$
7.81	$t^{-\frac{1}{2}}(b+t)^{-1}\cos(at^{\frac{1}{2}})$	$\frac{1}{2}\pi b^{-\frac{1}{2}}e^{bp}\{e^{-ab^{\frac{1}{2}}}\mathrm{Erfc}[(bp)^{\frac{1}{2}}-\frac{1}{2}a/p]$ $+\ e^{ab^{\frac{1}{2}}}\mathrm{Erfc}[(bp)^{\frac{1}{2}}+\frac{1}{2}a/p]\}$ Re $p > 0$

	$f(t)$	$g(p) = \int_0^\infty f(t)e^{-pt}dt$
7.82	$t^{-3/4}\sin(at^{1/2})$	$\pi(\frac{1}{2}a)^{1/2}p^{-1/2}\exp(-\frac{1}{8}a^2/p)I_{1/4}(\frac{1}{8}a^2/p)$ Re $p > 0$
7.83	$t^{-3/4}\cos(at^{1/2})$	$\pi(\frac{1}{2}a)^{1/2}p^{-1/2}\exp(-\frac{1}{8}a^2/p)I_{-1/4}(\frac{1}{8}a^2/p)$ Re $p > 0$
7.84	$t^{-1/4}\sin(at^{1/2})$	$\frac{1}{2}\pi(\frac{1}{2}a/p)^{3/2}\exp(-\frac{1}{8}a^2/p)$ $\cdot [I_{-1/4}(\frac{1}{8}a^2/p)-I_{3/4}(\frac{1}{8}a^2/p)]$ Re $p > 0$
7.85	$t^{-1/4}\cos(at^{1/2})$	$\frac{1}{2}\pi(\frac{1}{2}a/p)^{3/2}\exp(-\frac{1}{8}a^2/p)$ $\cdot [I_{-3/4}(\frac{1}{8}a^2/p)-I_{1/4}(\frac{1}{8}a^2/p)]$ Re $p > 0$
7.86	$t^n\sin(at^{1/2})$ $n = 0,1,2,\cdots$	$(-1)^n(2\pi)^{1/2}(2p)^{-n-1}\exp(-\frac{1}{4}a^2/p)$ $\cdot He_{2n+1}[(2p)^{-1/2}a]$ Re $p > 0$
7.87	$t^{n-1/2}\cos(at^{1/2})$ $n = 0,1,2,\cdots$	$(-1)^n(2\pi)^{1/2}(2p)^{-n-1/2}\exp(-\frac{1}{4}a^2/p)$ $\cdot He_{2n}[(2p)^{-1/2}a]$ Re $p > 0$

	$f(t)$	$g(p) = \int_0^\infty f(t)e^{-pt}dt$
7.88	$t^{\nu-1}\sin(at^{\frac{1}{2}})$ Re $\nu > -\frac{1}{2}$	$i(2p)^{-\nu}\Gamma(2\nu)\exp(-\frac{1}{8}a^2/p)$ $\cdot\{D_{-2\nu}[(2p)^{-\frac{1}{2}}ia]-D_{-2\nu}[-(2p)^{-\frac{1}{2}}ia]\}$ $= 2^{-\nu-\frac{1}{2}}\pi^{\frac{1}{2}}\sec(\pi\nu)p^{-\nu}\exp(-\frac{1}{8}a^2/p)$ $\cdot\{D_{2\nu-1}[-(2p)^{-\frac{1}{2}}a]-D_{2\nu-1}[(2p)^{-\frac{1}{2}}a]\}$ Re $p > 0$
7.89	$t^{\nu-1}\cos(at^{\frac{1}{2}})$ Re $\nu > 0$	$(2p)^{-\nu}\Gamma(2\nu)\exp(-\frac{1}{8}a^2/p)$ $\cdot\{D_{-2\nu}[(2p)^{-\frac{1}{2}}ia]+D_{-2\nu}[-(2p)^{-\frac{1}{2}}ia]\}$ $= 2^{-\nu-\frac{1}{2}}\pi^{\frac{1}{2}}\csc(\pi\nu)p^{-\nu}\exp(-\frac{1}{8}a^2/p)$ $\cdot\{D_{2\nu-1}[-(2p)^{-\frac{1}{2}}a]+D_{2\nu-1}[(2p)^{-\frac{1}{2}}a]\}$ Re $p > 0$
7.90	$t^{\nu-1}e^{-bt^{\frac{1}{2}}}\sin(at^{\frac{1}{2}})$ Re $\nu > -\frac{1}{2}$	$i(2p)^{-\nu}\Gamma(2\nu)\exp(\frac{b^2-a^2}{8p})$ $\cdot\{\exp(i\frac{1}{4}ab/p)D_{-2\nu}[(2p)^{-\frac{1}{2}}(b+ia)]$ $-\exp(-i\frac{1}{4}ab/p)D_{-2\nu}[(2p)^{-\frac{1}{2}}(b-ia)]\}$ Re $p > 0$

	$f(t)$	$g(p) = \int_0^\infty f(t)e^{-pt}dt$
7.91	$t^{\nu-1}e^{-bt^{\frac{1}{2}}}\cos(at^{\frac{1}{2}})$ $\mathrm{Re}\ \nu > 0$	$(2p)^{-\nu}\Gamma(2\nu)\exp(\frac{b^2-a^2}{8p})$ $\cdot\{\exp(i\frac{1}{4}ab/p)D_{-2\nu}[(2p)^{-\frac{1}{2}}(b+ia)]$ $+\exp(-i\frac{1}{4}ab/p)D_{-2\nu}[(2p)^{-\frac{1}{2}}(b-ia)]\}$ $\mathrm{Re}\ p > 0$
7.92	$t^{-\frac{1}{2}}(a^2+t^2)^{-\frac{1}{2}}$ $\cdot\sin[b(a^2+t^2)^{\frac{1}{2}}]$	$\frac{1}{4}\pi(\pi p)^{\frac{1}{2}}[Y_{\frac{1}{4}}(z_1)J_{\frac{1}{4}}(z_2)-J_{\frac{1}{4}}(z_1)Y_{\frac{1}{4}}(z_2)]$ $z_{\frac{1}{2}} = \frac{1}{2}a[(b^2+p^2)^{\frac{1}{2}}\pm b]$, $\mathrm{Re}\ p > \lvert\mathrm{Im} b\rvert$
7.93	$t^{-\frac{1}{2}}(a^2+t^2)^{-\frac{1}{2}}$ $\cdot\cos[b(a^2+t^2)^{\frac{1}{2}}]$	$\frac{1}{4}\pi(\pi p)^{\frac{1}{2}}[J_{\frac{1}{4}}(z_1)J_{\frac{1}{4}}(z_2)+Y_{\frac{1}{4}}(z_1)Y_{\frac{1}{4}}(z_2)]$ $z_{\frac{1}{2}} = \frac{1}{2}a[(b^2+p^2)^{\frac{1}{2}}\pm b]$, $\mathrm{Re}\ p > \lvert\mathrm{Im} b\rvert$
7.94	$\sin[b(t^2+at)^{\frac{1}{2}}]$	$\frac{1}{2}ab(p^2+b^2)^{-\frac{1}{2}}e^{\frac{1}{2}ap}K_1[\frac{1}{2}a(p^2+b^2)^{\frac{1}{2}}]$ $\mathrm{Re}\ p > \lvert\mathrm{Im}\ b\rvert$
7.95	$(a+t)^{-\frac{1}{2}}\sin[b(t^2+at)^{\frac{1}{2}}]$	$(\frac{1}{2}\pi)^{\frac{1}{2}}b[(p^2+b^2)^{\frac{1}{2}}+p]^{-\frac{1}{2}}e^{\frac{1}{2}ap}$ $\cdot(p^2+b^2)^{-\frac{1}{2}}\exp[-\frac{1}{2}a(p^2+b^2)^{\frac{1}{2}}]$ $\mathrm{Re}\ p > \lvert\mathrm{Im}\ b\rvert$
7.96	$(t+a)^{-\frac{1}{2}}\cos[b(t^2+at)^{\frac{1}{2}}]$	$(\frac{1}{2}\pi)^{\frac{1}{2}}[(p^2+b^2)^{\frac{1}{2}}+p]^{\frac{1}{2}}e^{\frac{1}{2}ap}$ $\cdot(p^2+b^2)^{-\frac{1}{2}}\exp[-\frac{1}{2}a(p^2+b^2)^{\frac{1}{2}}]$ $\mathrm{Re}\ p > \lvert\mathrm{Im}\ b\rvert$

	$f(t)$	$g(p) = \int_0^\infty f(t)e^{-pt}dt$
7.97	$(a+t)^{-\frac{1}{2}}(t^2+at)^{-\frac{1}{2}}$ $\cdot\sin[b(t^2+at)^{\frac{1}{2}}]$	$-i\pi a^{-\frac{1}{2}}e^{ap}\mathrm{Erfc}(z_1)\mathrm{Erf}(iz_2)$ $z_{\frac{1}{2}}=(\frac{1}{2}a)^{\frac{1}{2}}[(p^2+b^2)^{\frac{1}{2}}\pm p]^{\frac{1}{2}}$ $\mathrm{Re}\,p > \|\mathrm{Im}\,b\|$
7.98	$(a+t)^{-\frac{1}{2}}(t^2+at)^{-\frac{1}{2}}$ $\cdot\cos[b(t^2+at)^{\frac{1}{2}}]$	$\pi a^{-\frac{1}{2}}e^{ap}$ $\cdot\mathrm{Erfc}\{(\frac{1}{2}a)^{\frac{1}{2}}[(p^2+b^2)^{\frac{1}{2}}+p]^{\frac{1}{2}}\}$ $\mathrm{Re}\,p > \|\mathrm{Im}\,b\|$
7.99	$(t^2+at)^{-\frac{1}{2}}\cos[b(t^2+at)^{\frac{1}{2}}]$	$K_0[\frac{1}{2}a(p^2+b^2)^{\frac{1}{2}}]e^{\frac{1}{2}ap}$ $\mathrm{Re}\,p > \|\mathrm{Im}\,b\|$
7.100	$t^{-\frac{1}{2}}\cos[b(t^2+at)^{\frac{1}{2}}]$	$(\frac{1}{2}\pi)^{\frac{1}{2}}[p+(p^2+b^2)^{\frac{1}{2}}]^{\frac{1}{2}}e^{\frac{1}{2}ap}$ $\cdot(p^2+b^2)^{-\frac{1}{2}}\exp[-\frac{1}{2}a(p^2+b^2)^{\frac{1}{2}}]$ $\mathrm{Re}\,p > \|\mathrm{Im}\,b\|$
7.101	$(t^2+at)^{-3/4}\sin[b(t^2+at)^{\frac{1}{2}}]$	$(\frac{1}{2}\pi b)^{\frac{1}{2}}K_{\frac{1}{4}}(z_1)I_{\frac{1}{4}}(z_2)e^{\frac{1}{2}ap}$ $z_{\frac{1}{2}}=\frac{1}{4}a[(p^2+b^2)^{\frac{1}{2}}\pm p]$ $\mathrm{Re}\,p > \|\mathrm{Im}\,b\|$

	$f(t)$	$g(p) = \int_0^\infty f(t)e^{-pt}dt$
7.102	$(t^2+at)^{-3/4}$ $\cdot\cos[b(t^2+at)^{1/2}]$	$(\frac{1}{2}\pi b)^{1/2}K_{1/4}(z_1)I_{-1/4}(z_2)e^{\frac{1}{2}ap}$ $z_{\frac{1}{2}}=\frac{1}{4}a[(p^2+b^2)^{1/2}\pm p]$ $\mathrm{Re}\, p > \lvert\mathrm{Im}\, b\rvert$
7.103	$(t^2+at)^{-1/2}\sin[b(t^2+at)^{1/2}]$ $\cdot\{[(\frac{1}{2}a+t+(t^2+at)^{1/2}]^{\nu}$ $-[(\frac{1}{2}a+t-(t^2+at)^{1/2}]^{\nu}\}$	$2(\frac{1}{2}a)^{\nu}\sin[\nu\arctan(b/p)]$ $\cdot K_{\nu}[\frac{1}{2}a(p^2+b^2)^{1/2}]e^{\frac{1}{2}ap}$ $\mathrm{Re}\, p > \lvert\mathrm{Im}\, b\rvert$
7.104	$(t^2+at)^{-1/2}\cos[b(t^2+at)^{1/2}]$ $\cdot\{[\frac{1}{2}a+t+(t^2+at)^{1/2}]^{\nu}$ $+[\frac{1}{2}a-t-(t^2+at)^{1/2}]^{\nu}\}$	$2(\frac{1}{2}a)^{\nu}\cos[\nu\arctan(b/p)]$ $\cdot K_{\nu}[\frac{1}{2}a(p^2+b^2)^{1/2}]e^{\frac{1}{2}ap}$ $\mathrm{Re}\, p > \lvert\mathrm{Im}\, b\rvert$
7.105	$(a+t)^{1-2\nu}(t^2+at)^{\nu-5/4}$ $\cdot\sin[b(t^2+at)^{1/2}]$ $\mathrm{Re}\,\nu > -\frac{1}{4}$	$2a^{-1}(\frac{1}{2}b)^{-1/2}\Gamma(\frac{1}{4}+\nu)e^{\frac{1}{2}ap}$ $\cdot W_{\frac{1}{2}-\nu,\frac{1}{4}}(z_1)M_{\nu-\frac{1}{2},\frac{1}{4}}(z_2)$ $z_{\frac{1}{2}}=\frac{1}{2}a[(p^2+b^2)^{1/2}\pm p]$ $\mathrm{Re}\, p > \lvert\mathrm{Im}\, b\rvert$

	$f(t)$	$g(p) = \int_0^\infty f(t)e^{-pt}dt$
7.106	$(a+t)^{1-2\nu}(t^2+at)^{\nu-5/4}$ $\cdot\cos[b(t^2+at)^{1/2}]$ $\mathrm{Re}\,\nu > 1/4$	$2a^{-1}(2b)^{-1/2}\Gamma(\nu-1/4)e^{1/2 ap}$ $\cdot W_{1/2-\nu,1/4}(z_1)M_{\nu-1/2,-1/4}(z_2)$ $z_{1,2}=1/2 a[(p^2+b^2)^{1/2}\pm p]$ $\mathrm{Re}\,p > \lvert\mathrm{Im}\,b\rvert$
7.107	$0 \quad t > a$ $t^{-1/2}(a^2-t^2)^{-1/2}$ $\cdot\cos[b(a^2-t^2)^{1/2}]\, t<a$	$-1/2\pi(\pi p)^{1/2}$ $\cdot[J_{1/4}(z_1)J_{1/4}(z_2)-J_{-1/4}(z_1)J_{-1/4}(z_2)]$ $z_{1,2}=1/2 a[b\pm(b^2-p^2)^{1/2}]$
7.108	$(at-t^2)^{-1/2}\cos[b(at-t^2)^{1/2}]$ $t < a$ $0 \quad t > a$	$\pi e^{-1/2 ap}J_0[1/2 a(b^2-p^2)^{1/2}]$
7.109	$0 \quad t < a$ $\sin[b(t^2-a^2)^{1/2}]\, t > a$	$ab(p^2+b^2)^{-1/2}K_1[a(p^2+b^2)^{1/2}]$ $\mathrm{Re}\,p > \lvert\mathrm{Im}\,b\rvert$
7.110	$0 \quad t < a$ $t\sin[b(t^2-a^2)^{1/2}] \quad t > a$	$a^2bp(b^2+p^2)^{-1}K_2[a(p^2+b^2)^{1/2}]$ $\mathrm{Re}\,p > \lvert\mathrm{Im}\,b\rvert$

	$f(t)$	$g(p) = \int_0^\infty f(t)e^{-pt}dt$
7.111	$0 \quad t < a$ $(t+a)^{-\frac{1}{2}}\sin[b(t^2-a^2)^{\frac{1}{2}}]$ $t > a$	$(\frac{1}{2}\pi)^{\frac{1}{2}}b[(p^2+b^2)^{\frac{1}{2}}+p]^{-\frac{1}{2}}$ $\cdot(p^2+b^2)^{-\frac{1}{2}}\exp[-a(p^2+b^2)^{\frac{1}{2}}]$ $\mathrm{Re}\, p > \lvert \mathrm{Im} b\rvert$
7.112	$0 \quad t < a$ $(t+a)^{-\frac{1}{2}}\cos[b(t^2-a^2)^{\frac{1}{2}}]$ $t > a$	$(\frac{1}{2}\pi)^{\frac{1}{2}}[(p^2+b^2)^{\frac{1}{2}}+p]^{\frac{1}{2}}$ $\cdot(p^2+b^2)^{-\frac{1}{2}}\exp[-a(p^2+b^2)^{\frac{1}{2}}]$ $\mathrm{Re}\, p > \lvert \mathrm{Im}\, b\rvert$
7.113	$0 \quad t < a$ $(t+a)^{-\frac{1}{2}}(t^2-a^2)^{-\frac{1}{2}}$ $\cdot\sin[b(t^2-a^2)^{\frac{1}{2}}]$ $t > a$	$-i\pi(2a)^{-\frac{1}{2}}e^{ap}\mathrm{Erfc}(z_1)\mathrm{Erf}(iz_2)$ $z_{\frac{1}{2}} = a^{\frac{1}{2}}[(p^2+b^2)^{\frac{1}{2}}\pm p]^{\frac{1}{2}}$ $\mathrm{Re}\, p > \lvert \mathrm{Im}\, b\rvert$
7.114	$0 \quad t < a$ $(t+a)^{-\frac{1}{2}}(t^2-a^2)^{-\frac{1}{2}}$ $\cdot\cos[b(t^2-a^2)^{\frac{1}{2}}] \quad t > a$	$\pi(2a)^{-\frac{1}{2}}e^{ap}$ $\cdot\mathrm{Erfc}\{a^{\frac{1}{2}}[(p^2+b^2)^{\frac{1}{2}}+p]^{\frac{1}{2}}\}$ $\mathrm{Re}\, p > \lvert \mathrm{Im}\, b\rvert$
7.115	$0 \quad t < a$ $(t^2-a^2)^{-\frac{1}{2}}\cos[b(t^2-a^2)^{\frac{1}{2}}]$ $t > a$	$K_o[a(p^2+b^2)^{\frac{1}{2}}]$ $\mathrm{Re}\, p > \lvert \mathrm{Im}\, a\rvert$

	$f(t)$	$g(p) = \int_0^\infty f(t)e^{-pt}dt$
7.116	$0 \quad t < a$ $t^{-\frac{1}{2}}(t^2-a^2)^{-\frac{1}{2}}$ $\cdot\cos[b(t^2-a^2)^{\frac{1}{2}}]$ $t > a$	$(2\pi p)^{-\frac{1}{2}} K_{\frac{1}{4}}(z_1) K_{\frac{1}{4}}(z_2)$ $z_{\frac{1}{2}} = \frac{1}{2}a[(b^2+p^2)^{\frac{1}{2}} \pm p]$ $\mathrm{Re}\, p > \lvert \mathrm{Im} b \rvert$
7.117	$0 \quad t < a$ $t(t^2-a^2)^{-\frac{1}{2}}$ $\cdot\cos[b(t^2-a^2)^{\frac{1}{2}}]$ $t > a$	$ap(p^2+b^2)^{-\frac{1}{2}}$ $\cdot K_1[a(p^2+b^2)^{\frac{1}{2}}]$ $\mathrm{Re}\, p > \lvert \mathrm{Im}\, b \rvert$
7.118	$0 \quad t < a$ $(t-a)^{-\frac{1}{2}}\cos[b(t^2-a^2)^{\frac{1}{2}}$ $t > a$	$(\frac{1}{2}\pi)^{\frac{1}{2}}[p+(p^2+b^2)^{\frac{1}{2}}]^{\frac{1}{2}}$ $\cdot(p^2+b^2)^{-\frac{1}{2}}\exp[-a(p^2+b^2)^{\frac{1}{2}}]$ $\mathrm{Re}\, p > \lvert \mathrm{Im}\, b \rvert$
7.119	$0 \quad t < a$ $(t^2-a^2)^{-3/4}\sin[b(t^2-a^2)^{\frac{1}{2}}]$ $t > a$	$(\frac{1}{2}\pi b)^{\frac{1}{2}} K_{\frac{1}{4}}(z_1) I_{\frac{1}{4}}(z_2)$ $z_{\frac{1}{2}} = \frac{1}{2}a[(p^2+b^2)^{\frac{1}{2}} \pm p]$ $\mathrm{Re}\, p > \lvert \mathrm{Im}\, b \rvert$
7.120	$0 \quad t < a$ $(t^2-a^2)^{-3/4}\cos[b(t^2-a^2)^{\frac{1}{2}}]$ $t > a$	$(\frac{1}{2}\pi b)^{\frac{1}{2}} K_{\frac{1}{4}}(z_1) I_{-\frac{1}{4}}(z_2)$ $z_{\frac{1}{2}} = \frac{1}{2}a[(p^2+b^2)^{\frac{1}{2}} \pm p]$ $\mathrm{Re}\, p > \lvert \mathrm{Im}\, b \rvert$

	$f(t)$	$g(p) = \int_0^\infty f(t)e^{-pt}dt$
7.121	$0 \qquad t < a$ $(t^2-a^2)^{-\frac{1}{2}}\{[t+(t^2-a^2)^{\frac{1}{2}}]^\nu$ $-[t-(t^2-a^2)^{\frac{1}{2}}]^\nu\}\sin[b(t^2-a^2)^{\frac{1}{2}}]$	$2a^\nu\sin[\nu \arctan(b/p)]$ $\cdot K_\nu[a(p^2+b^2)^{\frac{1}{2}}]$ $\operatorname{Re} p > \lvert\operatorname{Im} b\rvert$
7.122	$0 \qquad t < a$ $(t^2-a^2)^{-\frac{1}{2}}\{[t+(t^2-a^2)^{\frac{1}{2}}]^\nu$ $+[t-(t^2-a^2)^{\frac{1}{2}}]^\nu\}\cos[b(t^2-a^2)^{\frac{1}{2}}]$	$2a^\nu\cos[\nu \arctan(b/p)]$ $\cdot K_\nu[a(p^2+b^2)^{\frac{1}{2}}]$ $\operatorname{Re} p > \lvert\operatorname{Im} b\rvert$
7.123	$0 \qquad t < a$ $(t^2-a^2)^{-\frac{1}{2}}\log[t/a+(t^2/a^2-1)^{\frac{1}{2}}]$ $\cdot\sin[b(t^2-a^2)^{\frac{1}{2}}]$ $t > a$	$\arctan(b/p)$ $\cdot K_o[a(p^2+b^2)^{\frac{1}{2}}]$ $\operatorname{Re} p > \lvert\operatorname{Im} b\rvert$
7.124	$0 \qquad t < a$ $(a+t)^{1-2\nu}(t^2-a^2)^{\nu-\frac{5}{4}}$ $\cdot\sin[b(t^2-a^2)^{\frac{1}{2}}]$ $t > a$ $\operatorname{Re} \nu > -\frac{1}{4}$	$a^{-1}(\frac{1}{2}b)^{-\frac{1}{2}}\Gamma(\nu+\frac{1}{4})$ $\cdot W_{\frac{1}{2}-\nu,\frac{1}{4}}(z_1)M_{\nu-\frac{1}{2},\frac{1}{4}}(z_2)$ $z_{\substack{1\\2}}=a[(p^2+b^2)^{\frac{1}{2}}\pm p]$ $\operatorname{Re} p > \lvert\operatorname{Im} b\rvert$

	$f(t)$	$g(p) = \int_0^\infty f(t)e^{-pt}dt$
7.125	$0 \quad t < a$ $(a+t)^{1-2\nu}(t^2-a^2)^{\nu-5/4}$ $\cdot\cos[b(t^2-a^2)^{1/2}]$ $t > a$ $\mathrm{Re}\,\nu > 1/4$	$a^{-1}(2b)^{-1/2}\Gamma(\nu-1/4)$ $\cdot W_{1/2-\nu,1/4}(z_1)M_{\nu-1/2,-1/4}(z_2)$ $z_{1,2} = a[(p^2+b^2)^{1/2}\pm p]$ $\mathrm{Re}\,p > \lvert\mathrm{Im}\,b\rvert$
7.126	$t^{-3/2}e^{-b/t}\sin(a/t)$	$\pi^{1/2}(a^2+b^2)^{-1/2}e^{-2p^{1/2}u}$ $[u\sin(2p^{1/2}v)+v\cos(2p^{1/2}v)]$ $u, v = 2^{-1/2}[(a^2+b^2)^{1/2}\pm b]^{1/2}$ $\mathrm{Re}\,p > 0$
7.127	$t^{-3/2}e^{-b/t}\cos(a/t)$	$\pi^{1/2}(a^2+b^2)^{-1/2}e^{-2p^{1/2}u}$ $\cdot[u\cos(2p^{1/2}v)-v\sin(2p^{1/2}v)]$ u,v as before, $\mathrm{Re}\,p > 0$
7.128	$t^{-1/2}e^{-b/t}\sin(a/t)$	$(\pi/p)^{1/2}e^{-2p^{1/2}u}\sin(2p^{1/2}v)$ u,v as before, $\mathrm{Re}\,p > 0$
7.129	$t^{-1/2}e^{-b/t}\cos(a/t)$	$(\pi/p)^{1/2}e^{-2p^{1/2}u}\cos(2p^{1/2}v)$ u,v as before, $\mathrm{Re}\,p > 0$

	$f(t)$	$g(p) = \int_0^\infty f(t)e^{-pt}dt$
7.130	$t^{\nu-1}e^{-b/t}\sin(a/t)$	$ip^{-\frac{1}{2}\nu}\{(b+ia)^{\frac{1}{2}\nu}K_\nu[2p^{\frac{1}{2}}(b+ia)^{\frac{1}{2}}]$ $-(b-ia)^{\frac{1}{2}\nu}K_\nu[2p^{\frac{1}{2}}(b-ia)^{\frac{1}{2}}]\}$ $\mathrm{Re}\, p > 0$
7.131	$t^{\nu-1}e^{-b/t}\cos(a/t)$	$p^{-\frac{1}{2}\nu}\{(b+ia)^{\frac{1}{2}\nu}K_\nu[2p^{\frac{1}{2}}(b+ia)^{\frac{1}{2}}]$ $+(b-ia)^{\frac{1}{2}\nu}K_\nu[2p^{\frac{1}{2}}(b-ia)^{\frac{1}{2}}]\}$ $\mathrm{Re}\, p > 0$
7.132	$\sin(at^2)$	$\frac{1}{2}(\frac{1}{2}a/\pi)^{-\frac{1}{2}}\{\cos z[\frac{1}{2}-C(z)]$ $+\sin z[\frac{1}{2}-S(z)]\}$ $z = \frac{1}{4}p^2/a,\quad \mathrm{Re}\, p > 0$
7.133	$\cos(at^2)$	$\frac{1}{2}(\frac{1}{2}a/\pi)^{-\frac{1}{2}}\{\cos z[\frac{1}{2}-S(z)]$ $-\sin z[\frac{1}{2}-C(z)]\}$ $z = \frac{1}{4}p^2/a),\quad \mathrm{Re}\, p > 0$
7.134	$t\sin(at^2)$	$\frac{1}{2}a^{-1}+\frac{1}{2}(\frac{1}{2}\pi)^{\frac{1}{2}}a^{-3/2}p$ $\cdot\{\sin z[\frac{1}{2}-C(z)]-\cos z[\frac{1}{2}-S(z)]\}$ $z = \frac{1}{2}p^2/a,\quad \mathrm{Re}\, p > 0$

	$f(t)$	$g(p) = \int_0^\infty f(t)e^{-pt}dt$
7.135	$t\cos(at^2)$	$\frac{1}{2}(\frac{1}{2}\pi)^{\frac{1}{2}}a^{-3/2}p$ $\cdot\{\cos z[\frac{1}{2}-C(z)]+\sin z[\frac{1}{2}-S(z)]\}$ $z=\frac{1}{4}p^2/a,$ Re $p>0$
7.136	$t^{-1}\sin(at^2)$	$\frac{1}{2}\pi[\frac{1}{2}-C(z)]^2+\frac{1}{2}\pi[\frac{1}{2}-S(z)]^2$ $z=\frac{1}{4}p^2/a,$ Re $p>0$
7.137	$t^{-\frac{1}{2}}\sin(at^2)$	$\frac{1}{4}\pi^{\frac{1}{2}}(\pi p/a)^{\frac{1}{2}}[J_{\frac{1}{4}}(z)\sin(z-3\pi/8)$ $-Y_{\frac{1}{4}}(z)\cos(z-3\pi/8)]$ $z=\frac{1}{8}p^2/a,$ Re $p>0$
7.138	$t^{-\frac{1}{2}}\cos(at^2)$	$\frac{1}{4}\pi^{\frac{1}{2}}(\pi p/a)^{\frac{1}{2}}[J_{\frac{1}{4}}(z)\cos(z-3\pi/8)$ $+Y_{\frac{1}{4}}(z)\sin(z-3\pi/8)]$ $z=\frac{1}{8}p^2/a,$ Re $p>0$
7.139	$t^{\nu-1}\sin(at^2)$ Re $\nu>-2$	$\frac{1}{2}i\Gamma(\nu)(2a)^{-\frac{1}{2}\nu}$ $\cdot\{\exp[-i(\frac{1}{4}\pi\nu+\frac{1}{8}p^2/a)]D_{-\nu}[(2a)^{-\frac{1}{2}}pe^{-i\frac{\pi}{4}}]$ $-\exp[i(\frac{1}{4}\pi\nu+\frac{1}{8}p^2/a)D_{-\nu}[(2a)^{-\frac{1}{2}}pe^{i\frac{\pi}{4}}]\}$ Re $p>0$

	$f(t)$	$g(p) = \int_0^\infty f(t)e^{-pt}dt$
7.140	$t^{\nu-1}\cos(at^2)$ $\mathrm{Re}\ \nu > -1$	$\frac{1}{2}(2a)^{-\frac{1}{2}\nu}\Gamma(\nu)$ $\cdot\{\exp[-i(\frac{1}{4}\pi\nu+\frac{1}{8}p^2/a)]D_{-\nu}[(2a)^{-\frac{1}{2}}pe^{-i\frac{\pi}{4}}]$ $+\exp[i(\frac{1}{4}\pi\nu+\frac{1}{8}p^2/a)]D_{-\nu}[(2a)^{-\frac{1}{2}}pe^{i\frac{\pi}{4}}]\}$ $\mathrm{Re}\ p > 0$
7.141	$t^{-\frac{1}{2}}\sin(at^{-\frac{1}{2}})$	$2a\sum_{n=0}^{\infty}[n!(2n+1)!]^{-1}(a^2p)^n$ $\cdot[\psi(2n+2)+\frac{1}{2}\psi(n+1)-\log(ap^{\frac{1}{2}})]$ $\mathrm{Re}\ p > 0$
7.142	$t^{-\frac{1}{2}}\cos(at^{-\frac{1}{2}})$	$\pi p^{-\frac{1}{2}}\sum_{n=0}^{\infty}[n!\Gamma(\frac{1}{2}+\frac{1}{2}n)]^{-1}(-1)^n(a^2p)^{\frac{1}{2}n}$ $\mathrm{Re}\ p > 0$
7.143	$t^{-2/3}\sin(at^{1/3})$	$\frac{3}{2}ia^{-1}z[e^{i\pi/4}\ S_{o,1/3}(ze^{i3\pi/4})$ $-e^{-i\pi/4}\ S_{o,1/3}(z\ e^{-i3\pi/4})]$ $z = 2(\frac{1}{3}ap^{-1/3})^{3/2}$, $\mathrm{Re}\ p > 0$

	$f(t)$	$g(p) = \int_0^\infty f(t)e^{-pt}dt$
7.144	$t^{-2/3}\cos(at^{1/3})$	$\frac{3}{2}a^{-1}z[e^{i\pi/4}S_{o,1/3}(ze^{i3\pi/4})$ $+ e^{-i\pi/4}S_{o,1/3}(ze^{-i3\pi/4})]$ $z = 2(\frac{1}{3}ap^{-1/3})^{3/2}$, Re $p > 0$
7.145	$\sin(ae^{-t})$	$a^{-p}\Gamma(p)[U_p(2a,0)\sin a$ $- U_{p+1}(2a,0)\cos a]$ Re $p > 0$
7.146	$\cos(ae^{-t})$	$a^{-p}\Gamma(p)[U_p(2a,0)\cos a+U_{p+1}(2a,0)\sin a]$ Re $p > 0$
7.147	$\sin[a(1-e^{-t})]$	$a^{-p}\Gamma(p)U_{p+1}(2a,0)$ Re $p > 0$
7.148	$\cos[a(1-e^{-t})]$	$a^{-p}\Gamma(p)U_p(2a,0)$ Re $p > 0$
7.149	$(e^t-1)^{-\frac{1}{2}}\sin[a(1-e^{-t})^{\frac{1}{2}}]$	$\pi^{\frac{1}{2}}\Gamma(\frac{1}{2}+p)(\frac{1}{2}a)^{-p}\mathbf{H}_p(a)$ Re $p > -\frac{1}{2}$
7.150	$(e^t-1)^{-\frac{1}{2}}\cos[a(1-e^{-t})^{\frac{1}{2}}]$	$\pi^{\frac{1}{2}}\Gamma(\frac{1}{2}+p)(\frac{1}{2}a)^{-p}J_p(a)$ Re $p > -\frac{1}{2}$

	$f(t)$	$g(p) = \int_0^\infty f(t)e^{-pt}dt$
7.151	$(1-e^{-t})^{-\frac{1}{2}}\sin[a(e^t-1)^{\frac{1}{2}}]$	$\pi^{\frac{1}{2}}\Gamma(\frac{1}{2}-p)(\frac{1}{2}a)^p[I_p(a)-\mathbf{L}_{-p}(a)]$ Re $p > -\frac{1}{2}$
7.152	$(1-e^{-t})^{-\frac{1}{2}}\cos[a(e^t-1)^{\frac{1}{2}}]$	$2\pi^{\frac{1}{2}}(\frac{1}{2}a)^p[\Gamma(\frac{1}{2}+p)]^{-1}K_p(a)$ Re $p > -\frac{1}{2}$
7.153	$\log[2\sin(\frac{1}{2}at)]$	$-p\sum_{n=1}^{\infty}[n(p^2+n^2a^2)]^{-1}$ Re $p > 0$
7.154	$\log[2\cos(\frac{1}{2}at)]$	$p\sum_{n=1}^{\infty}(-1)^{n-1}[n(p^2+n^2a^2)^{-1}]$ Re $p > 0$
7.155	$\log[\cot(\frac{1}{2}at)]$	$2p\sum_{n=1}^{\infty}\{(2n-1)[p^2+(2n-1)^2a^2]\}^{-1}$ Re $p > 0$

.8 Inverse Trigonometric Functions

.1	$\arcsin(t/a)$ $t < a$; 0 $t > a$	$\frac{1}{2}\pi p^{-1}[I_o(ap)-\mathbf{L}_o(ap)-e^{-ap}]$

	$f(t)$	$g(p) = \int_0^\infty f(t)e^{-pt}dt$
8.2	$\arccos(t/a)$ $t < a$ 0 $t > a$	$\frac{1}{2}\pi p^{-1}[1-I_o(ap)+\mathbf{L}_o(ap)]$
8.3	0 $t < a$ $\arccos(a/t)$ $t > a$	$p^{-1}\int_{ap}^\infty K_o(x)dx$ $\quad \mathrm{Re}\, p > 0$
8.4	$\arctan(at)$	$p^{-1}[\mathrm{Ci}(p/a)\sin(p/a)$ $-\mathrm{si}(p/a)\cos(p/a)]$ $\quad \mathrm{Re}\, p > 0$
8.5	$\mathrm{arccot}(at)$	$p^{-1}[\frac{1}{2}\pi-\mathrm{Ci}(p/a)\sin(p/a)$ $+\mathrm{si}(p/a)\cos(p/a)]$ $\quad \mathrm{Re}\, p > 0$
8.6	$t\arctan(at)$	$p^{-2}[-\mathrm{Ci}(p/a)\sin(p/a)$ $-\mathrm{si}(p/a)\cos(p/a)]$ $+(ap)^{-1}[\mathrm{Ci}(p/a)\cos(p/a)$ $-\mathrm{si}(p/a)\sin(p/a)]$ $\quad \mathrm{Re}\, p > 0$
8.7	$t\,\mathrm{arccot}(at)$	$p^{-2}[\frac{1}{2}\pi-\mathrm{Ci}(p/a)\sin(p/a)$ $+\mathrm{si}(p/a)\cos(p/a)]$ $+(ap)^{-1}[\mathrm{si}(p/a)\sin(p/a)$ $+\mathrm{Ci}(p/a)\cos(p/a)]$ $\quad \mathrm{Re}\, p > 0$

	$f(t)$	$g(p) = \int_0^\infty f(t)e^{-pt}dt$
8.8	$t^\nu(t^2+a^2)^{\frac{1}{2}\nu}$ $\cdot\sin[\nu \operatorname{arccot}(t/a)+b]$ $\operatorname{Re}\nu > -1$	$-\frac{1}{2}\pi^{\frac{1}{2}}(a/p)^{\nu+\frac{1}{2}}\Gamma(\nu+1)$ $\cdot[\cos(\frac{1}{2}ap+b)J_{\nu+\frac{1}{2}}(\frac{1}{2}ap)$ $+\sin(\frac{1}{2}ap+b)Y_{\nu+\frac{1}{2}}(\frac{1}{2}ap)]$ $\operatorname{Re}p > 0$
8.9	$t^\nu(t^2+a^2)^{\frac{1}{2}\nu}$ $\cdot\cos[\nu \operatorname{arccot}(t/a)+b]$ $\operatorname{Re}\nu > -1$	$\frac{1}{2}\pi^{\frac{1}{2}}(a/p)^{\nu+\frac{1}{2}}\Gamma(1+\nu)$ $\cdot[\sin(\frac{1}{2}ap+b)J_{\nu+\frac{1}{2}}(\frac{1}{2}ap)$ $-\cos(\frac{1}{2}ap+b)Y_{\nu+\frac{1}{2}}(\frac{1}{2}ap)]$ $\operatorname{Re}p > 0$
8.10	$t^{-\frac{1}{2}}(b+t)^{\frac{1}{2}\nu}\cos(at)$ $\cdot\cos\{\nu \arctan[(t/b)^{\frac{1}{2}}]\}$	$(\frac{1}{2}\pi)^{\frac{1}{2}}(2p)^{-\frac{1}{2}-\frac{1}{2}\nu}\exp(\frac{1}{2}pb-\frac{1}{8}a^2/p)$ $\cdot\{e^{\frac{1}{2}ab^{\frac{1}{2}}}D_\nu[(2p)^{\frac{1}{2}}(b^{\frac{1}{2}}+\frac{1}{2}a/p)]$ $+e^{-\frac{1}{2}ab^{\frac{1}{2}}}D_\nu[(2p)^{\frac{1}{2}}(b^{\frac{1}{2}}-\frac{1}{2}a/p)]\}$ $\operatorname{Re}p > 0$
8.11	$[t(t+1)(t+2)]^{-\frac{1}{2}}$ $\cdot\cos[\nu \arccos(1+t)^{-1}]$	$\pi^{\frac{1}{2}}e^pD_{\nu-\frac{1}{2}}[(2p)^{\frac{1}{2}}]D_{-\nu-\frac{1}{2}}[(2p)^{\frac{1}{2}}]$ $\operatorname{Re}p > 0$
8.12	$(1-e^{-2t})^{-\frac{1}{2}}$ $\cdot\cos[\nu \arccos(e^{-t})]$	$\pi 2^{-p}p^{-1}[B(\frac{1}{2}+\frac{1}{2}p+\frac{1}{2}\nu,\frac{1}{2}+\frac{1}{2}p-\frac{1}{2}\nu)]^{-1}$ $\operatorname{Re}p > 0$

	$f(t)$	$g(p) = \int_0^\infty f(t)\, e^{-pt} dt$
8.13	$t^{-\frac{1}{2}}(a^2-t^2)^{-\frac{1}{2}}$ $\cdot \cos[\nu \arccos(t/a)]$ $t<a$; 0 $t>a$	$(\frac{1}{2}\pi)^{3/2} p^{\frac{1}{2}} [I_{\frac{1}{2}\nu-\frac{1}{4}}(\frac{1}{2}ap) I_{-\frac{1}{2}\nu-\frac{1}{4}}(\frac{1}{2}ap) - I_{\frac{1}{2}\nu+\frac{1}{4}}(\frac{1}{2}ap) I_{-\frac{1}{2}\nu+\frac{1}{4}}(\frac{1}{2}ap)]$

1.9 Hyperbolic Functions

9.1	$\sin h(at)$	$a(p^2-a^2)^{-1}$ $\quad \mathrm{Re}\, p > \lvert \mathrm{Re}\, a \rvert$
9.2	$\cosh(at)$	$p(p^2-a^2)^{-1}$ $\quad \mathrm{Re}\, p > \lvert \mathrm{Re}\, a \rvert$
9.3	$t^{-1}\sinh(at)$	$\frac{1}{2}\log[(p+a)(p-a)^{-1}]$ $\quad \mathrm{Re}\, p > \lvert \mathrm{Re}\, a \rvert$
9.4	$t^{-1}[1-\cosh(at)]$	$\frac{1}{2}\log(1-a^2/p^2)$ $\quad \mathrm{Re}\, p > \lvert \mathrm{Re}\, a \rvert$
9.5	$t^{-1}\sinh^2(at)$	$-\frac{1}{4}\log(1-4a^2/p^2)$ $\quad \mathrm{Re}\, p > 2\lvert \mathrm{Re}\, a \rvert$
9.6	$t^{-1}[\cosh(at)-\cosh(bt)]$	$\frac{1}{2}\log[(p^2-b^2)(p^2-a^2)^{-1}]$ $\quad \mathrm{Re}\, p > \mathrm{Max}[\lvert \mathrm{Re}\, a \rvert, \lvert \mathrm{Re}\, b \rvert]$
9.7	$t^{-1}-\mathrm{csch}t$	$\psi(\frac{1}{2}+\frac{1}{2}p)-\log(\frac{1}{2}p)$ $\quad \mathrm{Re}\, p > 0$

	$f(t)$	$g(p) = \int_0^\infty f(t)e^{-pt}dt$
9.8	$\operatorname{sech}(at)$	$\frac{1}{2}a^{-1}[\psi(\frac{3}{4}+\frac{1}{4}p/a)-\psi(\frac{1}{4}+\frac{1}{4}p/a)$ $\operatorname{Re} p > \operatorname{Max}[-\operatorname{Re} a, -3\operatorname{Re} a]$
9.9	$t^{-1}(1-\operatorname{sech} t)$	$2\log\Gamma(\frac{3}{4}+\frac{1}{4}p)-2\log\Gamma(\frac{1}{4}+\frac{1}{4}p)$ $-\log(\frac{1}{4}p)$ $\operatorname{Re} p > 0$
9.10	$t^{\nu-1}\operatorname{csch}(at)$ $\operatorname{Re}\nu > 1$	$2(2a)^{-\nu}\Gamma(\nu)\zeta(\nu,\frac{1}{2}+\frac{1}{2}p/a)$ $\operatorname{Re} p > -\lvert\operatorname{Re} a\rvert$
9.11	$t^{\nu-1}\operatorname{sech}(at)$ $\operatorname{Re}\nu > 0$	$2^{1-2p}a^{-\nu}\Gamma(\nu)$ $\cdot[\zeta(\nu,\frac{1}{4}+\frac{1}{4}p/a)-\zeta(\nu,\frac{3}{4}+\frac{1}{4}p/a)]$ $\operatorname{Re} p > -\lvert\operatorname{Re} a\rvert$
9.12	$\operatorname{sech}^2 t$	$\frac{1}{2}p[\psi(\frac{1}{2}+\frac{1}{4}p)-\psi(\frac{1}{4}p)]-1$ $\operatorname{Re} p > -2$
9.13	$t^{-\frac{1}{2}}\sinh(at)$	$(\frac{1}{2}\pi)^{\frac{1}{2}}(p^2-a^2)^{-\frac{1}{2}}[p-(p^2-a^2)^{\frac{1}{2}}]^{\frac{1}{2}}$ $\operatorname{Re} p > \lvert\operatorname{Re} a\rvert$
9.14	$t^{-\frac{1}{2}}\cosh(at)$	$(\frac{1}{2}\pi)^{\frac{1}{2}}(p^2-a^2)^{-\frac{1}{2}}[p+(p^2-a^2)^{\frac{1}{2}}]^{\frac{1}{2}}$ $\operatorname{Re} p > \lvert\operatorname{Re} a\rvert$

	$f(t)$	$g(p) = \int_0^\infty f(t)e^{-pt}dt$
9.15	$t^{\nu-1}\sinh(at)$ $\mathrm{Re}\,\nu > -1$	$\frac{1}{2}\Gamma(\nu)[(p-a)^{-\nu}-(p+a)^{-\nu}]$ $\mathrm{Re}\,p > \lvert\mathrm{Re}\,a\rvert$
9.16	$t^{\nu-1}\cosh(at)$ $\mathrm{Re}\,\nu > 0$	$\frac{1}{2}\Gamma(\nu)[(p-a)^{-\nu}+(p+a)^{-\nu}]$ $\mathrm{Re}\,p > \lvert\mathrm{Re}\,a\rvert$
9.17	$\sinh^{2n}(at)$	$(2n)!a^{2n}p^{-1}(p^2-4a^2)^{-1}(p^2-4\cdot 4a^2)^{-1}$ $\cdots(p^2-n^2\cdot 4a^2)^{-1}$ $n=1,2,3,\cdots$ $\mathrm{Re}\,p > n\lvert\mathrm{Re}\,a\rvert$
9.18	$\sinh^{2n+1}(at)$	$(2n+1)!a^{2n+1}(p^2-a^2)^{-1}(p^2-9a^2)^{-1}$ $\cdots[p^2-(2n+1)^2a^2]^{-1}$ $n=0,1,2,\cdots;\mathrm{Re}\,p > (2n+1)\lvert\mathrm{Re}\,a\rvert$
9.19	$t^{-1}\sinh(at)$ $t < b$ 0 $t > b$	$\frac{1}{2}\log\left(\frac{p+a}{p-a}\right)+\frac{1}{2}\mathrm{Ei}[-b(p-a)]$ $-\frac{1}{2}\mathrm{Ei}[-b(p+a)]$ $= -\frac{1}{2}\log\left(\frac{a+p}{a-p}\right)+\frac{1}{2}\overline{\mathrm{Ei}}[b(a-p)]$ $-\frac{1}{2}\mathrm{Ei}[-b(a+p)]$
9.20	0 $t < b$ $t^{-1}\sinh(at)$ $t > b$	$-\frac{1}{2}\mathrm{Ei}[-b(p-a)]+\frac{1}{2}\mathrm{Ei}[-b(p+a)]$ $\mathrm{Re}\,p > \lvert\mathrm{Re}\,a\rvert$

	$f(t)$	$g(p) = \int_0^\infty f(t)e^{-pt}dt$
9.21	$0 \quad t < b$ $t^{-1}\cosh(at) \quad t > b$	$-\frac{1}{2}Ei[-b(p-a)]-\frac{1}{2}Ei[-b(p+a)]$ $\mathrm{Re}\, p > \lvert \mathrm{Re}\, a \rvert$
9.22	$0 \quad t < b$ $(\cosh t-\cosh b)^{\nu-1} \quad t > b$ $\mathrm{Re}\, \nu > 0$	$-i(\frac{1}{2}\pi)^{-\frac{1}{2}}e^{i\pi\nu}\Gamma(\nu)(\sinh b)^{\nu-\frac{1}{2}}$ $\cdot q_{p-\frac{1}{2}}^{\frac{1}{2}-\nu}(\cosh b)$ $= \Gamma(\nu)(\sinh b)^{\nu-1}\Gamma(1+p-\nu)$ $\cdot P_{\nu-1}^{-p}(\coth b)$ $\mathrm{Re}(1+p-\nu) > 0$
9.23	$\cosh^2(at)$	$p^{-1}(p^2-4a^2)^{-1}(p^2-2a^2)$
9.24	$\sinh^{\nu}(at)$ $\mathrm{Re}\, \nu > -1, \mathrm{Re}\, a > 0$	$2^{-\nu-1}a^{-1}B(\frac{1}{2}p/a-\frac{1}{2}\nu, \nu+1)$ $\mathrm{Re}\, p > \mathrm{Re}(a\nu)$
9.25	$[\cosh(at)-1]^{\nu}$ $\mathrm{Re}\, \nu > -\frac{1}{2}, \mathrm{Re}\, a > 0$	$2^{-\nu}a^{-1}B(p/a-\nu, 2\nu+1)$ $\mathrm{Re}\, p > \mathrm{Re}(a\nu)$
9.26	$(\cosh a+\cosh t)^{-1}$	$-p\,\mathrm{csch}\, a \sum_{n=0}^{\infty} (-1)^n \varepsilon_n (n^2-p^2)^{-1}e^{-na}$ $\mathrm{Re}\, p > -1$

	$f(t)$	$g(p) = \int_0^\infty f(t)e^{-pt}dt$
9.27	$\tanh(at)$	$(2a)^{-1}[\psi(\frac{1}{2}+\frac{1}{4}p/a)-\psi(\frac{1}{4}p/a)]-p^{-1}$ $\quad \mathrm{Re}\, p > 0$
9.28	$t^{-1}\tanh(at)$	$\log(\frac{1}{4}p/a)+2\log[\frac{\Gamma(\frac{1}{4}p/a)}{\Gamma(\frac{1}{2}+\frac{1}{4}p/a)}]$ $\quad \mathrm{Re}\, p > 0$
9.29	$t^{-1}-\coth t$	$\psi(\frac{1}{2}p)+p^{-1}-\log(\frac{1}{2}p)$ $\quad \mathrm{Re}\, p > 0$
9.30	$t^{\nu-1}\tanh(at)$ $\quad \mathrm{Re}\, \nu > -1$	$2^{1-2\nu}a^{-\nu}\Gamma(\nu)[\zeta(\nu,\frac{1}{4}p/a)-\zeta(\nu,\frac{1}{2}+\frac{1}{4}p/a)]$ $-p^{-\nu}\Gamma(\nu)$ $\quad \mathrm{Re}\, p > 0$
9.31	$t^{\nu-1}\coth(at)$ $\quad \mathrm{Re}\, \nu > 1$	$a^{-\nu}\Gamma(\nu)[2^{1-\nu}\zeta(\nu,\frac{1}{2}p/a)-(a/p)^{\nu}]$ $\quad \mathrm{Re}\, p > 0$
9.32	$t^{\nu-1}(\coth t-1)$ $\quad \mathrm{Re}\, \nu > 1$	$2^{1-\nu}\Gamma(\nu)\zeta(\nu,1+\frac{1}{2}p)$ $\quad \mathrm{Re}\, p > -2$
9.33	$\log(\sinh t)-\log t$ $= \log(\frac{\sinh t}{t})$	$p^{-1}[\log(\frac{1}{2}p)-p^{-1}-\psi(\frac{1}{2}p)]$ $\quad \mathrm{Re}\, p > 0$

	$f(t)$	$g(p) = \int_0^\infty f(t)e^{-pt}dt$
9.34	$\log(\cosh t)$	$(2p)^{-1}[\psi(\tfrac{1}{2}+p/4)-\psi(p/4)]-p^{-2}$ $\quad \mathrm{Re}\, p > 0$
9.35	$\sinh(at)\log t$	$(p^2-a^2)^{-1}\{\tfrac{1}{2}p \log[(p+a)(p-a)^{-1}]$ $-\gamma a-\tfrac{1}{2}a \log(p^2-a^2)]\}$ $\quad \mathrm{Re}\, p > \lvert\mathrm{Re}\, a\rvert$
9.36	$\cosh(at)\log t$	$(p^2-a^2)^{-1}\{\tfrac{1}{2}a \log[(p+a)(p-a)^{-1}]$ $-\gamma p-\tfrac{1}{2}p \log(p^2-a^2)\}$ $\quad \mathrm{Re}\, p > \lvert\mathrm{Re}\, a\rvert$
9.37	$\sinh(at^{\frac{1}{2}})$	$\tfrac{1}{2}a\pi^{\frac{1}{2}}p^{-3/2}\exp(\tfrac{1}{4}a^2/p)$ $\quad \mathrm{Re}\, p > 0$
9.38	$\cosh(at^{\frac{1}{2}})$	$p^{-1}+\tfrac{1}{2}a\pi^{\frac{1}{2}}p^{-3/2}\exp(\tfrac{1}{4}a^2/p)$ $\cdot\mathrm{Erf}(\tfrac{1}{2}ap^{-\frac{1}{2}})$ $\quad \mathrm{Re}\, p > 0$
9.39	$t^{\frac{1}{2}}\sinh(at^{\frac{1}{2}})$	$\pi^{\frac{1}{2}}p^{-5/2}(\tfrac{1}{2}p+\tfrac{1}{4}a^2)\exp(\tfrac{1}{4}a^2/p)$ $\cdot\mathrm{Erf}(\tfrac{1}{2}ap^{-\frac{1}{2}})-\tfrac{1}{2}ap^{-2}$ $\quad \mathrm{Re}\, p > 0$
9.40	$t^{\frac{1}{2}}\cosh(at^{\frac{1}{2}})$	$\pi^{\frac{1}{2}}p^{-5/2}(\tfrac{1}{2}p+\tfrac{1}{4}a^2)\exp(\tfrac{1}{4}a^2/p)$ $\quad \mathrm{Re}\, p > 0$

	$f(t)$	$g(p) = \int_0^\infty f(t)e^{-pt}dt$
9.41	$t^{-\frac{1}{2}}\sinh(at^{\frac{1}{2}})$	$(\pi/p)^{\frac{1}{2}}\exp(\frac{1}{4}a^2/p)$ $\cdot\mathrm{Erf}(\frac{1}{2}ap^{-\frac{1}{2}})$ Re $p > 0$
9.42	$t^{-\frac{1}{2}}\cosh(at^{\frac{1}{2}})$	$(\pi/p)^{\frac{1}{2}}\exp(\frac{1}{4}a^2/p)$ Re $p > 0$
9.43	$t^{-1}\sinh(at^{\frac{1}{2}})$	$\pi\mathrm{Erf}(\frac{1}{2}iap^{-\frac{1}{2}})$ Re $p > 0$
9.44	$t^{-3/4}\sinh(at^{\frac{1}{2}})$	$\pi(\frac{1}{2}a/p)^{\frac{1}{2}}\exp(\frac{1}{8}a^2/p)$ $\cdot I_{\frac{1}{4}}(\frac{1}{8}a^2/p)$ Re $p > 0$
9.45	$t^{-3/4}\cosh(at^{\frac{1}{2}})$	$\pi(\frac{1}{2}a/p)^{\frac{1}{2}}\exp(\frac{1}{8}a^2/p)$ $\cdot I_{-\frac{1}{4}}(\frac{1}{8}a^2/p)$ Re $p > 0$
9.46	$t^{-\frac{1}{4}}\sinh(at^{\frac{1}{2}})$	$\frac{1}{2}\pi(\frac{1}{2}a/p)^{3/2}\exp(\frac{1}{8}a^2/p)$ $\cdot[I_{-\frac{1}{4}}(\frac{1}{8}a^2/p)+I_{3/4}(\frac{1}{8}a^2/p)]$ Re $p > 0$
9.47	$t^{-\frac{1}{4}}\cosh(at^{\frac{1}{2}})$	$\frac{1}{2}\pi(\frac{1}{2}a/p)^{3/2}\exp(\frac{1}{8}a^2/p)$ $\cdot[I_{-3/4}(\frac{1}{8}a^2/p)+I_{\frac{1}{4}}(\frac{1}{8}a^2/p)]$ Re $p > 0$

	$f(t)$	$g(p) = \int_0^\infty f(t) e^{-pt} dt$
9.48	$t^n \sinh(at^{\frac{1}{2}})$ $n=0,1,2,\cdots$	$-i(-1)^n (2\pi)^{\frac{1}{2}} (2p)^{-n-1} \exp(\frac{1}{4}a^2/p)$ $\cdot He_{2n+1}[ia(2p)^{-\frac{1}{2}}]$ $\quad \mathrm{Re}\, p > 0$
9.49	$t^{n-\frac{1}{2}} \cosh(at^{\frac{1}{2}})$ $n=0,1,2,\cdots$	$(-1)^n (2\pi)^{\frac{1}{2}} (2p)^{-n-\frac{1}{2}} \exp(\frac{1}{4}a^2/p)$ $\cdot He_{2n}[ia(2p)^{-\frac{1}{2}}]$ $\quad \mathrm{Re}\, p > 0$
9.50	$t^{\nu-1} \sinh(at^{\frac{1}{2}})$ $\mathrm{Re}\, \nu > -\frac{1}{2}$	$\Gamma(2\nu) \exp(\frac{1}{8}a^2/p)(2p)^{-\nu}$ $\cdot \{D_{-2\nu}[-(2p)^{-\frac{1}{2}}a] - D_{-2\nu}[(2p)^{-\frac{1}{2}}a]\}$ $\mathrm{Re}\, p > 0$
9.51	$t^{\nu-1} \cosh(at^{\frac{1}{2}})$ $\mathrm{Re}\, \nu > 0$	$\Gamma(2\nu) \exp(\frac{1}{8}a^2/p)(2p)^{-\nu}$ $\cdot \{D_{-2\nu}[-(2p)^{-\frac{1}{2}}a] + D_{-2\nu}[(2p)^{-\frac{1}{2}}a]\}$ $\mathrm{Re}\, p > 0$
9.52	$t^{-\frac{1}{2}} \sinh^2(at^{\frac{1}{2}})$	$\frac{1}{2}(\pi/p)^{\frac{1}{2}}(e^{a^2/p}-1)$ $\quad \mathrm{Re}\, p > a$
9.53	$t^{-\frac{1}{2}} \cosh^2(at^{\frac{1}{2}})$	$\frac{1}{2}(\pi/p)^{\frac{1}{2}}(e^{a^2/p}+1)$ $\quad \mathrm{Re}\, p > a$

	$f(t)$	$g(p) = \int_0^\infty f(t)e^{-pt}dt$
9.54	$t^{\nu-1}e^{-bt^{1/2}}\sinh(at^{1/2})$ $\operatorname{Re}\nu > -\frac{1}{2}$	$(2p)^{-\nu}\Gamma(2\nu)\exp(\frac{b^2+a^2}{8p})$ $\cdot\{\exp(-\frac{1}{4}ab/p)D_{-2\nu}[(2p)^{-1/2}(b-a)]$ $-\exp(\frac{1}{4}ab/p)D_{-2\nu}[(2p)^{-1/2}(b+a)]\}$ $\operatorname{Re}p > 0$
9.55	$t^{\nu-1}e^{-bt^{1/2}}\cosh(at^{1/2})$ $\operatorname{Re}\nu > 0$	$(2p)^{-\nu}\Gamma(2\nu)\exp(\frac{b^2+a^2}{8p})$ $\cdot\{\exp(-\frac{1}{4}ab/p)D_{-2\nu}[(2p)^{-1/2}(b-a)]$ $+\exp(\frac{1}{4}ab/p)D_{-2\nu}[(2p)^{-1/2}(b+a)]\}$ $\operatorname{Re}p > 0$
9.56	$(a+t)^{-1/2}(t^2+at)^{-1/2}$ $\cdot\sinh[b(t^2+at)^{1/2}]$	$\pi a^{-1/2}e^{ap}\operatorname{Erfc}(z_1)\operatorname{Erf}(z_2)$ $z_{\frac{1}{2}} = (\frac{1}{2}a)^{1/2}[p\pm(p^2-b^2)^{1/2}]^{1/2}$ $\operatorname{Re}p > \lvert\operatorname{Re}b\rvert$
9.57	$(a+t)^{-1/2}(t^2+at)^{-1/2}$ $\cdot\cosh[b(t^2+at)^{1/2}]$	$\pi a^{-1/2}e^{ap}$ $\cdot\operatorname{Erfc}\{(\frac{1}{2}a)^{1/2}[p+(p^2-b^2)^{1/2}]^{1/2}\}$ $\operatorname{Re}p > \lvert\operatorname{Re}b\rvert$
9.58	$(t^2+at)^{-1/2}\cosh[b(t^2+at)^{1/2}]$	$K_o[\frac{1}{2}a(p^2-b^2)^{1/2}]e^{\frac{1}{2}ap}$ $\operatorname{Re}p > \lvert\operatorname{Re}b\rvert$

	$f(t)$	$g(p) = \int_0^\infty f(t)e^{-pt}dt$
9.59	$\sinh[b(t^2+at)^{\frac{1}{2}}]$	$\frac{1}{2}ab(p^2-b^2)^{-\frac{1}{2}}e^{\frac{1}{2}ap}K_1[\frac{1}{2}a(p^2-b^2)^{\frac{1}{2}}]$ $\operatorname{Re} p > \lvert\operatorname{Re} b\rvert$
9.60	$(a+t)^{-\frac{1}{2}}\sinh[b(t^2+at)^{\frac{1}{2}}]$	$(\frac{1}{2}\pi)^{\frac{1}{2}}b[p+(p^2-b^2)^{\frac{1}{2}}]^{-\frac{1}{2}}e^{\frac{1}{2}ap}$ $\cdot(p^2-b^2)^{-\frac{1}{2}}\exp[-\frac{1}{2}a(p^2-b^2)^{\frac{1}{2}}]$ $\operatorname{Re} p > \lvert\operatorname{Re} b\rvert$
9.61	$(t+a)^{-\frac{1}{2}}\cosh[b(t^2+at)^{\frac{1}{2}}]$	$(\frac{1}{2}\pi)^{\frac{1}{2}}[(p^2-b^2)^{\frac{1}{2}}+p]^{\frac{1}{2}}e^{\frac{1}{2}ap}$ $\cdot(p^2-b^2)^{-\frac{1}{2}}\exp[-\frac{1}{2}a(p^2-b^2)^{\frac{1}{2}}]$ $\operatorname{Re} p > \lvert\operatorname{Re} a\rvert$
9.62	$t^{-\frac{1}{2}}\cosh[b(t^2+at)^{\frac{1}{2}}]$	$(\frac{1}{2}\pi)^{\frac{1}{2}}[p+(p^2-b^2)^{\frac{1}{2}}]^{\frac{1}{2}}e^{\frac{1}{2}ap}$ $\cdot(p^2-b^2)^{-\frac{1}{2}}\exp[-\frac{1}{2}a(p^2+b^2)^{\frac{1}{2}}]$ $\operatorname{Re} p > \lvert\operatorname{Re} b\rvert$
9.63	$(t^2+at)^{-3/4}$ $\cdot\sinh[b(t^2+at)^{\frac{1}{2}}]$	$(\frac{1}{2}\pi b)^{\frac{1}{2}}e^{\frac{1}{2}ap}K_{\frac{1}{4}}(z_1)I_{\frac{1}{4}}(z_2)$ $z_{\frac{1}{2}} = \frac{1}{4}a[p\pm(p^2-b^2)^{\frac{1}{2}}]$ $\operatorname{Re} p > \lvert\operatorname{Re} b\rvert$
9.64	$(t^2+at)^{-3/4}$ $\cdot\cosh[b(t^2+at)^{\frac{1}{2}}]$	$(\frac{1}{2}\pi b)^{\frac{1}{2}}e^{\frac{1}{2}ap}K_{\frac{1}{4}}(z_1)I_{-\frac{1}{4}}(z_2)$ $z_{\frac{1}{2}} = \frac{1}{4}a[p\pm(p^2-b^2)^{\frac{1}{2}}]$ $\operatorname{Re} p > \lvert\operatorname{Re} a\rvert$

	$f(t)$	$g(p) = \int_0^\infty f(t)e^{-pt}dt$
9.65	$0 \quad t < a$ $\sinh[b(t^2-a^2)^{\frac{1}{2}}] \quad t > a$	$ab(p^2-b^2)^{-\frac{1}{2}}K_1[a(p^2-b^2)^{\frac{1}{2}}]$ $\mathrm{Re}\, p > \lvert\mathrm{Re}\, b\rvert$
9.66	$0 \quad t < a$ $t \sinh[b(t^2-a^2)^{\frac{1}{2}}] \quad t > a$	$a^2bp(p^2-b^2)^{-1}K_2[a(p^2-b^2)^{\frac{1}{2}}]$ $\mathrm{Re}\, p > \lvert\mathrm{Re}\, b\rvert$
9.67	$(at-t^2)^{-\frac{1}{2}}$ $\cdot\cosh[b(at-t^2)^{\frac{1}{2}}] \quad t < a$ $0 \quad t > a$	$\pi e^{-\frac{1}{2}ap}I_0[\frac{1}{2}a(p^2+b^2)^{\frac{1}{2}}]$
9.68	$0 \quad t < a$ $(t+a)^{-\frac{1}{2}}\sinh[b(t^2-a^2)^{\frac{1}{2}}] \quad t > a$	$(\frac{1}{2}\pi)^{\frac{1}{2}}b[(p^2-b^2)^{\frac{1}{2}}+p]^{-\frac{1}{2}}$ $\cdot(p^2-b^2)^{-\frac{1}{2}}\exp[-a(p^2-b^2)^{\frac{1}{2}}]$ $\mathrm{Re}\, p > \lvert\mathrm{Re}\, b\rvert$
9.69	$0 \quad t < a$ $(t+a)^{-\frac{1}{2}}\cosh[b(t^2-a^2)^{\frac{1}{2}}] \quad t > a$	$(\frac{1}{2}\pi)^{\frac{1}{2}}[p+(p^2-b^2)^{\frac{1}{2}}]^{\frac{1}{2}}$ $\cdot(p^2-b^2)^{-\frac{1}{2}}\exp[-a(p^2-b^2)^{\frac{1}{2}}]$ $\mathrm{Re}\, p > \lvert\mathrm{Re}\, a\rvert$

	$f(t)$	$g(p) = \int_0^\infty f(t)e^{-pt}dt$
9.70	$0 \quad t < a$ $(t+a)^{-\frac{1}{2}}(t^2-a^2)^{-\frac{1}{2}}$ $\cdot\sinh[b(t^2-a^2)^{\frac{1}{2}}]$ $t > a$	$\pi(2a)^{-\frac{1}{2}}e^{ap}\mathrm{Erfc}(z_1)\mathrm{Erf}(z_2)$ $z_{\frac{1}{2}} = a^{\frac{1}{2}}[p\pm(p^2-b^2)^{\frac{1}{2}}]^{\frac{1}{2}}$ $\mathrm{Re}\, p > \|\mathrm{Re}\, b\|$
9.71	$0 \quad t < a$ $(t+a)^{-\frac{1}{2}}(t^2-a^2)^{-\frac{1}{2}}$ $\cdot\cosh[b(t^2-a^2)^{\frac{1}{2}}]$ $t > a$	$\pi(2a)^{-\frac{1}{2}}e^{ap}$ $\cdot\mathrm{Erfc}\{a^{\frac{1}{2}}[p+(p^2-b^2)^{\frac{1}{2}}]^{\frac{1}{2}}\}$ $\mathrm{Re}\, p > \|\mathrm{Re}\, b\|$
9.72	$0 \quad t < a$ $(t-a)^{-\frac{1}{2}}\cosh[b(t^2-a^2)^{\frac{1}{2}}]$ $t > a$	$(\frac{1}{2}\pi)^{\frac{1}{2}}[p+(p^2-b^2)^{\frac{1}{2}}]^{\frac{1}{2}}$ $\cdot(p^2-b^2)^{-\frac{1}{2}}\exp[-a(p^2-b^2)^{\frac{1}{2}}]$ $\mathrm{Re}\, p > \|\mathrm{Re}\, b\|$
9.73	$0 \quad t < a$ $(t^2-a^2)^{-\frac{1}{2}}$ $\cdot\{[t+(t^2-a^2)^{\frac{1}{2}}]^{\nu}$ $-[t-(t^2-a^2)^{\frac{1}{2}}]^{\nu}\}$ $\cdot\sinh[b(t^2-a^2)^{\frac{1}{2}}]$ $t > a$	$a^{\nu}[(\frac{p+b}{p-b})^{\frac{1}{2}\nu} - (\frac{p-b}{p+b})^{\frac{1}{2}\nu}]$ $\cdot K_{\nu}[a(p^2-b^2)^{\frac{1}{2}}]$ $\mathrm{Re}\, p > \|\mathrm{Re}\, b\|$

	$f(t)$	$g(p) = \int_0^\infty f(t)e^{-pt}dt$
9.74	$0 \quad t < a$ $(t^2-a^2)^{-\frac{1}{2}}\{[t+(t^2-a^2)^{\frac{1}{2}}]^\nu$ $+[t-(t^2-a^2)^{\frac{1}{2}}]^\nu\}$ $\cdot\cosh[b(t^2-a^2)^{\frac{1}{2}}]$ $t > a$	$a^\nu[(\frac{p+b}{p-b})^{\frac{1}{2}\nu}+(\frac{p-b}{p+b})^{\frac{1}{2}\nu}]$ $\cdot K_\nu[a(p^2-b^2)^{\frac{1}{2}}]$ $\mathrm{Re}\, p > \lvert\mathrm{Re}\, b\rvert$
9.75	$0 \quad t < a$ $(t^2-a^2)^{-\frac{1}{2}}\log[t/a+(t^2/a^2-1)^{\frac{1}{2}}]$ $\cdot\sinh[b(t^2-b^2)^{\frac{1}{2}}]$ $t > a$	$\frac{1}{2}\log(\frac{p+b}{p-b})$ $\cdot K_o[a(p^2-b^2)^{\frac{1}{2}}]$ $\mathrm{Re}\, p > \lvert\mathrm{Re}\, b\rvert$
9.76	$t^{-\frac{1}{2}}e^{-b/t}\sinh(a/t)$	$\frac{1}{2}(\pi/p)^{\frac{1}{2}}\{\exp[-2p^{\frac{1}{2}}(b-a)^{\frac{1}{2}}]$ $-\exp[-2p^{\frac{1}{2}}(b+a)^{\frac{1}{2}}]\}$ $b > a \qquad \mathrm{Re}\, p > 0$
9.77	$t^{-\frac{1}{2}}e^{-b/t}\cosh(a/t)$	$\frac{1}{2}(\pi/p)^{\frac{1}{2}}\{\exp[-2p^{\frac{1}{2}}(b-a)^{\frac{1}{2}}]$ $+\exp[-2p^{\frac{1}{2}}(b+a)^{\frac{1}{2}}]\}$ $b > a \qquad \mathrm{Re}\, p > 0$

	$f(t)$	$g(p) = \int_0^\infty f(t)e^{-pt}dt$
9.78	$t^{-3/2}e^{-b/t}\sinh(a/t)$	$\frac{1}{2}\pi^{\frac{1}{2}}\{(b-a)^{-\frac{1}{2}}\exp[-2p^{\frac{1}{2}}(b-a)^{\frac{1}{2}}]$ $-(b+a)^{-\frac{1}{2}}\exp[-2p^{\frac{1}{2}}(b+a)^{\frac{1}{2}}]\}$ $b > a, \quad \mathrm{Re}\, p > 0$
9.79	$t^{-3/2}e^{-b/t}\cosh(a/t)$	$\frac{1}{2}\pi^{\frac{1}{2}}\{(b-a)^{-\frac{1}{2}}\exp[-2p^{\frac{1}{2}}(b-a)^{\frac{1}{2}}]$ $+(b+a)^{-\frac{1}{2}}\exp[-2p^{\frac{1}{2}}(b+a)^{\frac{1}{2}}]\}$ $b > a, \quad \mathrm{Re}\, p > 0$
9.80	$t^{\nu-1}e^{-b/t}\sinh(a/t)$	$p^{-\frac{1}{2}\nu}\{(b-a)^{\frac{1}{2}\nu}K_\nu[2p^{\frac{1}{2}}(b-a)^{\frac{1}{2}}]$ $-(b+a)^{\frac{1}{2}\nu}K_\nu[2p^{\frac{1}{2}}(b+a)^{\frac{1}{2}}]\}$ $b > a, \quad \mathrm{Re}\, p > 0$
9.81	$t^{\nu-1}e^{-b/t}\cosh(a/t)$	$p^{-\frac{1}{2}\nu}\{(b-a)^{\frac{1}{2}\nu}K_\nu[2p^{\frac{1}{2}}(b-a)^{\frac{1}{2}}]$ $+(b+a)^{\frac{1}{2}\nu}K_\nu[2p^{\frac{1}{2}}(b+a)^{\frac{1}{2}}]\}$ $b > a$
9.82	$e^{-a\sinh t}$ $\mathrm{Re}\, a > 0$	$\pi\csc(\pi p)[\mathbf{J}_p(a)-J_p(a)]$
9.83	$e^{-a\cosh t}$ $\mathrm{Re}\, a > 0$	$\csc(\pi p)[\int_0^\pi e^{a\cos t}\cos(pt)dt$ $-\pi I_p(a)]$

	$f(t)$	$g(p) = \int_0^\infty f(t)e^{-pt}dt$
9.84	$\sinh[a(1-e^{-t})^{\frac{1}{2}}]$ $\cdot(e^t-1)^{-\frac{1}{2}}$	$\pi^{\frac{1}{2}}\Gamma(\frac{1}{2}+p)(\frac{1}{2}a)^{-p}\mathbf{L}_p(a)$ Re $p > -\frac{1}{2}$
9.85	$\cosh[a(1-e^{-t})^{\frac{1}{2}}]$ $\cdot(e^t-1)^{-\frac{1}{2}}$	$\pi^{\frac{1}{2}}\Gamma(\frac{1}{2}+p)(\frac{1}{2}a)^{-p}I_p(a)$ Re $p > -\frac{1}{2}$
9.86	$\tanh[\frac{1}{2}\pi(e^{2t}-1)^{\frac{1}{2}}]$	$2^{-p}\zeta(p-1)$ Re $p > 0$
9.87	$[\sinh(\frac{1}{2}t)]^{2b}$ $\cdot\exp[-2a\coth(\frac{1}{2}t)]$	$\frac{1}{2}a^{\frac{1}{2}b-\frac{1}{2}}\Gamma(p-b)[W_{\frac{1}{2}-p,b}(4a)$ $-(p-b)W_{-\frac{1}{2}-p,b}(4a)]$ Re $a > 0$, Re $p >$ Re b
9.88	$\frac{1}{2}t^{-1}\mathrm{csch}t$ $\cdot(1+t-e^t)$	$\log\Gamma(\frac{1}{2}+\frac{1}{2}p)-\log\Gamma(\frac{1}{2}p)$ $-\frac{1}{2}\psi(\frac{1}{2}+\frac{1}{2}p)$ Re $p > 0$
9.89	$\begin{matrix}\sin\\ \cos\end{matrix}(bt)\begin{matrix}\tanh\\ \coth\end{matrix}(at^{\frac{1}{2}})$	See Mordell, L. J., 1920: Mess. of Math.49, 65-72
9.90	$(\pi t)^{-\frac{1}{2}}\mathrm{csch}[(\pi t)^{\frac{1}{2}}]$ $\cdot\sin[(2\nu+1)(\pi t)^{\frac{1}{2}}]$	For this and similar results see Mordell, L.J., 1933: Acta Math. 61, 323-360 and Quart. J. Math. 1920: 48, 339-342.

1.10 Inverse Hyperbolic Functions

	$f(t)$	$g(p) = \int_0^\infty f(t)e^{-pt}dt$
10.1	$\sinh^{-1}(at)$	$\frac{1}{2}\pi p^{-1}[\mathbf{H}_o(p/a)-Y_o(p/a)]$ Re $p > 0$
10.2	$\cosh^{-1}(1+at)$	$p^{-1}e^{p/a}K_o(p/a)$ Re $p > 0$
10.3	0 $t < a$ $\cosh^{-1}(t/a)$ $t > a$	$p^{-1}K_o(ap)$ Re $p > 0$
10.4	$t \sinh^{-1}t$	$\frac{1}{2}\pi p^{-2}[\mathbf{H}_o(p)-Y_o(p)$ $+p\mathbf{H}_1(p)-pY_1(p)]-p^{-1}$ Re $p > 0$
10.5	$(1+t^2)^{\frac{1}{2}}-t \sinh^{-1}t$	$p^{-2}S_{2,o}(p)$ Re $p > 0$
10.6	$\sinh^{-1}[(t/a)^{\frac{1}{2}}]$	$p^{-1}e^{ap}K_o(ap)$ Re $p > 0$
0.7	$(1+t^2)^{-\frac{1}{2}}\exp[n \sinh^{-1}t]$ $n = 0,1,2,\cdots$	$\frac{1}{2}[S_n(p)-\pi E_n(p)-\pi Y_n(p)]$ Re $p > 0$

	$f(t)$	$g(p) = \int_0^\infty f(t)e^{-pt}dt$
10.8	$(1+t^2)^{-\frac{1}{2}}\exp[-n \sinh^{-1}t]$ $n = 0,1,2,\cdots$	$\frac{1}{2}(-1)^{n+1}[S_n(p)+\Pi E_n(p)+\pi Y_n(p)]$ $\operatorname{Re} p > 0$
10.9	$(1+t^2)^{-\frac{1}{2}}$ $\cdot\exp(-\nu \sinh^{-1}(t)$	$\pi\csc(\pi\nu)[\mathbf{J}_\nu(p)-J_\nu(p)]$ $\operatorname{Re} p > 0$
10.10	$t^{-\frac{1}{2}}(a^2+t^2)^{-\frac{1}{2}}$ $\cdot\exp[\nu \sinh^{-1}(t/a)]$	$(\frac{1}{2}\pi)^{3/2}p^{\frac{1}{2}}[J_{\frac{1}{2}\nu+\frac{1}{4}}(\frac{1}{2}ap)J_{\frac{1}{2}\nu-\frac{1}{4}}(\frac{1}{2}ap)$ $+ Y_{\frac{1}{2}\nu+\frac{1}{4}}(\frac{1}{2}ap)Y_{\frac{1}{2}\nu-\frac{1}{4}}(\frac{1}{2}ap)]$ $\operatorname{Re} p > 0$
10.11	$t^{-\frac{1}{2}}(a^2+t^2)^{-\frac{1}{2}}$ $\cdot\exp[-\nu \sinh^{-1}(t/a)]$	$(\frac{1}{2}\pi)^{3/2}p^{\frac{1}{2}}[J_{\frac{1}{2}\nu+\frac{1}{4}}(\frac{1}{2}ap)Y_{\frac{1}{2}\nu-\frac{1}{4}}(\frac{1}{2}ap)$ $-J_{\frac{1}{2}\nu-\frac{1}{4}}(\frac{1}{2}ap)Y_{\frac{1}{2}\nu+\frac{1}{4}}(\frac{1}{2}ap)]$ $\operatorname{Re} p > 0$
10.12	$\sinh[(2n+1)\sinh^{-1}t]$ $n = 0,1,2,\cdots$	$O_{2n+1}(p)$ $\operatorname{Re} p > 0$
10.13	$\cosh(2n \sinh^{-1}t)$ $n = 0,1,2,\cdots$	$O_{2n}(p)$ $\operatorname{Re} p > 0$

	$f(t)$	$g(p) = \int_0^\infty f(t)e^{-pt}dt$
10.14	$\sinh(\nu \sinh^{-1}at)$	$\nu(\nu^2-1)^{-1}[p^{-1}S_{2,\nu}(p/a)-a^{-1}]$ $= \nu p^{-1}S_{o,\nu}(p/a)$ Re $p > 0$
10.15	$\cosh(\nu \sinh^{-1}t)$	$p^{-1}S_{1,\nu}(p)$ Re $p > 0$
10.16	$\sinh[\nu \cosh^{-1}(1+t/a)]$	$\nu p^{-1}e^{ap}K_\nu(ap)$ Re $p > 0$
10.17	$(1+t^2)^{-\frac{1}{2}}\sinh(\nu \sinh^{-1}t)$	$\nu S_{-1,\nu}(p)$ Re $p > 0$
10.18	$(1+t^2)^{-\frac{1}{2}}\cosh(\nu \sinh^{-1}t)$	$S_{o,\nu}(p)$ Re $p > 0$
10.19	$t^{-\frac{1}{2}}(t^2+a^2)^{-\frac{1}{2}}$ $\cdot\sinh[\nu\sinh^{-1}(t/a)]$	$-\frac{1}{8}i\pi^{3/2}p^{\frac{1}{2}}[e^{i\frac{1}{2}\pi\nu}H^{(1)}_{\frac{1}{2}+\frac{1}{2}\nu}(\tfrac{1}{2}ap)H^{(2)}_{\frac{1}{2}-\frac{1}{2}\nu}(\tfrac{1}{2}ap)$ $-e^{-i\frac{1}{2}\pi\nu}H^{(1)}_{\frac{1}{2}-\frac{1}{2}\nu}(\tfrac{1}{2}ap)H^{(2)}_{\frac{1}{2}+\frac{1}{2}\nu}(\tfrac{1}{2}ap)]$ Re $p > 0$
10.20	$t^{-\frac{1}{2}}(t^2+a^2)^{-\frac{1}{2}}$ $\cdot\cosh[\nu \sinh^{-1}(t/a)]$	$\frac{1}{8}\pi^{3/2}p^{\frac{1}{2}}[e^{i\frac{1}{2}\pi\nu}H^{(1)}_{\frac{1}{2}+\frac{1}{2}\nu}(\tfrac{1}{2}ap)H^{(2)}_{\frac{1}{2}-\frac{1}{2}\nu}(\tfrac{1}{2}ap)$ $+e^{-i\frac{1}{2}\pi\nu}H^{(1)}_{\frac{1}{2}-\frac{1}{2}\nu}(\tfrac{1}{2}ap)H^{(2)}_{\frac{1}{2}+\frac{1}{2}\nu}(\tfrac{1}{2}ap)]$ Re p 0

	$f(t)$	$g(p) = \int_0^\infty f(t)e^{-pt}dt$
10.21	$\sinh[\nu \sinh^{-1}(t/a)^{\frac{1}{2}}]$	$\nu p^{-1}e^{ap}K_\nu(ap)$ Re $p > 0$
10.22	$(t^2+2at)^{-\frac{1}{2}}$ $\cdot\cosh[\nu \sinh^{-1}(t/a)^{\frac{1}{2}}]$	$e^{ap}K_\nu(ap)$ Re $p > 0$
10.23	$(t^2+2at)^{-\frac{1}{2}}$ $\cdot\cosh[\nu\cosh^{-1}(1+t/a)]$	$e^{ap}K_\nu(ap)$ Re $p > 0$
10.24	$\cosh[\nu \sinh^{-1}(t/a)^{\frac{1}{2}}]$ $\cdot[t(t+a)(t+2a)]^{-\frac{1}{2}}$	$(\frac{1}{2}p/\pi)^{\frac{1}{2}}e^{ap}K_{\frac{1}{4}+\frac{1}{2}\nu}(\frac{1}{2}ap)K_{\frac{1}{4}-\frac{1}{2}\nu}(\frac{1}{2}ap)$ Re $p > 0$
10.25	$\cosh[\nu \cosh^{-1}(1+t/a)]$ $\cdot[t(t+a)(t+2a)]^{-\frac{1}{2}}$	$(\frac{1}{2}p/\pi)^{\frac{1}{2}}e^{ap}K_{\frac{1}{4}+\frac{1}{2}\nu}(\frac{1}{2}ap)K_{\frac{1}{4}-\frac{1}{2}\nu}(\frac{1}{2}ap)$ Re $p > 0$
10.26	$t^{-\frac{1}{2}}(a^2+t^2)^{-\frac{1}{2}}$ $\cdot[\cosh(\nu \sinh^{-1}t/a)$ $+i \sinh(\nu \sinh^{-1}t/a)]$	$\frac{1}{4}\pi^{3/2}e^{i\frac{1}{2}\pi\nu}p^{\frac{1}{2}}H^{(1)}_{\frac{1}{2}+\frac{1}{2}\nu}(\frac{1}{2}ap)H^{(2)}_{\frac{1}{2}-\frac{1}{2}\nu}(\frac{1}{2}ap)$ Re $p > 0$

1.11 Orthogonal Polynomials

	$f(t)$	$g(p) = \int_0^\infty f(t) e^{-pt} dt$
11.1	$t^{-\frac{1}{2}} T_n(1-2t)$	$(\pi/p)^{\frac{1}{2}} {}_2F_0(-n,n;;p^{-1})$ Re $p > 0$
11.2	$t^{\frac{1}{2}} U_n(2t-1)$	$(-1)^{n-1} \frac{1}{2} n \pi^{\frac{1}{2}} p^{-3/2} {}_2F_0(-n,n+2;;p^{-1})$ Re $p > 0$
11.3	$t^{-\frac{1}{2}}(1+t)^n T_n[(1-t)(1+t)^{-1}]$	$2^{-n} \pi^{\frac{1}{2}} p^{-n-\frac{1}{2}} He_{2n}[(2p)^{\frac{1}{2}}]$ Re $p > 0$
11.4	$t^{\frac{1}{2}}(1+t)^n U_n[(1-t)(1+t)^{-1}]$	$2^{-n-3/2} \pi^{\frac{1}{2}} (n+1) p^{-n-2} He_{2n+1}[(2p)^{\frac{1}{2}}]$ Re $p > 0$
11.5	0 $t < 1$; $(t^2-1)^{-\frac{1}{2}} T_n(t)$ $t > 1$	$K_n(p)$ Re $p > 0$
11.6	0 $t < 1$; $(t^2-1)^{\frac{1}{2}} U_n(t)$ $t > 1$	$\pi p^{-1} K_n(p)$ Re $p > 0$

	$f(t)$	$g(p) = \int_0^\infty f(t)e^{-pt}dt$
11.7	$t^{-\frac{1}{2}}(b^2-t^2)^{-\frac{1}{2}}T_n(t/b)$ $t < b$ $0 \quad t > b$	$(\frac{1}{2}\pi)^{3/2}p^{\frac{1}{2}}[I_{-\frac{1}{2}n-\frac{1}{4}}(\frac{1}{2}bp)$ $\cdot I_{\frac{1}{2}n-\frac{1}{4}}(\frac{1}{2}bp)-I_{\frac{1}{2}n+\frac{1}{4}}(\frac{1}{2}bp)$ $\cdot I_{\frac{1}{4}-\frac{1}{2}n}(\frac{1}{2}bp)]$ Re $p > 0$
11.8	$0 \quad t < a$ $(t-a)^{-\frac{1}{2}}(b-t)^{-\frac{1}{2}}$ $\cdot T_n[(2t-a-b)/(b-a)]$ $a < t < b$ $0 \quad t > b$	$\pi\exp[-\frac{1}{2}(a+b)p]$ $\cdot I_n[\frac{1}{2}(b-a)p]$
11.9	$0 \quad t < a$ $t^{-\frac{1}{2}}(t^2-a^2)^{-\frac{1}{2}}T_n(a/t) \quad t > a$	$(\pi/a)^{\frac{1}{2}}D_{n-\frac{1}{2}}[(2ap)^{\frac{1}{2}}]D_{-n-\frac{1}{2}}[(2ap)^{\frac{1}{2}}]$ Re $p > 0$
11.10	$(1-e^{-2t})^{-\frac{1}{2}}T_n(e^{-t})$	$\pi\Gamma(p)2^{-p}[\Gamma(\frac{1}{2}+\frac{1}{2}p-\frac{1}{2}n)$ $\cdot\Gamma(\frac{1}{2}+\frac{1}{2}p+\frac{1}{2}n)]^{-1}$ Re $p > 0$
11.11	$He_{2n}(t)$	$(2n-1)!(-2)^{-n}[(n-1)!]^{-1}p^{-1}$ $\cdot {}_2F_0(-n,1;;2p^{-2})$ Re $p > 0$

	$f(t)$	$g(p) = \int_0^\infty f(t) e^{-pt} dt$
11.12	$He_{2n+1}(t)$	$(2n+1)!(-2)^{-n}(n!)^{-1}p^{-2} \cdot {}_2F_0(-n,1;;2p^{-2})$ Re $p > 0$
11.13	$t^{\mu-1}He_n(t)$ Re $\mu > 0$ for n even Re $\mu > -1$ for n odd	$\sum_{m=0}^{[\frac{n}{2}]} [m!(n-2m)!]^{-1} n! \Gamma(\mu+n-2m) \cdot (-\frac{1}{2})^m p^{2m-\mu-n}$ Re $p > 0$
11.14	$t^{-\frac{1}{2}}He_{2n}(at^{\frac{1}{2}})$	$2^{-n}\pi^{\frac{1}{2}}(2n)!(n!)^{-1}p^{-n-\frac{1}{2}} \cdot (\frac{1}{2}a^2-p)^n$ Re $p > 0$
11.15	$He_{2n+1}(at^{\frac{1}{2}})$	$2^{-n-1}\pi^{\frac{1}{2}}a(2n+1)!(n!)^{-1}p^{-n-\frac{3}{2}} \cdot (\frac{1}{2}a^2-p)^n$ Re $p > 0$
11.16	$t^{\frac{1}{2}n-\frac{1}{2}}He_{2n}(t^{\frac{1}{2}})$	$2^{-\frac{1}{2}n}\pi^{\frac{1}{2}}n!\, p^{-\frac{1}{2}-\frac{1}{2}n}P_n[(2p)^{-\frac{1}{2}}]$ Re $p > 0$
11.17	$t^{n+\frac{1}{2}}He_{2n+1}(t^{\frac{1}{2}})$	$2^{-n}(2n+1)!p^{-n-2}U_{n+1}(p^{-1}-1)$ Re $p > 0$

	$f(t)$	$g(p) = \int_0^\infty f(t)e^{-pt}dt$
11.18	$t^{n-1}He_{2n}(t^{\frac{1}{2}})$	$2^{1-n}(2n-1)!(-p)^{-n}T_n(1-p^{-1})$ Re $p > 0$
11.19	$t^{\nu+\frac{1}{2}n}He_n[(2t)^{\frac{1}{2}}]$ Re $\nu > -\frac{1}{2}n$ if n even; Re $\nu > -\frac{1}{2}n-\frac{1}{2}$ if n odd	$2^{-\frac{1}{2}n}n!\Gamma(\nu+1)p^{-\nu-1-\frac{1}{2}n} \cdot C_n^{\nu+1}(p^{-\frac{1}{2}})$ Re $p > 0$
11.20	$t^{-\frac{1}{2}-\frac{1}{2}n}\exp(-\frac{1}{2}a/t) \cdot He_n[(a/t)^{\frac{1}{2}}]$	$2^{\frac{1}{2}n}\pi^{\frac{1}{2}}p^{\frac{1}{2}n-\frac{1}{2}}\exp[-(2ap)^{\frac{1}{2}}]$ Re $p > 0$
11.21	$t^{-\frac{1}{2}}\exp[-(a+b)t] \cdot He_n[(4at)^{\frac{1}{2}}] \cdot He_n[(4bt)^{\frac{1}{2}}]$	$\pi^{\frac{1}{2}}n!(a+b+p)^{-\frac{1}{2}n-\frac{1}{2}}(a+b-p)^{\frac{1}{2}n} \cdot P_n\{2(ab)^{\frac{1}{2}}[(a+b)^2-p^2]^{-\frac{1}{2}}\}$ Re $p > -$Re$(a+b)$
11.22	$t^{\nu-\frac{1}{2}n-1}He_n(t^{\frac{1}{2}})$ Re $\nu>\frac{1}{2}n$ if n odd; Re $\nu > \frac{1}{2}n-\frac{1}{2}$ if n even	$p^{-\nu}\Gamma(\nu)\,{}_2F_1(-\frac{1}{2}n,\frac{1}{2}-\frac{1}{2}n;1-\nu;2p)$ Re $p > 0$
11.23	$t^{-\frac{1}{2}}[(He_n(a+t^{\frac{1}{2}}) + He_n(a-t^{\frac{1}{2}})]$	$(2\pi/p)^{\frac{1}{2}}(1-\frac{1}{2}p^{-1})^{\frac{1}{2}n} \cdot He_n[a(1-\frac{1}{2}p^{-1})^{-\frac{1}{2}}]$ Re $p > 0$

	$f(t)$	$g(p) = \int_0^\infty f(t) e^{-pt} dt$
11.24	$t^{-\frac{1}{2}} He_n[(at)^{\frac{1}{2}}] \cdot He_n[(-at)^{\frac{1}{2}}]$	$n! \pi^{\frac{1}{2}} i^n p^{-\frac{1}{2}} P_n(\frac{1}{2}a/p)$ Re $p > 0$
11.25	$t^{-\frac{1}{2}} He_{2n}[(2at)^{\frac{1}{2}}] \cdot He_{2m}[(2bt)^{\frac{1}{2}}]$	$\pi^{\frac{1}{2}}(-2)^{-m-n}[(m+n)!]^{-1} p^{-m-n-\frac{1}{2}}(2m+2n)! \cdot (p-a)^n (p-b)^m {}_2F_1[-m,-n;-m-n+\frac{1}{2}; p(p-a-b)(p-a)^{-1}(p-b)^{-1}]$ Re $p > 0$
11.26	$t^{-\frac{1}{2}} He_{2n+1}[(2at)^{\frac{1}{2}}] \cdot He_{2m+1}[(2bt)^{\frac{1}{2}}]$	$\pi^{\frac{1}{2}} 2^{-m-n-1}(ab)^{\frac{1}{2}}[(m+n+1)!]^{-1}(2m+2n+2)! \cdot p^{-m-n-3/2}(p-a)^n (p-b)^m {}_2F_1[-m,-n; -m-n-\frac{1}{2}; p(p-a-b)(p-a)^{-1}(p-b)^{-1}]$ Re $p > 0$
11.27	$He_{2n+1}[a(1-e^{-t})^{\frac{1}{2}}]$	$(-2)^n a B(p, 3/2+n) n! L_n^{p+\frac{1}{2}}(\frac{1}{2}a^2)$ Re $p > 0$
11.28	$(e^t-1)^{-\frac{1}{2}} \cdot He_{2n}[a(1-e^{-t})^{\frac{1}{2}}]$	$(-2)^n B(p+\frac{1}{2}, n+\frac{1}{2}) n! L_n^p(\frac{1}{2}a^2)$ Re $p > 0$
11.29	$P_n(1+t)$	$(\frac{1}{2}\pi p)^{-\frac{1}{2}} e^p K_{n+\frac{1}{2}}(p)$ Re $p > 0$

	$f(t)$	$g(p) = \int_0^\infty f(t)e^{-pt}dt$
11.30	$P_n(1-t)$	$e^{-p}(p^{-1}\frac{d}{dp})^n(p^{-1}e^p)$ Re $p > 0$
11.31	$P_n(1-t)$ $t < 2$ 0 $t > 2$	$(\frac{1}{2}p/\pi)^{-\frac{1}{2}}e^{-p}I_{n+\frac{1}{2}}(p)$
11.32	$(t+1)^n P_n(\frac{t-1}{t+1})$	$n!p^{-n-1}L_n(p)$ Re $p > 0$
11.33	$P_{2n+1}(t^{\frac{1}{2}})$	$(-1)^n\frac{1}{2}\pi^{\frac{1}{2}}(2n+1)!p^{-3/2}$ $\cdot {}_2F_0(-n,\frac{1}{2}+n;;p^{-1})$ Re $p > 0$
11.34	$t^{-\frac{1}{2}}P_{2n}(t^{\frac{1}{2}})$	$(-1)^n(2n)!(\pi/p)^{\frac{1}{2}}{}_2F_0(-n,\frac{1}{2}+n;;p^{-1})$ Re $p > 0$
11.35	$P_n(e^{-t})$	$(p-1)(p-2)\cdots(p-n+1)$ $\cdot[(p+n)(p+n-2)\cdots(p-n+2)]^{-1}$ Re $p > 0$
11.36	$P_{2n}(\cos at)$	$[p^2+a^2][p^2+(3a)^2]\cdots[p^2+(2n-1)^2a^2]$ $\cdot\{p[p^2+(2a)^2][p^2+(4a)^2]\cdots$ $[p^2+(2na)^2]\}^{-1}$ Re $p > 0$

	$f(t)$	$g(p) = \int_0^\infty f(t)e^{-pt}dt$
11.37	$P_{2n+1}(\cos at)$	$p[p^2+(2a)^2]\cdots[p^2+(2na)^2]$ $\cdot\{(p^2+a^2)[p^2+(3a)^2]\cdots$ $[p^2+(2n+1)^2a^2]\}^{-1}$ $\quad \mathrm{Re}\, p > 0$
11.38	$P_{2n}(\cosh at)$	$[p^2-a^2][p^2-(3a)^2]\cdots[p^2-(2n-1)^2a^2]$ $\cdot\{p[p^2-(2a)^2][p^2-(4a)^2]\cdots$ $[p^2-(2na)^2]\}^{-1}$ $\quad \mathrm{Re}\, p > 2n\|a\|$
11.39	$P_{2n+1}(\cosh at)$	$p[p^2-(2a)^2]\cdots[p^2-(2na)^2]$ $\cdot\{(p^2-a^2)[p^2-(3a)^2]\cdots$ $[p^2-(2n+1)^2a^2]\}^{-1}$ $\quad \mathrm{Re}\, p > (2n+1)\|a\|$
11.40	$L_n(t)$	$p^{-n-1}(p-1)^n \quad \mathrm{Re}\, p > 0$
11.41	$L_n^\alpha(t)$	$\sum_{m=0}^{n} \binom{\alpha+m-1}{m} p^{m-n-1}(p-1)^{n-m} \quad \mathrm{Re}\, p > 0$
11.42	$t^n L_n(t)$	$n!p^{-n-1}P_n(1-2/p) \quad \mathrm{Re}\, p > 0$

	$f(t)$	$g(p) = \int_0^\infty f(t)e^{-pt}dt$
11.43	$t^{\alpha}L_n^{\alpha}(t)$	$(n!)^{-1}\Gamma(\alpha+n+1)p^{-\alpha-n-1}(p-1)^n$ $\mathrm{Re}\, p > 0$
11.44	$t^{\nu}L_n^{\alpha}(t)$ $\mathrm{Re}\, \nu > -1$	$(n!)^{-1}\Gamma(\nu+n+1)p^{-\nu-n-1}(p-1)^n$ $\cdot {}_2F_1[-n,\alpha-\nu;-\nu-n;p/(p-1)]$ $=[n!\Gamma(\alpha+1)]^{-1}\Gamma(\alpha+n+1)\Gamma(\nu+1)$ $\cdot p^{-\nu-1}{}_2F_1[-n,\nu+1;\alpha+1;p^{-1}]$ $\mathrm{Re}\, p > 0$
11.45	$t^{n+2\alpha}L_n^{\alpha}(t)$ $\mathrm{Re}\, \alpha > -\frac{1}{2}-\frac{1}{2}n$	$\pi^{-\frac{1}{2}}2^{2\alpha}\Gamma(\frac{1}{2}+\alpha)\Gamma(n+\alpha+1)$ $\cdot p^{-n-1-2\alpha}C_n^{\alpha+\frac{1}{2}}(1-2/p)$ $\mathrm{Re}\, p > 0$
11.46	$t^{\nu+n}L_n^{\alpha}(t)$ $\mathrm{Re}\, \nu > -1-n$	$\Gamma(\nu+n+1)p^{-\nu-n-1}P_n^{\alpha,\nu-\alpha}(1-2/p)$ $\mathrm{Re}\, p > 0$
11.47	$t^{2\alpha}[L_n^{\alpha}(t)]^2$ $\mathrm{Re}\, \alpha > -\frac{1}{2}$	$2^{2\alpha}\pi^{-1}(n!)^{-2}\Gamma(\frac{1}{2}+\alpha)\Gamma(\frac{1}{2}+n)p^{-2\alpha-1}$ $\cdot {}_2F_1[-n,\frac{1}{2}+\alpha;\frac{1}{2}-n;(1-2/p)^2]$ $\mathrm{Re}\, p > 0$

	$f(t)$	$g(p) = \int_0^\infty f(t)e^{-pt}dt$
11.48	$L_n(at)L_n(bt)$	$p^{-n-1}(p-a-b)^n \cdot P_n[\frac{p^2-(a+b)p+2ab}{p(p-a-b)}]$ Re $p > 0$
11.49	$L_n[(a-b)t] \cdot L_n[(a+b)t]$	$p^{-n-1}(p-2a)^n \cdot P_n[\frac{p^2+2(1-2a)(a^2-b^2)}{p(p-2a)}]$ Re $p > 0$
11.50	$t^{\alpha}L_n^{\alpha}(bt)L_m^{\alpha}(ct)$ Re $\alpha > -1$	$(m!n!)^{-1}\Gamma(m+n+\alpha+1)p^{-m-n-a-1} \cdot (p-b)^n(p-c)^m {}_2F_1[-m,-n;-m-n-a;p(p-b-c)(p-b)^{-1}(p-c)^{-1}]$ Re $p > 0$
11.51	$t^{-n}e^{-b/t}L_n^{\alpha}(b/t)$ Re $b > 0$	$2(-1)^n(n!)^{-1}b^{-\frac{1}{2}\alpha}p^{\frac{1}{2}\alpha+n}K_{\alpha}[2(bp)^{\frac{1}{2}}]$ Re $p > 0$
11.52	$t^{\nu-\frac{1}{2}}C_n^{\nu}(2t-1)$ Re $\nu > -\frac{1}{2}$	$2(-1)^n 2^{-2\nu}[n!\Gamma(\nu)]^{-1}\Gamma(n+2\nu) \cdot p^{-\nu-\frac{1}{2}} {}_2F_0[-n,n+2\nu;;p^{-1}]$ Re $p > 0$

	$f(t)$	$g(p) = \int_0^\infty f(t)e^{-pt}dt$
11.53	$t^{-\frac{1}{2}}C_{2n}^{\nu}(t^{\frac{1}{2}})$	$(-1)^n[n!\Gamma(\nu)]^{-1}(\pi/p)^{\frac{1}{2}}\Gamma(n+\nu)$ $\cdot {}_2F_0[-n,n+\nu;;p^{-1}]$ Re $p > 0$
11.54	$t^{-\frac{1}{2}}C_{2n+1}^{\nu}(t^{\frac{1}{2}})$	$(-1)^n[n!\Gamma(\nu)]^{-1}\pi^{\frac{1}{2}}\Gamma(n+\nu+1)p^{-3/2}$ $\cdot {}_2F_0[-n,n+\nu+1;;p^{-1}]$ Re $p > 0$
11.55	$t^{\nu}(1+t)^{\mu}P_n^{a,b}(1+2t)$ Re $\nu > -1$	$(n!)^{-1}\Gamma(n+\nu+1)p^{-1-\frac{1}{2}a-\frac{1}{2}b}$ $\cdot e^{\frac{1}{2}p}W_{\mu-\nu,n+\frac{1}{2}(\nu+\mu+1)}(a)$ Re $p > 0$
11.56	$t^{\nu}(1+t)^{n}$ $\cdot P_n^{a,b}[(t-1)/(t+1)]$ Re $\nu > -1$	$p^{-n-\nu-1}\Gamma(n+\nu+1)L_n^a(p)$ Re $p > 0$

1.12 Legendre Functions

	$f(t)$	$g(p) = \int_0^\infty f(t) e^{-pt} dt$
12.1	$P_{\nu-\frac{1}{2}}(1+2bt)$	$(\pi pb)^{-\frac{1}{2}} e^{\frac{1}{2}p/b} K_\nu(\frac{1}{2}p/b)$ Re $p > 0$
12.2	$P_{\nu-\frac{1}{2}}(1+at^2)$	$\frac{1}{4}\pi(\frac{1}{2}a)^{-\frac{1}{2}}\{J_\nu^2[p(2a)^{-\frac{1}{2}}]+Y_\nu^2[p(2a)^{-\frac{1}{2}}]\}$ Re $p > 0$
12.3	$P_{\nu-\frac{1}{2}}(1+at^2+2bt)$	$\pi^{-1}(\frac{1}{2}a)^{-\frac{1}{2}} e^{pb/a}$ $\cdot K_\nu(z_1) K_\nu(z_2)$ $z_{\frac{1}{2}} = \frac{1}{2}pa^{-1}[b \pm (b^2-2a)^{\frac{1}{2}}]$ Re $p > 0$
12.4	$0 \quad t < a+b$ $P_{\nu-\frac{1}{2}}[(2ab)^{-1}(t^2-a^2-b^2)]$ $t > a+b$	$2\pi^{-1}(ab)^{\frac{1}{2}} K_\nu(ap) K_\nu(bp)$
12.5	$0 \quad t < a-b$ $P_{\nu-\frac{1}{2}}[(2ab)^{-1}(a^2+b^2-t^2)]$ $a-b < t < a+b$ $2\pi^{-1}\cos(\pi\nu)$ $\cdot q_{\nu-\frac{1}{2}}[(2ab)^{-1}(t^2-a^2-b\ )]$ $t > a+b$	$2(ab)^{\frac{1}{2}} I_\nu(bp) K_\nu(ap)$ $a > b, \quad$ Re $p > 0$

	$f(t)$	$g(p) = \int_0^\infty f(t)e^{-pt}dt$
12.6	$(t^2+2at)^{-\frac{1}{2}\mu}$ $\cdot P_\nu^\mu(1+ta^{-1})$	$(2a/\pi)^{\frac{1}{2}} p^{\mu-\frac{1}{2}} e^{ap} K_{\nu+\frac{1}{2}}(ap)$ Re $\mu < 1$, Re $p > 0$
12.7	$(1+2at^{-1})^{\frac{1}{2}\mu}$ $\cdot P_\nu^\mu(1+ta^{-1})$	$p^{-1} e^{ap} W_{\mu,\frac{1}{2}+\nu}(2ap)$ Re $\nu < 1$, Re $p > 0$
12.8	$t^{-\frac{1}{2}\mu}(a^2+b^2+t)^{\frac{1}{2}\mu}$ $\cdot(a^2+t)^{\frac{1}{2}\nu}(b^2+t)^{-\frac{1}{2}\nu-\frac{1}{2}}$ $P_\nu^\mu[ab(a^2+t)^{-\frac{1}{2}}(b^2+t)^{-\frac{1}{2}}]$ Re $\mu < 1$	$(\frac{1}{2}p)^{-\frac{1}{2}} \exp[\frac{1}{2}p(a^2+b^2)]$ $\cdot D_{\mu+\nu}[a(2p)^{\frac{1}{2}}] D_{\mu-\nu-1}[b(2p)^{\frac{1}{2}}]$ Re $p > 0$
12.9	$t^{-\frac{1}{2}\mu}(1+at)^{-\frac{1}{2}}$ $\cdot P_\nu^\mu[(1+at)^{\frac{1}{2}}]$ Re $\mu < 1$	$2^\mu a^{-\frac{1}{4}} p^{\frac{1}{2}\mu-\frac{3}{4}} e^{\frac{1}{2}p/a}$ $\cdot W_{\frac{1}{2}\mu-\frac{1}{4},\frac{1}{2}\nu+\frac{1}{4}}(p/a)$ Re $p > 0$
12.10	$t^{-\frac{1}{2}\mu} P_\nu^\mu[(1+at)^{\frac{1}{2}}]$ Re $\mu < 1$	$2^\mu a^{\frac{1}{4}} p^{\frac{1}{2}\mu-\frac{5}{4}} e^{\frac{1}{2}p/a}$ $\cdot W_{\frac{1}{2}\mu+\frac{1}{4},\frac{1}{2}\nu+\frac{1}{4}}(p/a)$ Re $p > 0$

	$f(t)$	$g(p) = \int_0^\infty f(t)e^{-pt}dt$
12.11	$t^{-\frac{1}{2}\mu}(a^2+t)^{-\frac{1}{2}-\frac{1}{2}\nu}$ $\cdot P_\nu^\mu[a(a^2+t)^{-\frac{1}{2}}]$ Re $\mu < 1$	$2^{\frac{1}{2}(\nu+\mu+1)}p^{\frac{1}{2}(\nu+\mu-1)}e^{\frac{1}{2}a^2p}$ $\cdot D_{\mu-\nu-1}[a(2p)^{\frac{1}{2}}]$ Re $p > 0$
12.12	$t^{\frac{1}{2}\nu}q_\nu^\mu\{[1+(at)^{-1}]^{\frac{1}{2}}\}$ Re $\nu > -3/2$	$e^{i\pi\mu}\Gamma(\nu+\mu+1)2^{-\nu-1}$ $\cdot\pi^{\frac{1}{2}}a^{\frac{1}{2}}p^{-\frac{1}{2}\nu-3/2}e^{\frac{1}{2}p/a}$ $\cdot W_{-\frac{1}{2}\nu,-\frac{1}{2}\mu}(p/a)$ Re $p > 0$
12.13	$t^{\frac{1}{2}\nu}(1+at)^{-\frac{1}{2}}q_\nu^\mu\{[1+(at)^{-1}]^{\frac{1}{2}}\}$ Re $\nu > -3/2$	$e^{i\pi\mu}2^{-\nu-1}\Gamma(\nu+\mu+1)$ $\cdot\pi^{\frac{1}{2}}p^{-\frac{1}{2}\nu-1}e^{\frac{1}{2}p/a}$ $\cdot W_{-\frac{1}{2}\nu-\frac{1}{2},-\frac{1}{2}\mu}(p/a)$ Re $p > 0$
12.14	$(1+t^2)^{\frac{1}{2}\nu-1}\{2\pi^{-1}\sin[\frac{\pi}{2}(\nu+\mu)]$ $\cdot Q_{\nu-2}^\mu[t(1+t^2)^{-\frac{1}{2}}]$ $-\cos[\frac{\pi}{2}(\nu+\mu)]P_\nu^\mu[t(1+t^2)^{-\frac{1}{2}}]\}$ $\mu+\nu \neq 0,-1,-2,\cdots$	$2^\mu\pi^{-\frac{1}{2}}\Gamma(\frac{1}{2}\nu+\frac{1}{2}\mu-\frac{1}{2})$ $\cdot[\Gamma(\frac{1}{2}\nu-\frac{1}{2}\mu)]^{-1}$ $\cdot p^{1-\nu}S_{\nu-1,\mu}(p)$ Re $p > 0$

	$f(t)$	$g(p) = \int_0^\infty f(t)e^{-pt}dt$
12.15	$(1+t^2)^{-\frac{1}{2}\nu-\frac{1}{2}}\{\cos[\frac{\pi}{2}(\nu+\mu)]$ $\cdot P_\nu^\mu[t(1+t^2)^{-\frac{1}{2}}]$ $-2\pi^{-1}\sin[\frac{\pi}{2}(\nu+\mu)]Q_\nu^\mu[t(1+t^2)^{-\frac{1}{2}}]\}$ $\nu+\mu \neq -1,-2,-3,\cdots$	$2^\mu\pi^{-\frac{1}{2}}\Gamma(\frac{1}{2}+\frac{1}{2}\nu+\frac{1}{2}\mu)$ $\cdot[\Gamma(1+\frac{1}{2}\nu-\frac{1}{2}\mu)]^{-1}$ $\cdot p^\nu S_{-\nu,\mu}(p)$ $\mathrm{Re}\, p > 0$
12.16	$(1+at)^{-1}P_\nu[2(1+at)^{-2}-1]$	$p^{-1}e^{p/a}W_{\nu+\frac{1}{2},0}(p/a)$ $\cdot W_{-\nu-\frac{1}{2},0}(p/a)$, $\mathrm{Re}\, p > 0$
12.17	$t^{\frac{1}{2}}P_\nu^{\frac{1}{4}}[(1+a^2t^2)^{\frac{1}{2}}]$ $\cdot P_\nu^{-\frac{1}{4}}[(1+a^2t^2)^{\frac{1}{2}}]$	$2^{-3/2}\pi^{\frac{1}{2}}a^{-1}p^{-\frac{1}{2}}$ $\cdot[H_{\nu+\frac{1}{2}}^{(1)}(\frac{1}{2}p/a)H_{\nu+\frac{1}{2}}^{(2)}(\frac{1}{2}p/a)]$ $\mathrm{Re}\, p > 0$
12.18	$(1-e^{-2t})^{-\frac{1}{2}\mu}P_\nu^\mu(e^t)$ $\mathrm{Re}\,\mu < 1$	$\pi^{-\frac{1}{2}}2^{p-1}[\Gamma(p-\mu+1)]^{-1}$ $\cdot\Gamma(\frac{1}{2}+\frac{1}{2}\nu+\frac{1}{2}p)\Gamma(\frac{1}{2}p-\frac{1}{2}\nu)$ $\mathrm{Re}\, p > \mathrm{Max}[\mathrm{Re}\,\nu, -1-\mathrm{Re}\,\nu]$
12.19	$\sinh^\mu(\frac{1}{2}t)P_{2n}^{-\mu}(\cosh\frac{1}{2}t)$ $\mathrm{Re}\,\mu > -\frac{1}{2}$	$\pi^{-\frac{1}{2}}2^{-\mu}[\Gamma(1+n+\frac{1}{2}\mu+p)$ $\cdot\Gamma(\frac{1}{2}-n+\frac{1}{2}\mu+p)]^{-1}$ $\cdot\Gamma(\frac{1}{2}+\mu)\Gamma(p-n-\frac{1}{2}\mu)$ $\cdot\Gamma(p+n+\frac{1}{2}-\frac{1}{2}\mu)$ $n=0,1,2,\cdots;\mathrm{Re}\, p>n+\frac{1}{2}\mathrm{Re}\,\mu$

	$f(t)$	$g(p) = \int_0^\infty f(t)e^{-pt}dt$
12.20	$(1-e^{-t})^{-\frac{1}{2}}(1-a^2+a^2e^{-t})^{\mu}$ $\{P_{2\nu}^{2\mu}[a(1-e^{-t})^{\frac{1}{2}}]$ $+P_{2\nu}^{2\mu}[-a(1-e^{-t})^{\frac{1}{2}}]\}$ $\lvert a\rvert < 1$	$\pi\Gamma(p)[\Gamma(\frac{1}{2}-\mu-\nu)\Gamma(1-\mu+\nu)\Gamma(\frac{1}{2}+p)]^{-1}$ $\cdot 2^{2\mu+1}{}_2F_1(-\mu-\nu,\frac{1}{2}-\mu+\nu;\frac{1}{2}+p;a^2)$ $\mathrm{Re}\, p > 0$
12.21	$(1-a^2+a^2e^{-t})^{\mu}$ $\cdot\{P_{2\nu}^{2\mu}[a(1-e^{-t})^{\frac{1}{2}}]$ $-P_{2\nu}^{2\mu}[-a(1-e^{-t})^{\frac{1}{2}}]\}$ $\lvert a\rvert < 1$	$-\pi a\Gamma(p)[\Gamma(\frac{1}{2}-\mu+\nu)\Gamma(-\mu-\nu)\Gamma(\frac{3}{2}+p)]^{-1}$ $\cdot 2^{2\mu+1}{}_2F_1(\frac{1}{2}-\mu-\nu,\nu-\mu+1;\frac{3}{2}+p;a^2)$ $\mathrm{Re}\, p > 0$
12.22	$(e^t-1)[(a-2)^{-1}ae^t-1]^{\frac{1}{2}\mu}$ $\cdot P_\nu^{-\mu}(ae^t-a+1)$ $\mathrm{Re}\,\mu > -1$	$[\Gamma(p+1)]^{-1}\Gamma(p-\mu+\nu+1)\Gamma(p-\nu-\mu)$ $(\frac{1}{2}-2/a)^{-\frac{1}{2}p}P_\nu^{u-p}(a-1)$ $\mathrm{Re}\, a>0, \mathrm{Re}\, p>\mathrm{Max}[\mathrm{Re}(\mu-\nu-1),$ $\mathrm{Re}(\mu+\nu)]$
12.23	$0 \quad t < a$ $(t^2-a^2)^{\frac{1}{2}\mu}P_\nu^{-\mu}(t/a)$ $t > a$	$(2a/\pi)^{\frac{1}{2}}p^{-\mu-\frac{1}{2}}K_{\nu+\frac{1}{2}}(ap)$ $\mathrm{Re}\, p > 0$

	$f(t)$	$g(p) = \int_0^\infty f(t) e^{-pt} dt$
12.24	$0 \quad t < a$ $(t+a)^{\frac{1}{2}\mu}(t-a)^{-\frac{1}{2}\mu}$ $\cdot p_\nu^\mu(t/a) \quad t > a$ $\mathrm{Re}\ \mu > 1;\ \nu \neq 0, \pm 1, \pm 2, \cdots$	$p^{-1} W_{\mu, \nu+\frac{1}{2}}(2ap)$ $\mathrm{Re}\ p > 0$
12.25	$0 \quad t < a$ $p_\nu(2t^2/a^2-1)$ $t > a$	$\pi^{-1} a [K_{\nu+\frac{1}{2}}(\frac{1}{2}ap)]^2$
12.26	$t^{-1} q_\nu(1+2a^2/t^2)$ $\mathrm{Re}\ \nu > -\frac{3}{2}$	$\frac{1}{2} a^{-1} p^{-1} [\Gamma(1+\nu)]^2$ $\cdot W_{-\nu-\frac{1}{2}, 0}(iap) W_{-\nu-\frac{1}{2}, 0}(-iap)$

1.13 Bessel Functions of Order Zero and Unity*

	$f(t)$	$g(p) = \int_0^\infty f(t)e^{-pt}dt$
13.1	$t^{\frac{1}{2}}J_o(at)$	$\pi^{-\frac{1}{2}}r^{-3/2}\{2E[(\frac{1}{2}-\frac{1}{2}p/r)^{\frac{1}{2}}]$ $- K[(\frac{1}{2}-\frac{1}{2}p/r)^{\frac{1}{2}}]\}$ $\mathrm{Re}\, p > 0, \quad r = (p^2+a^2)^{\frac{1}{2}}$
13.2	$t^{-\frac{1}{2}}J_o(at)$	$2(\pi r)^{-\frac{1}{2}}K[(\frac{1}{2}-\frac{1}{2}p/r)^{\frac{1}{2}}]$ $\mathrm{Re}\, p > 0, \quad r = (p^2+a^2)^{\frac{1}{2}}$
13.3	$Y_o(at)$	$-2\pi^{-1}\,(p^2+a^2)^{-\frac{1}{2}}$ $\cdot \log[p/a+(1+p^2/a^2)^{\frac{1}{2}}]$ $\mathrm{Re}\, p > 0$
13.4	$t^{\frac{1}{2}}I_o(at)$	$\pi^{-\frac{1}{2}}(p-a)^{-\frac{1}{2}}E[2a(p+a)^{\frac{1}{2}}]$ $\mathrm{Re}\, p > a$
13.5	$t^{-\frac{1}{2}}I_o(at)$	$2\pi^{-\frac{1}{2}}(p+a)^{-\frac{1}{2}}K[2a(p+a)^{-\frac{1}{2}}]$ $\mathrm{Re}\, p > a$
13.6	$K_o(at)$	$(a^2-p^2)^{-\frac{1}{2}}\arccos(p/a) \quad p < a$ $(p^2-a^2)^{-\frac{1}{2}}\log[p/a+(p^2/a^2-1)^{\frac{1}{2}}] \quad p > a$

* Listed here are the cases where the results for arbitrary order are not available or where the general case simplifies considerably.

	$f(t)$	$g(p) = \int_0^\infty f(t) e^{-pt} dt$
13.7	$t^{\frac{1}{2}} K_0(at)$	$(\frac{1}{2}\pi)^{\frac{1}{2}} s^{-2} \{p(p+s)^{-\frac{1}{2}} K[(2s)^{\frac{1}{2}}(p+s)^{-\frac{1}{2}}]$ $- (p+s)^{\frac{1}{2}} E[(2s)^{\frac{1}{2}}(p+s)^{-\frac{1}{2}}]\}$ $\mathrm{Re}\, p > -a, \quad s = (p^2-a^2)^{\frac{1}{2}}$
13.8	$t^{-2} e^{-a/t} J_0(bt)$	$2b[z_1^{-1} J_0(z_1) K_1(z_2) + z_2^{-1} J_1(z_1) K_0(z_2)]$ $z_{\frac{1}{2}} = (2a)^{\frac{1}{2}}[(p^2+b^2)^{\frac{1}{2}} \mp p]$ $\mathrm{Re}\, p > 0$
13.9	$t^{-2} e^{-a/t} Y_0(bt)$	$2b[z_1^{-1} Y_0(z_1) K_1(z_2) + z_2^{-1} Y_1(z_1) K_0(z_2)]$ $z_{\frac{1}{2}} = (2a)^{\frac{1}{2}}[(p^2+b^2)^{\frac{1}{2}} \mp p]$ $\mathrm{Re}\, p > 0$
13.10	$t^{-2} e^{-a/t} I_0(bt)$	$2b[z_1^{-1} I_0(z_1) K_1(z_2) - z_2^{-1} I_1(z_1) K_0(z_2)]$ $z_{\frac{1}{2}} = a^{\frac{1}{2}}[(p+b)^{\frac{1}{2}} - (p-b)^{\frac{1}{2}}]$ $p > b, \quad \mathrm{Re}\, p > 0$
13.11	$t^{-2} e^{-a/t} K_0(bt)$	$2b[z_1^{-1} K_0(z_1) K_1(z_2) + z_2^{-1} K_1(z_1) K_0(z_2)]$ $z_{\frac{1}{2}} = a^{\frac{1}{2}}[(p+b)^{\frac{1}{2}} \pm (p-b)^{\frac{1}{2}}]$ $\mathrm{Re}\, p > 0$

	$f(t)$	$g(p) = \int_0^\infty f(t)e^{-pt}dt$
13.12	$\log t\, J_0(at)$	$(p^2+a^2)^{-\frac{1}{2}}[-\gamma-\log 2 + \log[p+(p^2+a^2)^{\frac{1}{2}}]-\log(p^2+a^2)]$ $\quad \mathrm{Re}\, p > 0$
13.13	$J_0(at)J_0(bt)$	$2\pi^{-1}[p^2+(a+b)^2]^{-\frac{1}{2}} \cdot K\{2(ab)^{\frac{1}{2}}[p^2+(a+b)^2]^{-\frac{1}{2}}\}$ $\quad \mathrm{Re}\, p > 0$
13.14	$t\, J_0(at)J_1(at)$	$\frac{1}{2}\pi^{-1}a^{-2}z\, [K(z) - E(z)]$ $\quad z = 2a(p^2+4a^2)^{-\frac{1}{2}}, \quad \mathrm{Re}\, p > 0$
13.15	$J_1(at)J_1(bt)$	$(\pi abz)^{-1}\{z^2(p^2+a^2+b^2) \cdot K[2(ab)^{\frac{1}{2}}z] - E[2(ab)^{\frac{1}{2}}z]\}$ $\quad z = [p^2+(a+b)^2]^{-\frac{1}{2}}, \quad \mathrm{Re}\, p > 0$
13.16	$J_0(at)\, J_1(at)$	$(2a)^{-1} - (\pi a)^{-1}p(p^2+4a^2)^{-\frac{1}{2}} \cdot K[2a(p^2+4a^2)^{-\frac{1}{2}}]$ $\quad \mathrm{Re}\, p > 0$
13.17	$t^{-1}\, J_0(at)J_1(at)$	$(\pi a)^{-1}\{(p^2+4a^2)^{\frac{1}{2}} \cdot E[2a(p^2+4a^2)^{-\frac{1}{2}}] - \frac{1}{2}\pi\, p\}$ $\quad \mathrm{Re}\, p > 0$

	$f(t)$	$g(p) = \int_0^\infty f(t)e^{-pt}dt$
13.18	$J_0(at)Y_0(at)$	$-2\pi^{-1}(p^2+4a^2)^{-\frac{1}{2}}$ $\cdot K[p(p^2+4a^2)^{-\frac{1}{2}}]$ Re $p > 0$
13.19	$J_0(at)Y_0(bt)$ $+ Y_0(at)J_0(bt)$	$-4\pi^{-1}[p^2+(a+b)^2]^{-\frac{1}{2}}$ $\cdot K\{[\frac{p^2+(a-b)^2}{p^2+(a+b)^2}]^{\frac{1}{2}}\}$ Re $p > 0$
13.20	$\log t\ I_0(at)$	$(p^2-a^2)^{-\frac{1}{2}}[-\gamma-\log 2$ $+\log[p+(p^2-a^2)^{\frac{1}{2}}]-\log(p^2-a^2)]$ Re $p > a$
13.21	$I_0(at)I_0(bt)$	$2\pi^{-1}[p^2-(a-b)^2]^{-\frac{1}{2}}$ $\cdot K\{2(ab)^{\frac{1}{2}}[p^2-(a-b)^2]^{-\frac{1}{2}}$ Re $p > a+b$
13.22	$I_0(at)I_1(at)$	$(\pi a)^{-1}K(2a/p) - (2a)^{-1}$ Re $p > 2a$
13.23	$t^{-1}I_0(at)I_1(at)$	$(\pi a)^{-1}p[\frac{1}{2}\pi-E(2a/p)]$ Re $p > 2a$

	$f(t)$	$g(p) = \int_0^\infty f(t)e^{-pt}dt$
13.24	$t\, I_0(at) I_1(at)$	$(\pi a)^{-1}[(p^2-4a^2)^{-1}pE(2a/p)$ $-p^{-1}K(2a/p)]$ Re $p > 2a$
13.25	$I_1(at) I_1(bt)$	$(\pi abz)^{-1}\{z^2(p^2-a^2-b^2)$ $\cdot K[2(ab)^{\frac{1}{2}}z] - E[2(ab)^{\frac{1}{2}}z]\}$ $z = [p^2-(a-b)^2]^{-\frac{1}{2}}$ Re $p > a+b$
13.26	$I_0(at) K_0(at)$	$p^{-1}K[(1-4a^2/p^2)^{\frac{1}{2}}]$ Re $p > 0$
13.27	$I_0(at) K_0(bt)$ $+I_0(bt) K_0(at)$	$2[p^2-(a-b)^2]^{-\frac{1}{2}}$ $\cdot K\{[\frac{p^2-(a+b)^2}{p^2-(a-b)^2}]^{\frac{1}{2}}\}$ Re $p > \|a-b\|$
13.28	$\log t\, J_0(at^{\frac{1}{2}})$	$-p^{-1}e^{-\frac{1}{4}a^2/p}\,[2\log(2p/a)$ $+ \overline{\mathrm{Ei}}(\frac{1}{4}a^2/p)$ Re $p > 0$
13.29	$t^{-\frac{1}{2}}\log t\, J_0(at^{\frac{1}{2}})$	$8p^{-1}\exp(-a^2/8p)$ $\cdot [\log(\frac{1}{2}a/p)-\frac{1}{2}\overline{\mathrm{Ei}}(a^2/4p)]$ Re $p > 0$

	$f(t)$	$g(p) = \int_0^\infty f(t)e^{-pt}dt$
13.30	$t^{-\frac{1}{2}}Y_0(at^{\frac{1}{2}})$	$-\pi^{-1}(\pi/p)^{\frac{1}{2}}\exp(-a^2/8p)K_0(a^2/8p)$ Re $p > 0$
13.31	$\log t\ I_0(at^{\frac{1}{2}})$	$-p^{-1}e^{-\frac{1}{4}a^2/p}[2\log(2p/a) + Ei(-\frac{1}{4}a^2/p)]$ Re $p > 0$
13.32	$t^{-\frac{1}{2}}\log t\ I_0(at^{\frac{1}{2}})$	$8p^{-1}\exp(a^2/8p) \cdot [\log(\frac{1}{2}a/p)-\frac{1}{2}Ei(-a^2/4p)]$ Re $p > 0$
13.33	$K_0(at^{\frac{1}{2}})$	$-\frac{1}{2}p^{-1}\exp(a^2/4p)Ei(-a^2/4p)$ Re $p > 0$
13.34	$K_1(at^{\frac{1}{2}})$	$\frac{1}{8}a\pi^{\frac{1}{2}}p^{-\frac{3}{2}}\exp(a^2/8p) \cdot [K_1(a^2/8p) - K_0(a^2/8p)]$ Re $p > 0$
13.35	$J_0(at^{\frac{1}{2}})Y_0(at^{\frac{1}{2}})$	$-(\pi p)^{-1}\exp(-\frac{1}{2}a^2/p) \cdot K_0(\frac{1}{2}a^2/p)$ Re $p > 0$

	$f(t)$	$g(p) = \int_0^\infty f(t)e^{-pt}dt$
13.36	$J_0(at^{\frac{1}{2}})Y_0(bt^{\frac{1}{2}}) + J_0(bt^{\frac{1}{2}})Y_0(at^{\frac{1}{2}})$	$-2(\pi p)^{-1}\exp[-(a^2+b^2)/(4p)] \cdot K_0(\frac{1}{2}ab/p)$ Re $p > 0$
13.37	$J_0(at^{\frac{1}{2}})Y_0(t) +2J_0(t)Y_0(at^{\frac{1}{2}})$	$(p^2+1)^{-\frac{1}{2}} \cdot \exp[-\frac{1}{4}a^2p(1+p^2)^{-1}] \cdot Y_0[\frac{1}{4}a^2(1+p^2)^{-1}]$ Re $p > 0$
13.38	$J_0(at^{\frac{1}{2}})\ I_0(at^{\frac{1}{2}})$	$p^{-1}\ J_0(\frac{1}{2}a^2/p)$ Re $p > 0$
13.39	$Y_0(at^{\frac{1}{2}})\ I_0(at^{\frac{1}{2}})$	$\frac{1}{2}p^{-1}[\mathbf{H}_0(\frac{1}{2}a^2/p) + Y_0(\frac{1}{2}a^2/p)]$ Re $p > 0$
13.40	$J_0(at^{\frac{1}{2}})K_0(at^{\frac{1}{2}})$	$\frac{1}{4}\pi p^{-1}[\mathbf{H}_0(\frac{1}{2}a^2/p) - Y_0(\frac{1}{2}a^2/p)]$
13.41	$tJ_1(at^{\frac{1}{2}})\ I_1(at^{\frac{1}{2}})$	$\frac{1}{2}p^{-3}a^2\ J_0(\frac{1}{2}a^2/p)$
13.42	$I_0(at^{\frac{1}{2}})K_0(at^{\frac{1}{2}})$	$(2p)^{-1}\exp(\frac{1}{2}a^2/p)K_0(\frac{1}{2}a^2/p)$ Re $p > 0$

	$f(t)$	$g(p) = \int_0^\infty f(t)e^{-pt}dt$
13.43	$I_o(at^{\frac{1}{2}})K_o(bt^{\frac{1}{2}})$ $+ I_o(bt^{\frac{1}{2}})K_o(at^{\frac{1}{2}})$	$p^{-1}\exp[(a^2+b^2)/(4p)]K_o(\frac{1}{2}ab/p)$ Re $p > 0$
13.44	$J_o(at^2)$	$\frac{1}{16}\pi a^{-1}p\{[J_{\frac{1}{4}}(p^2/8a)]^2+Y_{\frac{1}{4}}(p^2/8a)]^2\}$ Re $p > 0$
13.45	$\cos(at^2)J_o(at^2)$	$\frac{1}{4}(2a/\pi)^{-\frac{1}{2}}[J_o(\frac{p^2}{16a})\cos(\frac{p^2}{16a} - \frac{\pi}{4})$ $+ Y_o(\frac{p^2}{16a})\sin(\frac{p^2}{16a} - \frac{\pi}{4})]$ Re $p > 0$
13.46	$\sin(at^2)Y_o(at^2)$	$\frac{1}{4}(2a/\pi)^{-\frac{1}{2}}[J_o(\frac{p^2}{16a})\sin(\frac{p^2}{16a} - \frac{\pi}{4})$ $- Y_o(\frac{p^2}{16a})\cos(\frac{p^2}{16a} - \frac{\pi}{4})]$ Re $p > 0$
13.47	$e^{-at^2}I_o(at^2)$	$\frac{1}{4}(\frac{1}{2}\pi a)^{-\frac{1}{2}}\exp(\frac{p^2}{16a})K_o(\frac{p^2}{16a})$ Re $p > 0$

	$f(t)$	$g(p) = \int_0^\infty f(t)e^{-pt}dt$
13.48	$J_o[a(t^2+bt)^{\frac{1}{2}}]$	$e^{\frac{1}{2}pb}(p^2+a^2)^{-\frac{1}{2}}\exp[-\frac{1}{2}b(p^2+a^2)^{\frac{1}{2}}]$
13.49	$Y_o[a(t^2+bt)^{\frac{1}{2}}]$	$e^{\frac{1}{2}bp}(\pi r)^{-1}[e^{-\frac{1}{2}br}\overline{Ei}(\frac{1}{2}br-\frac{1}{2}bp)$ $-e^{\frac{1}{2}br}Ei(-\frac{1}{2}br-\frac{1}{2}bp)]$ $r=(p^2+a^2)^{\frac{1}{2}}$, Re $p > \|Im\ a\|$
13.50	$K_o[a(t^2+bt)^{\frac{1}{2}}]$	$-\frac{1}{2}s^{-1}e^{\frac{1}{2}bp}[e^{-\frac{1}{2}bs}Ei(-\frac{1}{2}bp+\frac{1}{2}bs)$ $-e^{\frac{1}{2}bs}Ei(-\frac{1}{2}bp-\frac{1}{2}bs)]$ $s=(p^2-a^2)^{\frac{1}{2}}$, Re $p > -\|Im\ a\|$
13.51	$(t^2+bt)^{\frac{1}{2}}J_1[a(t^2+bt)^{\frac{1}{2}}]$	$ar^{-3}(1+\frac{1}{2}br)e^{-\frac{1}{2}b(r-p)}$ $r=(p^2+a^2)^{\frac{1}{2}}$, Re $p > \|Im\ a\|$
13.52	$(t^2+bt)^{-\frac{1}{2}}J_1[a(t^2+bt)^{\frac{1}{2}}]$	$(\frac{1}{2}ab)^{-1}[1-e^{-\frac{1}{2}b(r-p)}]$ $r=(p^2+a^2)^{\frac{1}{2}}$, Re $p > \|Im\ a\|$
13.53	$(t^2+bt)^{-\frac{1}{2}}Y_o[a(t^2+bt)^{\frac{1}{2}}]$	$-\pi^{-1}e^{\frac{1}{2}bp}K_o(z_1)K_o(z_2)$ $z_{\frac{1}{2}} = \frac{1}{4}b[(p^2+a^2)^{\frac{1}{2}}\pm p]$ Re $p > \|Im\ a\|$

	$f(t)$	$g(p) = \int_0^\infty f(t)e^{-pt}dt$
13.54	$\log(t+b)$ $\cdot J_0[a(t^2+bt)^{\frac{1}{2}}]$	$e^{\frac{1}{2}bp}r^{-1}\{e^{-\frac{1}{2}br}[\log b/2+\log(1+p/r)]$ $-e^{\frac{1}{2}br}Ei(-br)\}$ $r=(p^2+a^2)^{\frac{1}{2}}, \quad \mathrm{Re}\, p > \lvert \mathrm{Im}\, a \rvert$
13.55	$\log t\, J_0[a(t^2+bt)^{\frac{1}{2}}]$	$r^{-1}e^{\frac{1}{2}bp}\{e^{-\frac{1}{2}br}[\log b/2+\log(1-p/r)]$ $-e^{-\frac{1}{2}br}\overline{Ei}(\frac{1}{2}br-\frac{1}{2}bp)$ $+ e^{\frac{1}{2}br}[Ei(-\frac{1}{2}br-\frac{1}{2}bp)$ $- Ei(-br)]\}$ $r=(p^2+a^2)^{\frac{1}{2}}, \quad \mathrm{Re}\, p > \lvert \mathrm{Im}\, a \rvert$
13.56	$\log(t+b)$ $\cdot I_0[a(t^2+bt)^{\frac{1}{2}}]$	$e^{\frac{1}{2}pb}s^{-1}\{e^{-\frac{1}{2}bs}[\log b/2+\log(1+p/s)]$ $-e^{\frac{1}{2}bs}Ei(-bs)\}$ $s=(p^2-a^2)^{\frac{1}{2}}, \quad \mathrm{Re}\, p > \mathrm{Re}\, a$
13.57	$\log t\, I_0[a(t^2+bt)^{\frac{1}{2}}]$	$s^{-1}e^{\frac{1}{2}bp}\{e^{-\frac{1}{2}bs}[\log b/2+\log(p/s-1)]$ $-e^{-\frac{1}{2}bs}Ei(-\frac{1}{2}bp+\frac{1}{2}bs)$ $+e^{\frac{1}{2}bs}[Ei(-\frac{1}{2}bp-\frac{1}{2}bs)$ $-Ei(-bs)]\}$ $s=(p^2-a^2)^{\frac{1}{2}}, \quad \mathrm{Re}\, p > \lvert \mathrm{Re}\, a \rvert$

	$f(t)$	$g(p) = \int_0^\infty f(t)e^{-pt}dt$
13.58	$0 \qquad t < b$ $J_0[a(t^2-b^2)^{1/2}] \quad t > b$	$(p^2+a^2)^{-1/2}\exp[-b(p^2+a^2)^{1/2}]$ Re $p > 0$
13.59	$0 \qquad t < b$ $Y_0[a(t^2-b^2)^{1/2}] \quad t > b$	$(\pi r)^{-1}[e^{-br}\overline{Ei}(br-bp)$ $-e^{br}Ei(-br-bp)]$ $r=(p^2+a^2)^{1/2}$, Re $p > 0$
13.60	$0 \qquad t < b$ $K_0[a(t^2-b^2)^{1/2}] \quad t > b$	$-\frac{1}{2}s^{-1}[e^{-bs}Ei(-bp+bs)$ $-e^{bs}Ei(-bp-bs)]$ $s = (p^2-a^2)^{1/2}$, Re $p > -a$
13.61	$0 \qquad t < b$ $(t^2-b^2)^{1/2}J_1[a(t^2-b^2)^{1/2}]$ $t > b$	$ar^{-3}(1+br)e^{-br}$ $r = (p^2+a^2)^{1/2}$, Re $p > 0$
13.62	$0 \qquad t < b$ $(t^2-b^2)^{-1/2}J_1[a(t^2-b^2)^{1/2}]$ $t > b$	$(ab)^{-1}(e^{-bp}-e^{-br})$ $r = (p^2+a^2)^{1/2}$, Re $p > 0$

	$f(t)$	$g(p) = \int_0^\infty f(t)e^{-pt}dt$
13.63	$0 \qquad t < b$ $(t^2-b^2)^{-\frac{1}{2}} Y_0[a(t^2-b^2)^{\frac{1}{2}}]$ $t > a$	$-\pi^{-1} K_0[\frac{1}{2}b(r+p)]$ $\cdot K_0[\frac{1}{2}b(r-p)]$ $r = (p^2+a^2)^{\frac{1}{2}}, \qquad \operatorname{Re} p > 0$
13.64	$0 \qquad t < b$ $\log(t+b) J_0[a(t^2-b^2)^{\frac{1}{2}}]$ $t > b$	$r^{-1}\{e^{-br}[\log b + \log(1+p/r)]$ $- e^{br} \operatorname{Ei}(-2br)\}$ $r = (p^2+a^2)^{\frac{1}{2}}, \qquad \operatorname{Re} p > 0$
13.65	$0 \qquad t < b$ $\log(t-b) J_0[a(t^2-b^2)^{\frac{1}{2}}]$ $t > b$	$r^{-1}\{e^{-br}[\log b + \log(1-p/r)]$ $- e^{-br} \overline{\operatorname{Ei}}(br-bp)$ $+ e^{br}[\operatorname{Ei}(-br-bp)$ $- \operatorname{Ei}(-2br)]\}$ $r = (p^2+a^2)^{\frac{1}{2}}, \qquad \operatorname{Re} p > 0$
13.66	$0 \qquad t < b$ $\log(t+b) I_0[a(t^2-b^2)^{\frac{1}{2}}]$ $t > b$	$s^{-1}\{e^{-bs}[\log b + \log(1+p/s)]$ $- e^{bs} \operatorname{Ei}(-2bs)\}$ $s = (p^2-a^2)^{\frac{1}{2}}, \qquad \operatorname{Re} p > a$

	$f(t)$	$g(p) = \int_0^\infty f(t)e^{-pt}dt$
13.67	$0 \quad t < b$ $\log(t-b)I_0[a(t^2-b^2)^{\frac{1}{2}}] \quad t > b$	$s^{-1}\{e^{-bs}[\log b+\log(p/s-1)]$ $- e^{-bs}Ei(-bp+bs)$ $+ e^{bs}[Ei(-bp-bs)$ $- Ei(-2bs)]\}$ $s = (p^2-a^2)^{\frac{1}{2}}, \quad \text{Re } p > a$
13.68	$J_0(b/t)$	$(2b/p)^{\frac{1}{2}}\{J_0[(2bp)^{\frac{1}{2}}]K_1[(2bp)^{\frac{1}{2}}]$ $+ J_1[(2bp)^{\frac{1}{2}}]K_0[(2bp)^{\frac{1}{2}}]\}$ $\text{Re } p > 0$
13.69	$Y_0(b/t)$	$(2b/p)^{\frac{1}{2}}[Y_0[(2bp)^{\frac{1}{2}}]K_1[(2bp)^{\frac{1}{2}}]$ $+ Y_1[(2bp)^{\frac{1}{2}}]K_0[(2bp)^{\frac{1}{2}}]$ $\text{Re } p > 0$
13.70	$K_0(b/t)$	$2(2b/\pi)^{\frac{1}{2}}\{\text{ker}(z)[\text{ker}_1(z)-\text{kei}_1(z)]$ $+ \text{kei}(z)[\text{ker}_1(z)+\text{kei}_1(z)]\}$ $z = (2pb)^{\frac{1}{2}}, \quad \text{Re } p > 0$

	$f(t)$	$g(p) = \int_0^\infty f(t)e^{-pt}dt$
13.71	$e^{-a/t}J_0(b/t)$	$2b[z_1^{-1}J_0(z_1)K_1(z_2)+z_2^{-1}J_1(z_1)K_o(z_2)]$ $z_{\frac{1}{2}} = (2p)^{\frac{1}{2}}[(a^2+b^2)^{\frac{1}{2}}\mp a]^{\frac{1}{2}}$ $\mathrm{Re}\, p > 0$
13.72	$e^{-a/t}Y_0(b/t)$	$2b[z_1^{-1}Y_0(z_1)K_1(z_2)$ $+ z_2^{-1}Y_1(z_1)K_0(z_2)]$ $z_{\frac{1}{2}} = (2p)^{\frac{1}{2}}[(a^2+b^2)^{\frac{1}{2}}\mp a]^{\frac{1}{2}}$ $\mathrm{Re}\, p > 0$
13.73	$e^{-a/t}I_0(b/t)$	$2b[z_1^{-1}I_0(z_1)K_1(z_2)$ $- z_2^{-1}I_1(z_1)K_0(z_2)]$ $z_{\frac{1}{2}} = p^{\frac{1}{2}}[(a+b)^{\frac{1}{2}}\mp(a-b)^{\frac{1}{2}}]$ $a > b,$ $\mathrm{Re}\, p > 0$
13.74	$e^{-a/t}K_0(b/t)$	$2b[z_1^{-1}K_0(z_1)K_1(z_2)$ $+ z_2^{-1}K_1(z_1)K_0(z_2)]$ $z_{\frac{1}{2}} = p^{\frac{1}{2}}[(a+b)^{\frac{1}{2}}\pm(a-b)^{\frac{1}{2}}]$ $\mathrm{Re}\, p > 0$

	$f(t)$	$g(p) = \int_0^\infty f(t)e^{-pt}dt$
13.75	$Y_0(at^{\frac{1}{2}})$	$(\pi p)^{-1}\exp(-\frac{1}{4}a^2/p)\overline{E}i(\frac{1}{4}a^2/p)$ Re $p > 0$
13.76	$t^{-\frac{1}{2}}e^{a/t}K_0(a/t)$	$\pi^{3/2}p^{-\frac{1}{2}}\{\mathbf{H}_0[(8ap)^{\frac{1}{2}}]-Y_0[(8ap)^{\frac{1}{2}}]\}$ Re $p > 0$
13.77	$t^{-\frac{1}{2}}e^{-a/t)}I_0(a/t)$	$(\pi/p)^{\frac{1}{2}}\{I_0[(8ap)^{\frac{1}{2}}]-\mathbf{L}_0[(8ap)^{\frac{1}{2}}]\}$ Re $p > 0$
13.78	$t^{-\frac{1}{2}}K_0(at^{\frac{1}{2}})$	$\frac{1}{2}(\pi/p)^{\frac{1}{2}}\exp(\frac{1}{8}a^2/p)K_0(\frac{1}{8}a^2/p)$ Re $p > 0$
13.79	$t^{-3/2}e^{a/t}$ $\cdot[K_1(a/t)-K_0(a/t)]$	$-(\frac{1}{2}a)^{-\frac{1}{2}}\pi^{3/2}\{Y_1[(8ap)^{\frac{1}{2}}]$ $-\mathbf{H}_{-1}[(8ap)^{\frac{1}{2}}]\}$ Re $p > 0$

1.14 Bessel Functions

	$f(t)$	$g(p) = \int_0^\infty f(t) e^{-pt} dt$
14.1	$J_\nu(at)$ $\mathrm{Re}\ \nu > -1$	$a^\nu (p^2+a^2)^{-\frac{1}{2}} [p+(p^2+a^2)^{\frac{1}{2}}]^{-\nu}$ $\mathrm{Re}\ p > \lvert \mathrm{Im}\ a \rvert$
14.2	$t\ J_\nu(at)$ $\mathrm{Re}\ \nu > -2$	$a^\nu (p^2+a^2)^{-3/2} [p+(p^2+a^2)^{\frac{1}{2}}]^{-\nu}$ $\cdot [p+\nu(p^2+a^2)^{\frac{1}{2}}]$ $\quad \mathrm{Re}\ p > \lvert \mathrm{Im}\ a \rvert$
14.3	$t^n J_n(at)$ $n = 0,1,2,\cdots$	$1\cdot 3\cdot 5 \cdots (2n-1) a^n (p^2+a^2)^{-n-\frac{1}{2}}$ $\mathrm{Re}\ p > \lvert \mathrm{Im}\ a \rvert$
14.4	$t^2 J_\nu(at)$ $\mathrm{Re}\ \nu > -3$	$a^\nu [p+(p^2+a^2)^{\frac{1}{2}}]^{-\nu} (p^2+a^2)^{-3/2}$ $\cdot \{\nu^2-1+3p(p^2+a^2)^{-1}[p+\nu(p^2+a^2)^{\frac{1}{2}}]\}$ $\mathrm{Re}\ p > \lvert \mathrm{Im}\ a \rvert$
14.5	$t^{-1} J_\nu(at)$ $\mathrm{Re}\ \nu > 0$	$\nu^{-1} a^\nu [p+(p^2+a^2)^{\frac{1}{2}}]^{-\nu}$ $\mathrm{Re}\ p > \lvert \mathrm{Im}\ a \rvert$
14.6	$t^\nu J_\nu(at)$ $\mathrm{Re}\ \nu > -\frac{1}{2}$	$2^\nu \pi^{-\frac{1}{2}} \Gamma(\tfrac{1}{2}+\nu) a^\nu p (p^2+a^2)^{-\nu-\frac{1}{2}}$ $\mathrm{Re}\ p > \lvert \mathrm{Im}\ a \rvert$
14.7	$t^{\nu+1} J_\nu(at)$ $\mathrm{Re}\ \nu > -1$	$2^{\nu+1} \pi^{-\frac{1}{2}} \Gamma(3/2+\nu) a^\nu (p^2+a^2)^{-\nu-3/2}$ $\times p$ $\mathrm{Re}\ p > \lvert \mathrm{Im}\ a \rvert$

	$f(t)$	$g(p) = \int_0^\infty f(t) d^{-pt} dt$
14.8	$t^{\mu} J_{\nu}(at)$ $\mathrm{Re}(\mu+\nu) > -1$	$\Gamma(\nu+\mu+1)(p^2+a^2)^{-\frac{1}{2}\mu-\frac{1}{2}} P_{\mu}^{-\nu}[p(p^2+a^2)^{-\frac{1}{2}}]$ $\mathrm{Re}\, p > \lvert \mathrm{Im}\, a \rvert$
14.9	$t^{-1} e^{-b/t} J_{\nu}(at)$ $\mathrm{Re}\, b > 0$	$2J_{\nu}(z_1) K_{\nu}(z_2)$ $z_{\frac{1}{2}} = (2b)^{\frac{1}{2}}[(p^2+a^2)^{\frac{1}{2}} \mp p]^{\frac{1}{2}}$ $\mathrm{Re}\, p > \lvert \mathrm{Im}\, a \rvert$
14.10	$J_{\nu}(at) J_{\nu}(bt)$ $\mathrm{Re}\, \nu > -\frac{1}{2}$	$\pi^{-1}(ab)^{-\frac{1}{2}} q_{\nu-\frac{1}{2}}[(2ab)^{-1}(p^2+a^2+b^2)]$ $\mathrm{Re}\, p > \lvert \mathrm{Im}\, a \rvert + \lvert \mathrm{Im}\, b \rvert$
14.11	$t\, J_{\nu}(at) J_{\nu}(bt)$ $\mathrm{Re}\, \nu > -1$	$-\pi^{-1}(ab)^{-3/2} p(z^2-1)^{-\frac{1}{2}} q_{\nu-\frac{1}{2}}^{1}(z)$ $z = (2ab)^{-1}(a^2+b^2+p^2)$ $\mathrm{Re}\, p > \lvert \mathrm{Im}\, a \rvert + \lvert \mathrm{Im}\, b \rvert$
14.12	$t^{\frac{1}{2}} J_{\nu}^{2}(at)$ $\mathrm{Re}\, \nu > -3/4$	$-2^{\frac{1}{2}} \pi^{-3/2} a^{-1}(\nu^2 - 1/16) p^{-\frac{1}{2}}(1+4a^2p^{-2})^{-\frac{1}{2}}$ $\cdot q_{\nu-\frac{1}{2}}^{-3/4}[(1+\frac{1}{4}p^2a^{-2})^{\frac{1}{2}}] q_{\nu-\frac{1}{2}}^{-\frac{1}{4}}[(1+\frac{1}{4}p^2a^{-2})^{\frac{1}{2}}]$ $\mathrm{Re}\, p > 2\lvert \mathrm{Im}\, a \rvert$
14.13	$t^{-\frac{1}{2}} J_{\nu}(at) J_{-\nu}(at)$	$(\pi/p)^{\frac{1}{2}} P_{-\frac{1}{4}}^{\nu}[(1+4a^2p^{-2})^{\frac{1}{2}}] P_{-\frac{1}{4}}^{-\nu}[(1+4a^2p^{-2})^{\frac{1}{2}}]$ $\mathrm{Re}\, p > 2\lvert \mathrm{Im}\, a \rvert$

	$f(t)$	$g(p) = \int_0^\infty f(t)e^{-pt}dt$
14.14	$t^{-\frac{1}{2}}J_\nu^2(at)$ $\mathrm{Re}\,\nu > -\frac{1}{4}$	$2^{-2\nu}p^{-\frac{1}{2}}\Gamma(\frac{1}{2}+2\nu)\{P_{-\frac{1}{4}}^{-\nu}[(1+4a^2p^{-2})^{\frac{1}{2}}]\}^2$ $\mathrm{Re}\,p > 2\lvert\mathrm{Im}\,a\rvert$
14.15	$t^{\frac{1}{2}}J_\nu(at)J_{\nu+1}(at)$ $\mathrm{Re}\,\nu > -5/4$	$(\pi a)^{-1}(\frac{1}{2}\pi p)^{-\frac{1}{2}}(3/4+\nu)(1+4a^2p^{-2})^{\frac{1}{2}}$ $\cdot q_{\nu-\frac{1}{2}}^{\frac{1}{4}}[(1+\frac{1}{4}p^2a^{-2})^{\frac{1}{2}}]q_{\nu+\frac{1}{2}}^{-\frac{1}{4}}[(1+\frac{1}{4}p^2a^{-2})^{\frac{1}{2}}]$ $\mathrm{Re}\,p > 2\lvert\mathrm{Im}\,a\rvert$
14.16	$t^{2\nu}J_\nu^2(at)$ $\mathrm{Re}\,\nu > -\frac{1}{4}$	$[\pi p\Gamma(1+\nu)]^{-1}(4ap^{-2})^{2\nu}\Gamma(\frac{1}{2}+\nu)\Gamma(\frac{1}{2}+2\nu)$ $\cdot {}_2F_1(\frac{1}{2}+\nu,\frac{1}{2}+2\nu;1+\nu;-4a^2p^{-2})$ $\mathrm{Re}\,p > 2\lvert\mathrm{Im}\,a\rvert$
14.17	$t^{\mu+\nu}J_\mu(at)J_\nu(at)$ $\mathrm{Re}(\mu+\nu) > -\frac{1}{2}$	$2\pi^{-3/2}(4a)^{\mu+\nu}\Gamma(\frac{1}{2}+\nu+\mu)$ $\cdot\int_0^{\frac{1}{2}\pi}(p^2+4a^2\cos^2 t)^{-\mu-\nu-\frac{1}{2}}$ $\cdot(\cos t)^{\mu+\nu}\cos[(\mu-\nu)t]dt$ $\mathrm{Re}\,p > 2\lvert\mathrm{Im}\,a\rvert$
14.18	$J_\nu(bt)Y_{-\nu}(at)$ $+J_{-\nu}(at)Y_\nu(bt)$ $-\frac{1}{2} < \mathrm{Re}\,\nu < \frac{1}{2}$	$-\sec(\pi\nu)(ab)^{-\frac{1}{2}}$ $P_{\nu-\frac{1}{2}}[(2ab)^{-1}(a^2+b^2+p^2)]$ $\mathrm{Re}\,p > \lvert\mathrm{Im}\,a\rvert+\lvert\mathrm{Im}\,b\rvert$

	$f(t)$	$g(p) = \int_0^\infty f(t)e^{-pt}dt$
14.19	$Y_\nu(at)$ $-1 < \mathrm{Re}\,\nu < 1$	$a^\nu(p^2+a^2)^{-\frac{1}{2}}\{\cot(\pi\nu)[(p^2+a^2)^{\frac{1}{2}}-p]^\nu$ $-\csc(\pi\nu)[(p^2+a^2)^{\frac{1}{2}}+p]^\nu\}$ $\mathrm{Re}\,p > \lvert\mathrm{Im}\,a\rvert$
14.20	$t^\mu Y_\nu(at)$	$(p^2+a^2)^{-\frac{1}{2}\mu-\frac{1}{2}}\{\Gamma(\mu+\nu+1)\cot(\pi\nu)P_\mu^{-\nu}[p(p^2+a^2)^{-\frac{1}{2}}]$ $-\Gamma(\mu-\nu+1)\csc(\pi\nu)P_\mu^\nu[p(p^2+a^2)^{-\frac{1}{2}}]\}$ $\mathrm{Re}\,p > \lvert\mathrm{Im}\,a\rvert$
14.21	$t^{-1}e^{-b/t}Y_\nu(at)$ $\mathrm{Re}\,b > 0$	$2Y_\nu(z_1)K_\nu(z_2)$ $z_{\frac{1}{2}} = (2b)^{\frac{1}{2}}[(p^2-a^2)^{\frac{1}{2}}\mp p]$ $\mathrm{Re}\,p > \lvert\mathrm{Im}\,a\rvert$
14.22	$J_\nu(at^{\frac{1}{2}})$ $\mathrm{Re}\,\nu > -2$	$\frac{1}{4}a\pi^{\frac{1}{2}}p^{-3/2}\exp(-\frac{1}{8}a^2/p)$ $\cdot[I_{\frac{1}{2}\nu-\frac{1}{2}}(\frac{1}{8}a^2/p)-I_{\frac{1}{2}\nu+\frac{1}{2}}(\frac{1}{8}a^2/p)]$ $\mathrm{Re}\,p > 0$
14.23	$t^{-\frac{1}{2}}J_\nu(at^{\frac{1}{2}})$ $\mathrm{Re}\,\nu > -1$	$(\pi/p)^{\frac{1}{2}}\exp(-\frac{1}{8}a^2/p)I_{\frac{1}{2}\nu}(\frac{1}{8}a^2/p)$ $\mathrm{Re}\,p > 0$

	$f(t)$	$g(p) = \int_0^\infty f(t)e^{-pt}dt$
14.24	$t^{\frac{1}{2}\nu}J_\nu(at^{\frac{1}{2}})$ $\mathrm{Re}\,\nu > -1$	$(\frac{1}{2}a)^\nu p^{-\nu-1}\exp(-\frac{1}{4}a^2/p)$ $\mathrm{Re}\,p > 0$
14.25	$t^{-\frac{1}{2}\nu}J_\nu(at^{\frac{1}{2}})$	$e^{i\pi\nu}(2/a)^\nu[\Gamma(\nu)]^{-1}p^{\nu-1}\exp(-\frac{1}{4}a^2/p)$ $\cdot\gamma(\nu,\frac{1}{4}a^2e^{-i\pi}p^{-1})$ $\mathrm{Re}\,p > 0$
14.26	$t^{\frac{1}{2}\nu-1}J_\nu(at^{\frac{1}{2}})$ $\mathrm{Re}\,\nu > 0$	$(2/a)^\nu\gamma(\nu,\frac{1}{4}a^2/p)$ $\mathrm{Re}\,p > 0$
14.27	$t^{\frac{1}{2}\nu+n}J_\nu(at^{\frac{1}{2}})$ $\mathrm{Re}(\nu+n) > -1$ $n = 0,1,2,\cdots$	$n!(\frac{1}{2}a)^\nu p^{-n-\nu-1}\exp(-\frac{1}{4}a^2/p)$ $\cdot L_n^\nu(\frac{1}{4}a^2/p)$ $\mathrm{Re}\,p > 0$
14.28	$t^\mu J_\nu(at^{\frac{1}{2}})$ $\mathrm{Re}(\mu+\frac{1}{2}\nu) > -1$	$2a^{-1}[\Gamma(1+\nu)]^{-1}\Gamma(1+\mu+\frac{1}{2}\nu)p^{-\mu-\frac{1}{2}}$ $\cdot\exp(-\frac{1}{8}a^2/p)M_{\mu+\frac{1}{2},\frac{1}{2}\nu}(\frac{1}{4}a^2/p)$ $=(\frac{1}{2}a)^\nu\Gamma(1+\mu+\frac{1}{2}\nu)[\Gamma(1+\nu)]^{-1}p^{-\mu-\frac{1}{2}\nu-1}$ $\cdot{}_1F_1(1+\mu+\frac{1}{2}\nu;1+\nu;-\frac{1}{4}a^2/p)$ $\mathrm{Re}\,p > 0$

	$f(t)$	$g(p) = \int_0^\infty f(t)e^{-pt}dt$
14.29	$t^{\frac{1}{2}\nu-\mu-1}\int_{at}^{\infty}u^{\mu}J_{\nu}(2u^{\frac{1}{2}})du$ $\mathrm{Re}\,\mu < \frac{3}{4}, \mathrm{Re}(\mu-\frac{1}{2}\nu) > 0$	$p^{\mu-\frac{1}{2}\nu}\Gamma(\mu+1+\frac{1}{2}\nu, ap^{-1})$ $\mathrm{Re}\,p > 0$
14.30	$t^{\frac{1}{2}\nu}L_n^{\nu}(t)J_{\nu}(at^{\frac{1}{2}})$ $n = 0,1,2,\cdots$	$(\frac{1}{2}a)^{\nu}(p-1)^n\exp(-\frac{1}{4}a^2/p)p^{-\nu-n-1}$ $\cdot L_n^{\nu}[(1-p)^{-1}ap^{-1}]$ $\mathrm{Re}\,p > 0$
14.31	$J_{\nu}(t)J_{2\nu}(at^{\frac{1}{2}})$ $\mathrm{Re}\,\nu > -\frac{1}{2}$	$(p^2+1)^{-\frac{1}{2}}\exp[-\frac{1}{4}a^2p(1+p^2)^{-1}]$ $\cdot J_{\nu}[\frac{1}{4}a^2(1+p^2)^{-1}]$ $\mathrm{Re}\,p > 0$
14.32	$J_{\nu}^2(at^{\frac{1}{2}})$ $\mathrm{Re}\,\nu > -1$	$p^{-1}e^{-\frac{1}{2}a^2/p}I_{\nu}(\frac{1}{2}a^2/p)$ $\mathrm{Re}\,p > 0$
14.33	$t^{-1}J_{\nu}^2(at^{\frac{1}{2}})$ $\mathrm{Re}\,\nu > 0$	$\nu^{-1}e^{-2a/p}[I_{\nu}(\frac{1}{2}a^2/p)+2\sum_{n=1}^{\infty}I_{\nu+n}(\frac{1}{2}a^2/p)]$ $\mathrm{Re}\,p > 0$
14.34	$J_{\nu}(at^{\frac{1}{2}})J_{\nu}(bt^{\frac{1}{2}})$ $\mathrm{Re}\,\nu > -1$	$p^{-1}\exp[-\frac{1}{4}p^{-1}(a^2+b^2)]I_{\nu}(\frac{1}{2}ab/p)$ $\mathrm{Re}\,p > 0$

	$f(t)$	$g(p) = \int_0^\infty f(t)e^{-pt}dt$
14.35	$t^{\lambda-1}J_\mu(at^{\frac{1}{2}})J_\nu(at^{\frac{1}{2}})$ $\mathrm{Re}(\nu+\mu+2\lambda)>0$	$2^{-\nu-\mu}a^{\nu+\mu}\Gamma(\lambda+\frac{1}{2}\mu+\frac{1}{2}\nu)$ $\cdot[\Gamma(\mu+1)\Gamma(\nu+1)]^{-1}p^{-\lambda-\frac{1}{2}\nu-\frac{1}{2}\mu}$ $\cdot {}_3F_3\left[\begin{matrix}\frac{1}{2}\nu+\frac{1}{2}\mu+\frac{1}{2}, \frac{1}{2}\nu+\frac{1}{2}\mu+1, \frac{1}{2}\nu+\frac{1}{2}\mu+\lambda;\\ \mu+1, \quad \nu+1, \quad \mu+\nu+1;\end{matrix} -a^2/p\right]$ $\mathrm{Re}\, p > 0$
14.36	$J_\nu[(-it)^{\frac{1}{2}}]J_\nu[(it)^{\frac{1}{2}}]$ $\mathrm{Re}\,\nu > -1$	$p^{-1}I_\nu(\frac{1}{2}p^{-1})$ $\mathrm{Re}\, p > 0$
14.37	$Y_\nu(at^{\frac{1}{2}})$ $-2 < \mathrm{Re}\,\nu < 2$	$\frac{1}{4}a\pi^{\frac{1}{2}}p^{-3/2}\exp(-\frac{1}{8}a^2/p)\{\cot\pi\nu[I_{\frac{1}{2}\nu-\frac{1}{2}}(\frac{1}{8}a^2/p)$ $-I_{\frac{1}{2}\nu+\frac{1}{2}}(\frac{1}{8}a^2/p)]-\csc\pi\nu[I_{-\frac{1}{2}\nu-\frac{1}{2}}(\frac{1}{8}a^2/p)$ $+ I_{-\frac{1}{2}\nu+\frac{1}{2}}(\frac{1}{8}a^2/p)]\}$ $\mathrm{Re}\, p > 0$
14.38	$t^{-\frac{1}{2}}Y_\nu(at^{\frac{1}{2}})$ $-1 < \mathrm{Re}\,\nu < 1$	$-(\pi/p)^{\frac{1}{2}}\exp(-\frac{1}{8}a^2/p)[\tan(\frac{1}{2}\pi\nu)I_{\frac{1}{2}\nu}(\frac{1}{8}a^2/p)$ $+\pi^{-1}\sec(\frac{1}{2}\pi\nu)K_{\frac{1}{2}\nu}(\frac{1}{8}a^2/p)]$, $\mathrm{Re}\, p > 0$
14.39	$t^\mu Y_\nu(at^{\frac{1}{2}})$ $\mathrm{Re}(\mu\pm\frac{1}{2}\nu) > -1$	$-2a^{-1}p^{-\mu-\frac{1}{2}}\exp(-\frac{1}{8}a^2/p)\{[\Gamma(\nu+1)]^{-1}$ $\cdot\cot(\pi\mu-\frac{1}{2}\pi\nu)\Gamma(1+\mu+\frac{1}{2}\nu)M_{\frac{1}{2}+\mu,\frac{1}{2}\nu}(\frac{1}{4}a^2/p)$ $+ \csc(\pi\mu-\frac{1}{2}\pi\nu)\}$ $\mathrm{Re}\, p > 0$

	$f(t)$	$g(p) = \int_0^\infty f(t)e^{-pt}dt$
14.40	$t^{-\frac{1}{2}}[\sin(\frac{1}{2}\pi\nu)J_\nu(at^{\frac{1}{2}})$ $+\cos(\frac{1}{2}\pi\nu)Y_\nu(at^{\frac{1}{2}})]$ $-1 < \operatorname{Re}\nu < 1$	$-(\pi p)^{-\frac{1}{2}}\exp(-\frac{1}{8}a^2/p)K_{\frac{1}{2}\nu}(\frac{1}{8}a^2/p)$ $\operatorname{Re} p > 0$
14.41	$t^{\frac{1}{2}\nu-\mu-1}\int_0^\infty u^\nu J_\nu(2u^{\frac{1}{2}})du$ $\operatorname{Re}(\mu+\frac{1}{2}\nu)>-1,\ \operatorname{Re}\nu > -1$	$p^{\mu-\frac{1}{2}\nu}\gamma(\mu+1+\frac{1}{2}\nu, ap^{-1})$ $\operatorname{Re} p > 0$
14.42	$t^{\nu-\frac{1}{2}}[J_{2\mu}(at^{\frac{1}{2}})\cos(\pi\nu+\pi\mu)$ $-J_{-2\mu}(at^{\frac{1}{2}})\cos(\pi\nu-\pi\mu)]$ $\operatorname{Re}(\nu\pm\mu) > -\frac{1}{2}$	$-2a^{-1}\sin(2\pi\mu)p^{-\nu}\exp(-\frac{1}{8}a^2/p)$ $\cdot W_{\nu,\mu}(\frac{1}{4}a^2/p)$, $\operatorname{Re} p > 0$
14.43	$(t^2+bt)^{-\frac{1}{2}}J_\nu[a(t^2+bt)^{\frac{1}{2}}]$ $\operatorname{Re}\nu > -1$	$e^{\frac{1}{2}bp}I_{\frac{1}{2}\nu}(z_1)K_{\frac{1}{2}\nu}(z_2)$ $z_{\frac{1}{2}} = \frac{1}{4}b[(p^2+a^2)^{\frac{1}{2}}\mp p]$, $\operatorname{Re} p>\lvert\operatorname{Im} a\rvert$
14.44	$(t^2+bt)^{\frac{1}{2}\nu}J_\nu[a(t^2+bt)^{\frac{1}{2}}]$ $\operatorname{Re}\nu > -1$	$(b/\pi)^{\frac{1}{2}}(\frac{1}{2}ab)^\nu(p^2+a^2)^{-\frac{1}{2}\nu-\frac{1}{4}}$ $\cdot e^{\frac{1}{2}bp}K_{\nu+\frac{1}{2}}[\frac{1}{2}b(p^2+a^2)^{\frac{1}{2}}]$ $\operatorname{Re} p>\lvert\operatorname{Im} a\rvert$

	$f(t)$	$g(p) = \int_0^\infty f(t)e^{-pt}dt$
14.45	$t^{\frac{1}{2}\nu}(t+b)^{-\frac{1}{2}\nu}J_\nu[a(t^2+bt)^{\frac{1}{2}}]$ $\mathrm{Re}\,\nu>-1,\ \lvert\arg b\rvert<\pi$	$a^\nu(p^2+a^2)^{-\frac{1}{2}}[p+(p^2+a^2)^{\frac{1}{2}}]^{-\nu}$ $\cdot\exp\{-\frac{1}{2}b[(p^2+a^2)^{\frac{1}{2}}-p]\}$ $\mathrm{Re}\,p>\lvert\mathrm{Im}\,a\rvert$
14.46	$t^{\mu-1}(t+b)^{-\mu}J_\nu[a(t^2+bt)^{\frac{1}{2}}]$ $\mathrm{Re}(\nu+2\mu)>0$	$2(ab)^{-1}\Gamma(\frac{1}{2}\nu+\mu)[\Gamma(1+\nu)]^{-1}$ $\cdot e^{\frac{1}{2}bp}W_{\frac{1}{2}-\mu,\frac{1}{2}\nu}(z_1)M_{\mu-\frac{1}{2},\frac{1}{2}\nu}(z_2)$ $z_{\frac{1}{2}} = \frac{1}{2}b[(p^2+a^2)^{\frac{1}{2}}\pm p]$ $\mathrm{Re}\,p>\lvert\mathrm{Im}\,a\rvert$
14.47	$t^{\frac{1}{2}\nu-1}(1+t)^{-\frac{1}{2}\nu}J_\nu[a(t^2+t)^{\frac{1}{2}}]$ $\mathrm{Re}\,\nu>0$	$(\frac{1}{2}a)^{-\nu}\gamma[\nu,\frac{1}{2}(p^2+a^2)^{\frac{1}{2}}-\frac{1}{2}p]$ $\mathrm{Re}\,p>\lvert\mathrm{Im}\,a\rvert$
14.48	$(t^2+2it)^{\frac{1}{2}\nu}J_\nu[a(t^2+2it)^{\frac{1}{2}}]$ $\mathrm{Re}\,\nu>-1$	$-i(\frac{1}{2}\pi)^{\frac{1}{2}}a^\nu(p^2+a^2)^{-\frac{1}{2}\nu-\frac{1}{4}}$ $\cdot e^{ip}H^{(2)}_{\nu+\frac{1}{2}}[(p^2+a^2)^{\frac{1}{2}}]$ $\mathrm{Re}\,p>\lvert\mathrm{Im}\,a\rvert$
14.49	$(t^2+2it)^{\lambda-\frac{1}{2}\nu}J_\nu[a(t^2+2it)^{\frac{1}{2}}]$ $\mathrm{Re}\,\lambda>-1$	$-i\pi^{\frac{1}{2}}2^{\lambda-\nu-\frac{1}{2}}\Gamma(\lambda+1)[\Gamma(\nu-\lambda)]^{-1}$ $\cdot(p^2+a^2)^{-\frac{1}{2}\lambda-\frac{1}{4}}e^{ip}$ $\cdot\sum_{n=0}^{\infty}2^{-n}\frac{\Gamma(\nu-\lambda+n)(p^2+a^2)^{-\frac{1}{2}n}}{n!\Gamma(\nu+n+1)}$ $\cdot H^{(2)}_{\lambda+n+\frac{1}{2}}[(p^2+a^2)^{\frac{1}{2}}]$ $\mathrm{Re}\,p>\lvert\mathrm{Im}\,a\rvert$

	$f(t)$	$g(p) = \int_0^\infty f(t)e^{-pt}dt$
14.50	$0 \quad t < b$ $(t^2-b^2)^{-\frac{1}{2}}J_\nu[a(t^2-b^2)^{\frac{1}{2}}]$ $t > b$ $\mathrm{Re}\,\nu > -1$	$I_{\frac{1}{2}\nu}(z_1)K_{\frac{1}{2}\nu}(z_2)$ $z_{\frac{1}{2}} = \frac{1}{2}b[(p^2+a^2)^{\frac{1}{2}} \mp p]$ $\mathrm{Re}\,p > \lvert\mathrm{Im}\,a\rvert$
14.51	$0 \quad t < b$ $(t^2-b^2)^{\frac{1}{2}\nu}J_\nu[a(t^2-b^2)^{\frac{1}{2}}]$ $t > b$ $\mathrm{Re}\,\nu > -1$	$(2b/\pi)^{\frac{1}{2}}(ab)^\nu(p^2+a^2)^{-\frac{1}{2}\nu-\frac{1}{4}}$ $\cdot K_{\nu+\frac{1}{2}}[b(p^2+a^2)^{\frac{1}{2}}]$ $\mathrm{Re}\,p > \lvert\mathrm{Im}\,a\rvert$
14.52	$0 \quad t < b$ $(t-b)^{\frac{1}{2}\nu}(t+b)^{-\frac{1}{2}\nu}$ $\cdot J_\nu[a(t^2-b^2)]$ $t > b$ $\mathrm{Re}\,\nu > -1$	$a^\nu(p^2+a^2)^{-\frac{1}{2}}[p+(p^2+a^2)^{\frac{1}{2}}]^{-\nu}$ $\cdot\exp[-b(p^2+a^2)^{\frac{1}{2}}]$ $\mathrm{Re}\,p > \lvert\mathrm{Im}\,a\rvert$
14.53	$0 \quad t < b$ $(t-b)^{\frac{1}{2}\nu-1}(t+b)^{-\frac{1}{2}\nu}$ $\cdot J_\nu[a(t^2-b^2)^{\frac{1}{2}}]$ $t > b$ $\mathrm{Re}\,\nu > 0$	$(ab)^{-\nu}e^{-bp}\gamma[\nu,b(p^2+a^2)^{\frac{1}{2}}-bp]$ $\mathrm{Re}\,p > \lvert\mathrm{Im}\,a\rvert$

	$f(t)$	$g(p) = \int_0^\infty f(t)e^{-pt}dt$
14.54	$0 \quad t < b$ $(t-b)^{\mu-1}(t+b)^{-\mu}$ $\cdot J_\nu[a(t^2-b^2)^{\frac{1}{2}}]$ $t > b$ $Re(\nu+2\mu) > 0$	$[ab\Gamma(1+\nu)]^{-1}\Gamma(\mu+\frac{1}{2}\nu)$ $\cdot W_{\frac{1}{2}-\mu,\frac{1}{2}\nu}(z_1)M_{\mu-\frac{1}{2},\frac{1}{2}\nu}(z_2)$ $z_{\frac{1}{2}} = b[(p^2+a^2)^{\frac{1}{2}}\pm p]$ $Re\ p > \|Im\ a\|$
14.55	$t^{-1}J_\nu(a/t)$ $a > 0$	$2J_\nu[(2ap)^{\frac{1}{2}}]K_\nu[(2ap)^{\frac{1}{2}}]$ $Re\ p > 0$
14.56	$t^{-1}Y_\nu(a/t)$ $a > 0$	$2Y_\nu[(2ap)^{\frac{1}{2}}]K_\nu[(2ap)^{\frac{1}{2}}]$ $Re\ p > 0$
14.57	$t^{-1}e^{-a/t}J_\nu(b/t)$ $Re\ a>0,\ b > 0$	$2J_\nu(z_1)K_\nu(z_2)$ $z_{\frac{1}{2}} = (2p)^{\frac{1}{2}}[(a^2+b^2)^{\frac{1}{2}}\mp a]^{\frac{1}{2}}$ $Re\ p > 0$
14.58	$t^{-1}e^{-a/t}Y_\nu(b/t)$ $Re\ a>0,\ b > 0$	$2Y_\nu(z_1)K_\nu(z_2)$ $z_{\frac{1}{2}} = (2p)^{\frac{1}{2}}[(a^2+b^2)^{\frac{1}{2}}\mp a]^{\frac{1}{2}}$ $Re\ p > 0$

	$f(t)$	$g(p) = \int_0^\infty f(t)e^{-pt}dt$
14.59	$t^{-\frac{1}{2}}[\cos(a/t)J_\nu(a/t)$ $-\sin(a/t)Y_\nu(a/t)]$	$-4(\pi p)^{-\frac{1}{2}}\{[\sin(\frac{1}{2}\pi\nu)\mathrm{ker}_{2\nu}[(8ap)^{\frac{1}{2}}]$ $+\cos(\frac{1}{2}\pi\nu)\mathrm{kei}_{2\nu}[(8ap)^{\frac{1}{2}}]\}$ $\mathrm{Re}\ p > 0$
14.60	$t^{-\frac{1}{2}}[\cos(a/t)Y_\nu(a/t)$ $+\sin(a/t)J_\nu(a/t)]$	$4(\pi p)^{-\frac{1}{2}}\{\sin(\frac{1}{2}\pi\nu)\mathrm{kei}_{2\nu}[(8ap)^{\frac{1}{2}}]$ $-\cos(\frac{1}{2}\pi\nu)\mathrm{ker}_{2\nu}[(8ap)^{\frac{1}{2}}]\}$ $\mathrm{Re}\ p > 0$
14.61	$t^{-1}[\cos(\frac{1}{2}a^2/t)J_\nu(\frac{1}{2}a^2/t)$ $-\sin(\frac{1}{2}a^2/t)Y_\nu(\frac{1}{2}a^2/t)]$	$4\pi^{-1}\{\sin(\frac{1}{2}\pi\nu)$ $\cdot[\mathrm{kei}_\nu^2(ap^{\frac{1}{2}})-\mathrm{ker}_\nu^2(ap^{\frac{1}{2}})]$ $-2\cos(\frac{1}{2}\pi\nu)\mathrm{ker}_\nu(ap^{\frac{1}{2}})\mathrm{kei}_\nu(ap^{\frac{1}{2}})\}$ $\mathrm{Re}\ p > 0$
14.62	$t^{-1}[\sin(\frac{1}{2}a^2/t)J_\nu(\frac{1}{2}a^2/t)$ $+\cos(\frac{1}{2}a^2/t)Y_\nu(\frac{1}{2}a^2/t)]$	$4\pi^{-1}\{2\sin(\frac{1}{2}\pi\nu)\mathrm{ker}_\nu(ap^{\frac{1}{2}})\mathrm{kei}_\nu(ap^{\frac{1}{2}})$ $-\cos(\frac{1}{2}\pi\nu)[\mathrm{ker}_\nu^2(ap^{\frac{1}{2}})-\mathrm{kei}_\nu^2(ap^{\frac{1}{2}})]\}$ $\mathrm{Re}\ p > 0$

	$f(t)$	$g(p) = \int_0^\infty f(t)e^{-pt}dt$
14.63	$t^{\frac{1}{2}}J_{\frac{1}{4}}(at^2)$ $a > 0$	$\frac{1}{4}a^{-1}(\pi p)^{\frac{1}{2}}[\mathbf{H}_{-\frac{1}{4}}(\frac{1}{4}p^2/a)-Y_{-\frac{1}{4}}(\frac{1}{4}p^2/a)]$ $\mathrm{Re}\, p > 0$
14.64	$t^{\frac{1}{2}}J_{-\frac{1}{4}}(at^2)$ $a > 0$	$\frac{1}{4}a^{-1}(\pi p)^{\frac{1}{2}}[\mathbf{H}_{\frac{1}{4}}(\frac{1}{4}p^2/a)-Y_{\frac{1}{4}}(\frac{1}{4}p^2/a)]$ $\mathrm{Re}\, p > 0$
14.65	$t^{3/2}J_{-\frac{1}{4}}(at^2)$ $a > 0$	$-\frac{1}{8}a^{-2}p(\pi p)^{\frac{1}{2}}[\mathbf{H}_{-3/4}(\frac{1}{4}p^2/a)-Y_{-3/4}(\frac{1}{4}p^2/a)]$ $\mathrm{Re}\, p > 0$
14.66	$t^{3/2}J_{-3/4}(at^2)$ $a > 0$	$\frac{1}{8}a^{-2}p(\pi p)^{\frac{1}{2}}[\mathbf{H}_{-\frac{1}{4}}(\frac{1}{4}p^2/a)-Y_{-\frac{1}{4}}(\frac{1}{4}p^2/a)]$ $\mathrm{Re}\, p > 0$
14.67	$t^{\frac{1}{2}}J_{1/8}(at^2)J_{-1/8}(at^2)$ $a > 0$	$(16a)^{-1}(\frac{1}{2}\pi p)^{\frac{1}{2}}\sec(\pi/8)$ $\cdot H_{1/8}^{(1)}(\frac{1}{16}p^2/a)H_{1/8}^{(2)}(\frac{1}{16}p^2/a)$ $\mathrm{Re}\, p > 0$
14.68	$J_{\nu+\frac{1}{2}}(at^2)$ $a > 0$, $\mathrm{Re}\,\nu>-1$	$(2a\pi)^{-\frac{1}{2}}\Gamma(1+\nu)D_{-\nu-1}[p(2ai)^{-\frac{1}{2}}]$ $\cdot D_{-\nu-1}[p(-2ai)^{-\frac{1}{2}}]$, $\mathrm{Re}\, p > 0$

	$f(t)$	$g(p) = \int_0^\infty f(t) e^{-pt} dt$
14.69	$(1-e^{-t})^{\frac{1}{2}\nu} J_\nu[a(1-e^{-t})^{\frac{1}{2}}]$ $\mathrm{Re}\ \nu > -1$	$\Gamma(p)(2/a)^p J_{\nu+p}(a)$ $\mathrm{Re}\ p > 0$
14.70	$(1-e^{-t})^{-\frac{1}{2}\nu} J_\nu[a(1-e^{-t})^{\frac{1}{2}}]$	$2^{-\nu}[\Gamma(\nu)]^{-1} a^{-p} s_{\nu+p-1,p-\nu}(a)$ $\mathrm{Re}\ p > 0$
14.71	$(e^t-1)^{\frac{1}{2}\nu} J_\nu[2a(e^t-1)^{\frac{1}{2}}]$ $a > 0$	$2[\Gamma(1+p)]^{-1} a^p K_{\nu-p}(2a)$ $\mathrm{Re}\ p > \frac{1}{2}\ \mathrm{Re}(\nu-\frac{3}{2})$
14.72	$(e^t-1)^\mu J_{2\nu}[2a(e^t-1)^{\frac{1}{2}}]$ $a>0,\ \mathrm{Re}(\mu+\nu)>-1$	$[\Gamma(2\nu+1)]^{-1} a^{2\nu} B(\mu+\nu+1,\ p-\mu-\nu)$ $\cdot\ {}_1F_2[{\mu+\nu+1; \atop \mu+\nu+1-p; 2\nu+1;}\ a^2]$ $+[\Gamma(\nu-\mu+p+1)]^{-1} a^{2p-2\mu} \Gamma(\mu+\nu-p)$ $\cdot\ {}_1F_2[{p+1; \atop p+1+\nu-\mu,\ p+1-\mu-\nu;}\ a^2]$ $\mathrm{Re}\ p > -\frac{7}{4}+\mathrm{Re}\ \mu$
14.73	$J_\nu(a\ \sinh t)$ $\mathrm{Re}\ a > 0,\ \mathrm{Re}\ \nu>-1$	$I_{\frac{1}{2}\nu+\frac{1}{2}p}(\frac{1}{2}a) K_{\frac{1}{2}\nu-\frac{1}{2}p}(\frac{1}{2}a)$ $\mathrm{Re}\ p > -\frac{1}{2}$
14.74	$J_0[a(\sinh t)^{\frac{1}{2}}]$ $\mathrm{Re}\ a > 0$	$2J_p(2^{-\frac{1}{2}}a) K_p(2^{-\frac{1}{2}}a)$ $\mathrm{Re}\ p > -\frac{1}{4}$

	$f(t)$	$g(p) = \int_0^\infty f(t)e^{-pt}dt$
14.75	$\operatorname{csch}t\, J_\nu(a \operatorname{csch}t)$ $a > 0$	$[a\Gamma(1+\nu)]^{-1}\Gamma(\frac{1}{2}+\frac{1}{2}\nu+\frac{1}{2}p)$ $\cdot W_{-\frac{1}{2}p,\frac{1}{2}\nu}(a)M_{\frac{1}{2}p,\frac{1}{2}\nu}(a)$ $\operatorname{Re} p > -\operatorname{Re}(\nu+1)$
14.76	$\operatorname{csch}t \exp[(a-b)\coth t]$ $\cdot J_\nu[(ab)^{\frac{1}{2}}\operatorname{csch}t]$ $\operatorname{Re} a > 0,\ \operatorname{Re} b > 0$	$(ab)^{-\frac{1}{2}}[\Gamma(\nu+1)]^{-1}\Gamma(\frac{1}{2}p+\frac{1}{2}+\frac{1}{2}\nu)$ $\cdot M_{\frac{1}{2}p,\frac{1}{2}\nu}(a)W_{-\frac{1}{2}p,\frac{1}{2}\nu}(b)$ $\operatorname{Re} p > \frac{1}{2}-\operatorname{Re}(\frac{1}{2}\nu)$

1.15 Modified Bessel Functions

15.1	$I_\nu(at)$ $\operatorname{Re}\nu > -1$	$a^{-\nu}(p^2-a^2)^{-\frac{1}{2}}[p-(p^2-a^2)^{\frac{1}{2}}]^\nu$ $\operatorname{Re} p > \lvert\operatorname{Re} a\rvert$
15.2	$t\, I_\nu(at)$ $\operatorname{Re}\nu > -2$	$a^{-\nu}(p^2-a^2)^{-3/2}[p+\nu(p^2-a^2)^{\frac{1}{2}}]$ $\cdot[p-(p^2-a^2)^{\frac{1}{2}}]^\nu,$ $\operatorname{Re} p > \lvert\operatorname{Re} a\rvert$
15.3	$t^n I_n(at)$ $n=1,2,3,\cdots$	$(p^2-a^2)^{-n-\frac{1}{2}}1\cdot 3\cdots(2n-1)a^n$ $\operatorname{Re} p > \lvert\operatorname{Re} a\rvert$

	$f(t)$	$g(p) = \int_0^\infty f(t)e^{-pt}dt$
15.4	$t^{-1}I_\nu(at)$ $\mathrm{Re}\ \nu > 0$	$\nu^{-1}a^{-\nu}[p-(p^2-a^2)^{\frac{1}{2}}]^\nu$ $\mathrm{Re}\ p > \lvert\mathrm{Re}\ a\rvert$
15.5	$t^{-2}I_\nu(at)$ $\mathrm{Re}\ \nu > 1$	$\nu^{-1}a^{-\nu}(\nu^2-1)^{-1}[p+\nu(p^2-a^2)^{\frac{1}{2}}]$ $\cdot[p-(p^2-a^2)^{\frac{1}{2}}]^\nu$, $\mathrm{Re}\ p > \lvert\mathrm{Re}\ a\rvert$
15.6	$t^\nu I_\nu(at)$ $\mathrm{Re}\ \nu > -\frac{1}{2}$	$2^\nu\pi^{-\frac{1}{2}}\Gamma(\frac{1}{2}+\nu)a^\nu(p^2-a^2)^{-\nu-\frac{1}{2}}$ $\mathrm{Re}\ p > \lvert\mathrm{Re}\ a\rvert$
15.7	$t^{\nu+1}I_\nu(at)$ $\mathrm{Re}\ \nu > -1$	$2^{\nu+1}\pi^{-\frac{1}{2}}\Gamma(\frac{3}{2}+\nu)a^\nu p(p^2-a^2)^{-\nu-\frac{3}{2}}$ $\mathrm{Re}\ p > \lvert\mathrm{Re}\ a\rvert$
15.8	$t^\mu I_\nu(at)$ $\mathrm{Re}(\nu+\mu) > -1$	$-i(\frac{1}{2}\pi a)^{-\frac{1}{2}}(p^2-a^2)^{-\frac{1}{2}\mu-\frac{1}{4}}q_{\nu-\frac{1}{2}}^{\mu+\frac{1}{2}}(p/a)e^{-i\pi\mu}$ $=\Gamma(\nu+\mu+1)(p^2-a^2)^{-\frac{1}{2}\mu-\frac{1}{2}}P_\mu^{-\nu}[p(p^2-a^2)^{-\frac{1}{2}}]$ $\mathrm{Re}\ p > \lvert\mathrm{Re}\ a\rvert$
15.9	$t^{-1}e^{-b/t}I_\nu(at)$ $\mathrm{Re}\ b > 0$	$2I_\nu(z_1)K_\nu(z_2)$ $z_{\frac{1}{2}} = b^{\frac{1}{2}}[(p+a)^{\frac{1}{2}} \mp (p-a)^{\frac{1}{2}}]$ $\mathrm{Re}\ p > \lvert\mathrm{Re}\ a\rvert$

	$f(t)$	$g(p) = \int_0^\infty f(t)e^{-pt}dt$
15.10	$I_\nu(at)I_\nu(bt)$ $\mathrm{Re}\,\nu > -\frac{1}{2}$	$\pi^{-1}(ab)^{-\frac{1}{2}}q_{\nu-\frac{1}{2}}[(2ab)^{-1}(p^2-a^2-b^2)]$ $\mathrm{Re}\,p > \lvert\mathrm{Re}\,a\rvert+\lvert\mathrm{Re}\,b\rvert$
15.11	$t^{-\frac{1}{2}}I_\mu(at)I_\nu(bt)$ $\mathrm{Re}(\mu+\nu) > -\frac{1}{2}$	$c^{\frac{1}{2}}\Gamma(\frac{1}{2}+\mu+\nu)P_{\nu-\frac{1}{2}}^{-\mu}(\cosh\alpha)$ $\cdot P_{\mu-\frac{1}{2}}^{-\nu}(\cosh\beta)$ $\mathrm{Re}(p\pm a\pm b) > 0$ $\sinh\alpha=ac, \sinh\beta=bc$ $\cosh\alpha\,\cosh\beta = pc$ $\lvert\mathrm{Im}\,a\rvert, \lvert\mathrm{Im}\,b\rvert < \frac{1}{2}\pi$
15.12	$t^{-\frac{1}{2}}I_\nu(at)I_{-\nu}(at)$ $a > 0$	$(p/\pi)^{-\frac{1}{2}}P_{-\frac{1}{4}}^{\nu}(z)P_{-\frac{1}{4}}^{-\nu}(z)$ $z = (1-4a^2p^{-2})^{\frac{1}{2}}$, $\mathrm{Re}\,p > 2a$
15.13	$t^{-\frac{1}{2}}I_\nu^2(at)$ $\mathrm{Re}\,\nu > -\frac{1}{4}$, $a > 0$	$2^{-2\nu}p^{-\frac{1}{2}}\Gamma(\frac{1}{2}+2\nu)[P_{-\frac{1}{4}}^{-\nu}(z)]^2$ $z = (1-4a^2p^{-2})$, $\mathrm{Re}\,p > 2a$
15.14	$K_\nu(at)$ $-1 < \mathrm{Re}\,\nu < 1$	$\frac{1}{2}\pi a^{-\nu}\csc(\pi\nu)(p^2-a^2)^{-\frac{1}{2}}$ $\cdot\{[p+(p^2-a^2)^{\frac{1}{2}}]^{\nu}-[p-(p^2-a^2)^{\frac{1}{2}}]^{\nu}\}$ $\mathrm{Re}\,p > -\mathrm{Re}\,a$

	$f(t)$	$g(p) = \int_0^\infty f(t)e^{-pt}dt$
15.15	$t^{\mu}K_{\nu}(at)$ $\text{Re}(\mu\pm\nu)>-1$	$(p^2-a^2)^{-\frac{1}{2}\mu-\frac{1}{2}}\Gamma(1+\mu-\nu)e^{-i\pi\nu}$ $\cdot q_{\mu}^{\nu}[p(p^2-a^2)^{-\frac{1}{2}}]$ $\quad \text{Re } p > -\text{Re } a$ $=(\frac{1}{2}\pi/a)^{\frac{1}{2}}\Gamma(\mu-\nu+1)\Gamma(\mu+\nu+1)$ $\cdot(p^2-a^2)^{-\frac{1}{2}\mu-\frac{1}{4}}P_{\nu-\frac{1}{2}}^{-\mu-\frac{1}{2}}(p/a)$ $\text{Re } p > -\text{Re } a$
15.16	$t^{\mu}K_{\nu}(at)$ $\text{Re}(\mu\pm\nu)>-1,\ a>0$	$(\frac{1}{2}\pi/a)^{\frac{1}{2}}\Gamma(\mu-\nu+1)\Gamma(\mu+\nu+1)$ $\cdot(a^2-p^2)^{-\frac{1}{2}\mu-\frac{1}{4}}\mathrm{P}_{\nu-\frac{1}{2}}^{-\mu-\frac{1}{2}}(p/a)$ for $-a < p < a$
15.17	$t^{-1}e^{-b/t}K_{\nu}(at)$	$2K_{\nu}(z_1)K_{\nu}(z_2)$ $z_{\frac{1}{2}} = b^{\frac{1}{2}}[(p+a)^{\frac{1}{2}}\pm(p-a)^{\frac{1}{2}}]$
15.18	$[I_{\nu}(at)+I_{-\nu}(at)]$ $\cdot K_{\nu}(at)$ $-\frac{1}{2} < \text{Re } \nu < \frac{1}{2}$	$\frac{1}{2}\pi a^{-1}\sec(\pi\nu)P_{\nu-\frac{1}{2}}(\frac{1}{2}p^2a^{-2}-1)$ $\text{Re } p > 0$
15.19	$t^{-\frac{1}{2}}I_{\nu}(at)K_{\nu}(at)$ $\text{Re } \nu > -\frac{1}{4}$	$-(2a)^{-1}(\frac{1}{2}\pi p)^{\frac{1}{2}}[P_{\nu-\frac{1}{2}}^{-\frac{1}{4}}(z)Q_{\nu-\frac{1}{2}}^{\frac{1}{4}}(z)$ $-P_{\nu-\frac{1}{2}}^{\frac{1}{4}}(z)Q_{\nu-\frac{1}{2}}^{-\frac{1}{4}}(z)]$, $\text{Re } p > 0$ $z = (1-\frac{1}{4}p^2a^{-2})^{\frac{1}{2}}$, $\quad p < 2a$

	$f(t)$	$g(p) = \int_0^\infty f(t)e^{-pt}dt$
15.20	$t^{2\nu}I_\nu(at)K_\nu(at)$ $0 > \mathrm{Re}\ \nu > -\frac{1}{2}$	$\frac{1}{4}\pi a^{-2\nu-1}\sec^2(\pi\nu)P_{\nu-\frac{1}{2}}(\frac{1}{2}p^2a^{-2}-1)$ $p < 2a$ $\frac{1}{4}\pi a^{-2\nu-1}\sec^2(\pi\nu)P_{\nu-\frac{1}{2}}(\frac{1}{2}p^2a^{-2}-1)$ $\mathrm{Re}\ p > 0$
15.21	$t^{-\frac{1}{2}}I_\mu(at)K_\nu(bt)$ $\mathrm{Re}(\mu\pm\nu) > -\frac{1}{2}$	$c^{\frac{1}{2}}\cos(\pi\mu)\sec[\pi(\mu+\nu)]\Gamma(\frac{1}{2}+\mu-\nu)$ $\cdot p^{-\mu}_{-\nu-\frac{1}{2}}(\cosh A)q^{-\nu}_{-\mu-\frac{1}{2}}\cosh B)$ $\mathrm{Re}(p\pm a+b) > 0$ $\sinh A = ac,\ \sinh B = bc$ $\cosh A\ \cosh B = pc$ $\lvert\mathrm{Im}\ A\rvert,\ \lvert\mathrm{Im}\ B\rvert < \frac{1}{2}\pi$
15.22	$I_\nu(at^{\frac{1}{2}})$ $\mathrm{Re}\ \nu > -2$	$\frac{1}{4}a\pi^{\frac{1}{2}}p^{-\frac{3}{2}}\exp(\frac{1}{8}a^2/p)$ $\cdot[I_{\frac{1}{2}\nu-\frac{1}{2}}(\frac{1}{8}a^2/p)-I_{\frac{1}{2}\nu+\frac{1}{2}}(\frac{1}{8}a^2/p)]$ $\mathrm{Re}\ p > 0$
15.23	$t^{-\frac{1}{2}}I_\nu(at^{\frac{1}{2}})$ $\mathrm{Re}\ \nu > -1$	$(\pi/p)^{\frac{1}{2}}\exp(\frac{1}{8}a^2/p)I_{\frac{1}{2}\nu}(\frac{1}{8}a^2/p)$ $\mathrm{Re}\ p > 0$

	$f(t)$	$g(p) = \int_0^\infty f(t) e^{-pt} dt$
15.24	$t^{\frac{1}{2}\nu} I_\nu(at^{\frac{1}{2}})$ $\mathrm{Re}\,\nu > -1$	$(\frac{1}{2}a)^\nu p^{-\nu-1} \exp(\frac{1}{4}a^2/p)$ $\mathrm{Re}\,p > 0$
15.25	$t^{-\frac{1}{2}\nu} I_\nu(at^{\frac{1}{2}})$ $\mathrm{Re}\,\nu > 0$	$(2/a)^\nu [\Gamma(\nu)]^{-1} p^{\nu-1} \gamma(\nu, \frac{1}{4}a^2/p)$
15.26	$t^{\frac{1}{2}\nu-1} I_\nu(at^{\frac{1}{2}})$ $\mathrm{Re}\,\nu > 0$	$(\frac{1}{2}ae^{i\pi})^{-\nu} \gamma(\nu, \frac{1}{4}a^2 e^{i\pi} p^{-1})$ $\mathrm{Re}\,p > 0$
15.27	$t^{\frac{1}{2}\nu+n} I_\nu(at^{\frac{1}{2}})$ $\mathrm{Re}(\nu+n) > -1$ $n=0,1,2,\cdots$	$n!(\frac{1}{2}a)^\nu p^{-n-\nu-1} \exp(\frac{1}{4}a^2/p)$ $\cdot L_n^\nu(-\frac{1}{4}a^2/p)$ $\quad \mathrm{Re}\,p > 0$
15.28	$t^\mu I_\nu(at^{\frac{1}{2}})$ $\mathrm{Re}(\mu+\frac{1}{2}\nu) > -1$	$2a^{-1}[\Gamma(1+\nu)]^{-1}\Gamma(1+\mu+\frac{1}{2}\nu) p^{-\mu-\frac{1}{2}}$ $\cdot \exp(\frac{1}{8}a^2/p) M_{-\mu-\frac{1}{2}, \frac{1}{2}\nu}(\frac{1}{4}a^2/p)$ $= (\frac{1}{2}a)^\nu \Gamma(1+\mu+\frac{1}{2}\nu) [\Gamma(1+\nu)]^{-1} p^{-\mu-1-\frac{1}{2}\nu}$ $\cdot {}_1F_1(1+\mu+\frac{1}{2}\nu; 1+\nu; \frac{1}{4}a^2/p)$ $\mathrm{Re}\,p > 0$

	$f(t)$	$g(p) = \int_0^\infty f(t) e^{-pt} dt$
15.29	$I_\nu^2(at^{\frac{1}{2}})$ $\mathrm{Re}\,\nu > -1$	$p^{-1} e^{\frac{1}{2}a^2/p} I_\nu(\tfrac{1}{2}a^2/p)$ $\mathrm{Re}\,p > 0$
15.30	$t^{-1} I_\nu^2(at^{\frac{1}{2}})$ $\mathrm{Re}\,\nu > -1$	$\nu^{-1} e^{2a/p} [J_\nu(\tfrac{1}{2}a^2/p) + 2 \sum_{n=1}^{\infty} (-1)^n J_{\nu+n}(\tfrac{1}{2}a^2/p)]$
15.31	$I_\nu(at^{\frac{1}{2}}) I_\nu(bt^{\frac{1}{2}})$ $\mathrm{Re}\,\nu > -1$	$p^{-1} \exp[\tfrac{1}{4}p^{-1}(a^2+b^2)] I_\nu(\tfrac{1}{2}ab/p)$ $\mathrm{Re}\,p > 0$
15.32	$I_\nu(at^{\frac{1}{2}}) J_\nu(bt^{\frac{1}{2}})$ $\mathrm{Re}\,\nu > -1$	$p^{-1} \exp[\tfrac{1}{4}p^{-1}(a^2-b^2)] J_\nu(\tfrac{1}{2}ab/p)$ $\mathrm{Re}\,p > 0$
15.33	$t^{\lambda-1} I_\mu(at^{\frac{1}{2}}) I_\nu(at^{\frac{1}{2}})$ $\mathrm{Re}(\nu+\mu+2\lambda) > 0$	$2^{-\nu-\mu} a^{\nu+\mu} \Gamma(\lambda+\tfrac{1}{2}\mu+\tfrac{1}{2}\nu)$ $\cdot [\Gamma(\mu+1)\Gamma(\nu+1)]^{-1} p^{-\lambda-\frac{1}{2}\nu-\frac{1}{2}\mu}$ $\cdot {}_3F_3\left[\begin{matrix} \tfrac{1}{2}\nu+\tfrac{1}{2}\mu+\tfrac{1}{2}, \tfrac{1}{2}\nu+\tfrac{1}{2}\mu+1, \tfrac{1}{2}\nu+\tfrac{1}{2}\mu+\lambda; \\ \mu+1, \quad \nu+1, \quad \mu+\nu+1; \end{matrix} a^2/p\right]$ $\mathrm{Re}\,p > 0$
15.34	$t^{-\frac{1}{2}} K_{1/3}(at^{-\frac{1}{2}})$	$3^{-\frac{1}{2}} \pi (\tfrac{1}{2}ap^2)^{-1/3} \exp[-3(\tfrac{1}{4}a^2p)^{1/3}]$ $\mathrm{Re}\,p > 0$

	$f(t)$	$g(p) = \int_0^\infty f(t)e^{-pt}dt$
15.35	$t^{-3/2}K_{1/3}(at^{-1/2})$	$2\pi 3^{-1/2}a^{-1}\exp[-3(\frac{1}{4}a^2p)^{1/3}]$ Re $p > 0$
15.36	$t^{-1}K_{2/3}(at^{-1/2})$	$2^{2/3}\pi 3^{-1/2}a^{-2/3}p^{-1/3}\exp[-3(\frac{1}{4}a^2p)^{1/3}]$ Re $p > 0$
15.37	$t^{-1/2}K_\nu(at^{1/2})$ $-1 < \text{Re}\,\nu < 1$	$\frac{1}{2}(\pi/p)^{1/2}\sec(\frac{1}{2}\pi\nu)\exp(\frac{1}{8}a^2/p)$ $\cdot K_{\frac{1}{2}\nu}(\frac{1}{8}a^2/p)$ Re $p > 0$
15.38	$t^{\frac{1}{2}\nu}K_\nu(at^{1/2})$ Re $\nu > -1$	$a^\nu\Gamma(1+\nu)(2p)^{-\nu-1}\exp(\frac{1}{4}a^2/p)$ $\cdot\Gamma(-\nu,\frac{1}{4}a^2/p)$ Re $p > 0$
15.39	$t^\mu K_\nu(at^{1/2})$ $\text{Re}(\mu\pm\frac{1}{2}\nu) > -1$	$a^{-1}\Gamma(1+\mu+\frac{1}{2}\nu)\Gamma(1+\mu-\frac{1}{2}\nu)$ $\cdot p^{-\mu-\frac{1}{2}}\exp(\frac{1}{8}a^2/p)W_{-\mu-\frac{1}{2},\frac{1}{2}\nu}(\frac{1}{4}a^2/p)$ Re $p > 0$
15.40	$t^\nu K_\nu(at^{1/2})I_\nu(at^{1/2})$ Re $\nu > -\frac{1}{2}$	$\frac{1}{2}\Gamma(\frac{1}{2}+\nu)a^{\nu-1}p^{-\frac{1}{2}-\frac{3}{2}\nu}$ $\cdot\exp(\frac{1}{2}a^2/p)W_{-\frac{1}{2}\nu,\frac{1}{2}\nu}(a^2/p)$

	$f(t)$	$g(p) = \int_0^\infty f(t)e^{-pt}dt$
15.41	$(t^2+bt)^{-\frac{1}{2}} I_\nu[a(t^2+bt)^{\frac{1}{2}}]$ $\mathrm{Re}\,\nu > -1$	$e^{\frac{1}{2}bp} I_{\frac{1}{2}\nu}(z_1) K_{\frac{1}{2}\nu}(z_2)$ $z_{\frac{1}{2}} = \frac{1}{4}b[p \mp (p^2-a^2)^{\frac{1}{2}}]$, $\mathrm{Re}\,p > \lvert\mathrm{Re}\,a\rvert$
15.42	$(t^2+b^2)^{\frac{1}{2}\nu} I_\nu[a(t^2+bt)^{\frac{1}{2}}]$ $\mathrm{Re}\,\nu > -1$	$(b/\pi)^{\frac{1}{2}}(\frac{1}{2}ab)^\nu (p^2-a^2)^{-\frac{1}{2}\nu-\frac{1}{4}} e^{\frac{1}{2}bp}$ $\cdot K_{\nu+\frac{1}{4}}[\frac{1}{2}b(p^2-a^2)^{\frac{1}{2}}]$, $\mathrm{Re}\,p > \lvert\mathrm{Re}\,a\rvert$
15.43	$t^{\frac{1}{2}\nu}(t+b)^{-\frac{1}{2}\nu} I_\nu[a(t^2+bt)^{\frac{1}{2}}]$ $\mathrm{Re}\,\nu > -1$, $\lvert\arg b\rvert < \pi$	$a^\nu (p^2-a^2)^{-\frac{1}{2}} [p+(p^2-a^2)^{\frac{1}{2}}]^{-\nu}$ $\cdot \exp\{\frac{1}{2}b[p-(p^2-a^2)^{\frac{1}{2}}]\}$ $\mathrm{Re}\,p > \lvert\mathrm{Re}\,a\rvert$
15.44	$t^{\mu-1}(t+b)^{-\mu} I_\nu[a(t^2+bt)^{\frac{1}{2}}]$ $\mathrm{Re}(\nu+2\mu) > 0$	$2(ab)^{-1}\Gamma(\frac{1}{2}\nu+\mu)[\Gamma(1+\nu)]^{-1} e^{\frac{1}{2}bp}$ $\cdot W_{\frac{1}{2}-\mu,\frac{1}{2}\nu}(z_1) M_{\frac{1}{2}-\mu,\frac{1}{2}\nu}(z_2)$ $z_{\frac{1}{2}} = \frac{1}{2}b[p \pm (p^2-a^2)^{\frac{1}{2}}]$ $\mathrm{Re}\,p > \lvert\mathrm{Re}\,a\rvert$
15.45	$(t^2+bt)^{-\frac{1}{2}} K_\nu[a(t^2+bt)^{\frac{1}{2}}]$ $-1 < \mathrm{Re}\,\nu < 1$	$\frac{1}{2}\sec(\frac{1}{2}\pi\nu) e^{\frac{1}{2}pb} K_{\frac{1}{2}\nu}(z_1) K_{\frac{1}{2}\nu}(z_2)$ $z_{\frac{1}{2}} = \frac{1}{4}b[p \mp (p^2-a^2)^{\frac{1}{2}}]$

	$f(t)$	$g(p) = \int_0^\infty f(t)e^{-pt}dt$
15.46	$t^{\frac{1}{2}\nu}(t+b)^{-\frac{1}{2}\nu}K_\nu[a(t^2+bt)^{\frac{1}{2}}]$ $\mathrm{Re}\,\nu > -1$	$\frac{1}{2}\Gamma(1+\nu)a^{-\nu}s^{-1}e^{\frac{1}{2}bp}$ $\cdot\{(p-s)^\nu e^{-\frac{1}{2}bs}\Gamma[-\nu,\frac{1}{2}b(p-s)]$ $-(p+s)^\nu e^{\frac{1}{2}bs}\Gamma[-\nu,\frac{1}{2}(p+s)]\}$ $s = (p^2-a^2)^{\frac{1}{2}}$, $\mathrm{Re}\,p > \lvert\mathrm{Re}\,a\rvert$
15.47	$t^{\mu-1}(t+b)^{-\mu}K_\nu[a(t^2+bt)^{\frac{1}{2}}]$ $\mathrm{Re}(2\mu\pm\nu) > 0$	$(ab)^{-1}\Gamma(\mu+\frac{1}{2}\nu)\Gamma(\mu-\frac{1}{2}\nu)e^{\frac{1}{2}bp}$ $\cdot W_{\frac{1}{2}-\mu,\frac{1}{2}\nu}(z_1)W_{\frac{1}{2}-\mu,\frac{1}{2}\nu}(z_2)$ $z_{\frac{1}{2}} = \frac{1}{2}b[p\pm(p^2-a^2)^{\frac{1}{2}}]$ $\mathrm{Re}\,p > \lvert\mathrm{Re}\,a\rvert$
15.48	$0 \quad t < b$ $(t^2-b^2)^{-\frac{1}{2}}I_\nu[a(t^2-b^2)^{\frac{1}{2}}]$ $t > b$ $\mathrm{Re}\,\nu > -1$	$I_{\frac{1}{2}\nu}(z_1)K_{\frac{1}{2}\nu}(z_2)$ $z_{\frac{1}{2}} = \frac{1}{2}b[p\mp(p^2-a^2)^{\frac{1}{2}}]$ $\mathrm{Re}\,p > \lvert\mathrm{Re}\,a\rvert$
15.49	$0 \quad t < b$ $(t^2-b^2)^{\frac{1}{2}\nu}I_\nu[a(t^2-b^2)^{\frac{1}{2}}]$ $t > b$ $\mathrm{Re}\,\nu > -1$	$(2b/\pi)^{\frac{1}{2}}(ab)^\nu(p^2-a^2)^{-\frac{1}{2}\nu-\frac{1}{4}}$ $\cdot K_{\nu+\frac{1}{2}}[b(p^2-a^2)^{\frac{1}{2}}]$ $\mathrm{Re}\,p > \lvert\mathrm{Re}\,a\rvert$

	$f(t)$	$g(p) = \int_0^\infty f(t)e^{-pt}dt$
15.50	$(t-b)^{\frac{1}{2}\nu}(t+b)^{-\frac{1}{2}\nu}$ $\cdot I_\nu[a(t^2-b^2)^{\frac{1}{2}}]$ $t > b$ $\text{Re}\ \nu > -1$	$a^\nu(p^2-a^2)^{-\frac{1}{2}}[p+(p^2-a^2)^{\frac{1}{2}}]^{-\nu}$ $\cdot\exp[-b(p^2-a^2)^{\frac{1}{2}}]$ $\text{Re}\ p > \|\text{Re}\ a\|$
15.51	$0 \quad t < b$ $(t-b)^{\mu-1}(t+b)^{-\mu}$ $I_\nu[a(t^2-b^2)^{\frac{1}{2}}] \quad t > b$ $\text{Re}\ (\nu+2\mu) > 0$	$(ab)^{-1}\Gamma(\frac{1}{2}\nu+\mu)[\Gamma(1+\nu)]^{-1}$ $W_{\frac{1}{2}-\mu,\frac{1}{2}\nu}(z_1)M_{\frac{1}{2}-\mu,\frac{1}{2}\nu}(z_2)$ $z_{\frac{1}{2}} = b[p\pm(p^2-a^2)^{\frac{1}{2}}]$ $\text{Re}\ p > \|\text{Re}\ a\|$
15.52	$0 \quad t < b$ $(t^2-b^2)^{-\frac{1}{2}}K_\nu[a(t^2-b^2)^{\frac{1}{2}}]$ $-1 < \text{Re}\ \nu < 1$	$\frac{1}{2}\sec(\frac{1}{2}\pi\nu)K_{\frac{1}{2}\nu}(z_1)K_{\frac{1}{2}\nu}(z_2)$ $z_{\frac{1}{2}} = \frac{1}{2}b[p\mp(p^2-a^2)^{\frac{1}{2}}]$
15.53	$0 \quad t < b$ $(t-b)^{\frac{1}{2}\nu}(t+b)^{-\frac{1}{2}\nu}$ $\cdot K_\nu[a(t^2-b^2)^{\frac{1}{2}}] \quad t > b$ $\text{Re}\ \nu > -1$	$\frac{1}{2}\Gamma(1+\nu)a^{-\nu}s^{-1}$ $\{(p-s)^\nu e^{-bs}\Gamma[-\nu,b(p-s)]$ $-(p+s)^\nu e^{bs}\Gamma[-\nu,b(p+s)]\}$ $s = (p^2-a^2)^{\frac{1}{2}}, \quad \text{Re}\ p > \|\text{Re}\ a\|$

	$f(t)$	$g(p) = \int_0^\infty f(t)e^{-pt}dt$
15.54	$0 \quad t < b$ $(t-b)^{\mu-1}(t+b)^{-\mu}$ $\cdot K_\nu[a(t^2-b^2)^{\frac{1}{2}}] \quad t > b$ $\mathrm{Re}[2\mu\pm\nu) > 0$	$\frac{1}{2}(ab)^{-1}\Gamma(\mu+\frac{1}{2}\nu)\Gamma(\mu-\frac{1}{2}\nu)$ $\cdot W_{\frac{1}{2}-\mu,\frac{1}{2}\nu}(z_1)W_{\frac{1}{2}-\mu,\frac{1}{2}\nu}(z_2)$ $z_{\frac{1}{2}} = b[p\pm(p^2-a^2)^{\frac{1}{2}}]$ $\mathrm{Re}\, p > \lvert\mathrm{Re}\, a\rvert$
15.55	$t^{-1}e^{-a/t}I_\nu(b/t)$ $a \geqq b$	$2I_\nu(z_1)K_\nu(z_2)$ $z_{\frac{1}{2}} = p^{\frac{1}{2}}[(a+b)^{\frac{1}{2}}\mp(a-b)^{\frac{1}{2}}]$ $\mathrm{Re}\, p > 0$
15.56	$t^{-\frac{1}{2}}e^{-a/t}K_\nu(a/t)$ $\mathrm{Re}\, a > 0$	$2(\pi/p)^{\frac{1}{2}}K_{2\nu}[(8ap)^{\frac{1}{2}}]$ $\mathrm{Re}\, p > 0$
15.57	$t^{-1}e^{a/t}K_\nu(a/t)$ $\mathrm{Re}\, a > 0$	$\frac{1}{2}\pi^2\{J_\nu^2[(2ap)^{\frac{1}{2}}]+Y_\nu^2[(2ap)^{\frac{1}{2}}]\}$ $\mathrm{Re}\, p > 0$
15.58	$t^{-1}e^{-a/t}K_\nu(b/t)$	$2K_\nu(z_1)K_\nu(z_2)$ $z_{\frac{1}{2}} = p^{\frac{1}{2}}[(a+b)^{\frac{1}{2}}\pm(a-b)^{\frac{1}{2}}]$ $\mathrm{Re}\, p > 0$

	$f(t)$	$g(p) = \int_0^\infty f(t)e^{-pt}dt$
15.59	$t^{\frac{1}{2}}e^{-at^2}I_{\frac{1}{4}}(at^2)$ $\operatorname{Re} a > 0$	$(2pa)^{-\frac{1}{2}}\exp(\frac{1}{8}p^2/a)[\Gamma(\frac{1}{4})]^{-1}$ $\cdot\Gamma(\frac{1}{4},\frac{1}{8}p^2/a)$ $\operatorname{Re} p > 0$
15.60	$t^{2\nu}e^{-at^2}I_\nu(at^2)$ $\operatorname{Re} a > 0,\ \operatorname{Re}\nu > -\frac{1}{4}$	$2^{-\frac{11}{2}\nu}[\Gamma(\nu+1)]^{-1}\Gamma(1+4\nu)a^{-\frac{1}{2}\nu}p^{-\nu-1}$ $\cdot\exp(\frac{1}{16}p^2/a)W_{-\frac{3}{2}\nu,\frac{1}{2}\nu}(\frac{1}{8}p^2/a)$ $\operatorname{Re} p > 0$
15.61	$K_\nu(2a\sinh t)$ $\operatorname{Re} a > 0, -1<\operatorname{Re}\nu < 1$	$\frac{1}{4}\pi^2\csc(\pi\nu)[J_{\frac{1}{2}\nu-\frac{1}{2}p}(a)Y_{-\frac{1}{2}\nu-\frac{1}{2}p}(a)$ $-J_{-\frac{1}{2}\nu-\frac{1}{2}p}(a)Y_{\frac{1}{2}\nu-\frac{1}{2}p}(a)]$
15.62	$\operatorname{csch}(\frac{1}{2}t)K_\nu[a\operatorname{csch}(\frac{1}{2}t)]$ $\operatorname{Re} a > 0$	$a^{-1}\Gamma(p+\frac{1}{2}\nu+\frac{1}{2})\Gamma(p-\frac{1}{2}\nu+\frac{1}{2})$ $\cdot W_{-p,\frac{1}{2}\nu}(ia)W_{-p,\frac{1}{2}\nu}(-ia)$ $\operatorname{Re}(p\pm\frac{1}{2}\nu) > -1$
15.63	$\operatorname{csch}(\frac{1}{2}t)$ $\cdot\exp[-(e^t-1)^{-1}(ae^t+b)]$ $\cdot K_\nu[(ab)^{\frac{1}{2}}\operatorname{csch}(\frac{1}{2}t)]$ $\operatorname{Re} a > 0, \operatorname{Re} b > 0$	$(ab)^{-\frac{1}{2}}\Gamma(p+\frac{1}{2}+\frac{1}{2}\nu)\Gamma(p+\frac{1}{2}-\frac{1}{2}\nu)$ $\cdot W_{-p,\frac{1}{2}\nu}(a)W_{-p,\frac{1}{2}\nu}(b)\exp[\frac{1}{2}(b-a)]$ $\operatorname{Re}(p\pm\frac{1}{2}\nu) > -\frac{1}{2}$

	$f(t)$	$g(p) = \int_0^\infty f(t)e^{-pt}dt$
15.64	$\operatorname{csch} t \exp[(a+b)\coth t]$ $\cdot K_\nu[2(ab)^{\frac{1}{2}}\operatorname{csch} t]$ $\operatorname{Re} a > 0, \operatorname{Re} b > 0$	$\frac{1}{2}(ab)^{-\frac{1}{2}}\Gamma(\frac{1}{2}+\frac{1}{2}p+\frac{1}{2}\nu)\Gamma(\frac{1}{2}+\frac{1}{2}p-\frac{1}{2}\nu)$ $\cdot W_{-\frac{1}{2}p,\frac{1}{2}\nu}(2a)W_{\frac{1}{2}p,\frac{1}{2}\nu}(2b)$ $\operatorname{Re}(p\pm\nu) > -1$

1.16 Functions Related to Bessel Functions and Kelvin Functions

16.1	$\mathbf{H}_{\frac{1}{2}}(at)$	$(\frac{1}{2}ap)^{-\frac{1}{2}} - a^{\frac{1}{2}}(p^2+a^2)^{-\frac{1}{2}}[p+(p^2+a^2)^{\frac{1}{2}}]^{\frac{1}{2}}$ $\operatorname{Re} p > \lvert\operatorname{Im} a\rvert$
16.2	$\mathbf{L}_{\frac{1}{2}}(at)$	$a^{-\frac{1}{2}}(p^2-a^2)^{-\frac{1}{2}}[p+(p^2-a^2)^{\frac{1}{2}}]^{\frac{1}{2}} - (\frac{1}{2}ap)^{-\frac{1}{2}}$ $\operatorname{Re} p > \lvert\operatorname{Re} a\rvert$
16.3	$\mathbf{H}_{-n-\frac{1}{2}}(at)$ $n=0,1,2,\cdots$	$(-1)^n a^{-n-\frac{1}{2}}(p^2+a^2)^{-\frac{1}{2}}[(p^2+a^2)^{\frac{1}{2}}-p]^{n+\frac{1}{2}}$ $\operatorname{Re} p > \lvert\operatorname{Im} a\rvert$
16.4	$\mathbf{L}_{-n-\frac{1}{2}}(at)$ $n=0,1,2,\cdots$	$a^{-n-\frac{1}{2}}(p^2-a^2)^{-\frac{1}{2}}[p-(p^2-a^2)^{\frac{1}{2}}]^{n+\frac{1}{2}}$ $\operatorname{Re} p > \lvert\operatorname{Re} a\rvert$
16.5	$t^{\frac{1}{2}}\mathbf{H}_{-\frac{1}{2}}(at)$	$(2a/\pi)^{\frac{1}{2}}(p^2+a^2)^{-1}$ $\operatorname{Re} p > \lvert\operatorname{Im} a\rvert$

	$f(t)$	$g(p) = \int_0^\infty f(t)e^{-pt}dt$
16.6	$t^{\frac{1}{2}}\mathbf{L}_{-\frac{1}{2}}(at)$	$(2a/\pi)^{\frac{1}{2}}(p^2-a^2)^{-1}$ $\qquad$ Re $p > \lvert$Re $a\rvert$
16.7	$t^{-\frac{1}{2}}\mathbf{H}_{-\frac{1}{2}}(at)$	$(\frac{1}{2}\pi a)^{-\frac{1}{2}}\arctan(a/p)$ $\qquad$ Re $p > \lvert$Im $a\rvert$
16.8	$t^{-\frac{1}{2}}\mathbf{L}_{-\frac{1}{2}}(at)$	$(2\pi a)^{-\frac{1}{2}}\log[(p-a)/(p+a)]$ $\quad$ Re $p > \lvert$Re $a\rvert$
16.9	$t^{\frac{1}{2}}\mathbf{H}_{-3/2}(at)$	$(\frac{1}{2}a\pi)^{-\frac{1}{2}}[p(p^2+a^2)^{-1}-a^{-1}\arctan(a/p)]$ Re $p > \lvert$Im $a\rvert$
16.10	$t^{\frac{1}{2}}\mathbf{L}_{-3/2}(at)$	$(\frac{1}{2}\pi a)^{-\frac{1}{2}}\{p(p^2-a^2)^{-1}-\frac{1}{2}a^{-1}\log[(p-a)/(p+a)]\}$ Re $p > \lvert$Re $a\rvert$
16.11	$t^{\frac{1}{2}}\mathbf{H}_{\frac{1}{2}}(at)$	$p^{-1}(p^2+a^2)^{-1}(\frac{1}{2}\pi)^{-\frac{1}{2}}a^{3/2}$ $\qquad$ Re $p > \lvert$Im $a\rvert$
16.12	$t^{\frac{1}{2}}\mathbf{L}_{\frac{1}{2}}(at)$	$p^{-1}(p^2-a^2)^{-1}(\frac{1}{2}\pi)^{-\frac{1}{2}}a^{3/2}$ $\qquad$ Re $p > \lvert$Re $a\rvert$
16.13	$t^{-\frac{1}{2}}\mathbf{H}_{\frac{1}{2}}(at)$	$(2\pi a)^{-\frac{1}{2}}\log(1+a^2/p^2)$ $\qquad$ Re $p > \lvert$Im $a\rvert$
16.14	$t^{-\frac{1}{2}}\mathbf{L}_{-\frac{1}{2}}(at)$	$-(2\pi a)^{-\frac{1}{2}}\log(1-a^2/p^2)$ $\qquad$ Re $p > \lvert$Re $a\rvert$

	$f(t)$	$g(p) = \int_0^\infty f(t)e^{-pt}dt$
16.15	$t^{\frac{1}{2}}\mathbf{H}_{3/2}(at)$	$(2a/\pi)^{\frac{1}{2}}[\frac{1}{2}p^{-2}-(p^2+a^2)^{-1}$ $+\frac{1}{2}a^{-2}\log(1+a^2/p^2)]$ $\quad \text{Re } p > \lvert\text{Im } a\rvert$
16.16	$t^{\frac{1}{2}}\mathbf{L}_{3/2}(at)$	$(2a/\pi)^{\frac{1}{2}}[(p^2-a^2)^{-1}-\frac{1}{2}p^{-2}$ $-\frac{1}{2}a^{-2}\log(1-a^2/p^2)]$ $\quad \text{Re } p > \lvert\text{Re } a\rvert$
16.17	$t^{-\frac{1}{2}}\mathbf{H}_{3/2}(at)$	$(2\pi a)^{-\frac{1}{2}}[a/p-p/a \log(1+a^2/p^2)]$ $\quad \text{Re } p > \lvert\text{Im } a\rvert$
16.18	$t^{-\frac{1}{2}}\mathbf{L}_{3/2}(at)$	$(2\pi a)^{-\frac{1}{2}}[p/a \log(1-a^2/p^2)-a/p]$ $\quad \text{Re } p > \lvert\text{Re } a\rvert$
16.19	$t^{3/2}\mathbf{H}_{3/2}(at)$	$(2/\pi)^{\frac{1}{2}}a^{5/2}p^{-2}(p^2+a^2)^{-2}(3p^2+a^2)$ $\quad \text{Re } p > \lvert\text{Im } a\rvert$
16.20	$t^{3/2}\mathbf{L}_{3/2}(at)$	$(2/\pi)^{\frac{1}{2}}a^{5/2}p^{-2}(p^2-a^2)^{-2}(3p^2-a^2)$ $\quad \text{Re } p > \lvert\text{Re } a\rvert$
16.21	$\mathbf{H}_0(at)$	$2\pi^{-1}(p^2+a^2)^{-\frac{1}{2}}\log[(1+a^2/p^2)^{\frac{1}{2}}+a/p]$ $\quad \text{Re } p > \lvert\text{Im } a\rvert$
16.22	$\mathbf{L}_0(at)$	$2\pi^{-1}(p^2-a^2)^{-\frac{1}{2}}\arcsin(a/p)$ $\quad \text{Re } p > \lvert\text{Re } a\rvert$

	$f(t)$	$g(p) = \int_0^\infty f(t)e^{-pt}dt$
16.23	$\mathbf{H}_1(at)$	$-2\pi^{-1}\{pa^{-1}(p^2+a^2)^{-\frac{1}{2}}\log[(1+a^2/p^2)+a/p]$ $-p^{-1}\}$ Re $p > \lvert \text{Im } a \rvert$
16.24	$\mathbf{L}_1(at)$	$2\pi^{-1}[pa^{-1}(p^2-a^2)^{-\frac{1}{2}}\arcsin(a/p)$ $-p^{-1}]$ Re $p > \lvert \text{Re } a \rvert$
16.25	$t^{-1}\mathbf{H}_1(at)$	$2\pi^{-1}\{a^{-1}(p^2+a^2)^{\frac{1}{2}}\log[(1+a^2/p^2)^{\frac{1}{2}}+a/p]-1\}$ Re $p > \lvert \text{Im } a \rvert$
16.26	$t^{-1}\mathbf{L}_1(at)$	$2\pi^{-1}[1-(p^2/a^2-1)^{\frac{1}{2}}\arcsin(a/p)]$ Re $p > \lvert \text{Re } a \rvert$
16.27	$t^{\frac{1}{2}\nu}\mathbf{H}_\nu(at^{\frac{1}{2}})$ Re $\nu > -\frac{3}{2}$	$-ip^{-1}(\frac{1}{2}a/p)^\nu\exp(-\frac{1}{4}a^2/p)\mathrm{Erf}(\frac{1}{2}iap^{-\frac{1}{2}})$ Re $p > 0$
16.28	$t^{\frac{1}{2}\nu}\mathbf{L}_\nu(at^{\frac{1}{2}})$ Re $\nu > -\frac{3}{2}$	$p^{-1}(\frac{1}{2}a/p)^\nu\exp(\frac{1}{4}a^2/p)\mathrm{Erf}(\frac{1}{2}ap^{-\frac{1}{2}})$ Re $p > 0$
16.29	$t^{\frac{1}{2}\nu}\mathbf{L}_{-\nu}(at^{\frac{1}{2}})$	$(\frac{1}{2}a/p)^\nu[p\Gamma(\frac{1}{2}-\nu)]^{-1}\exp(\frac{1}{4}a^2/p)$ $\cdot\gamma(\frac{1}{2}-\nu,\frac{1}{4}a^2/p)$ Re $p > 0$

	$f(t)$	$g(p) = \int_0^\infty f(t)e^{-pt}dt$
6.30	$t^{\frac{1}{2}\nu}[I_\nu(at^{\frac{1}{2}})-\mathbf{L}_\nu(at^{\frac{1}{2}})]$ $\mathrm{Re}\,\nu > -1$	$(\frac{1}{2}a)^\nu p^{-\nu-1}\exp(\frac{1}{4}a^2/p)$ $\cdot\mathrm{Erfc}(\frac{1}{2}ap^{-\frac{1}{2}})$ $\mathrm{Re}\,p > 0$
6.31	$t^{\frac{1}{2}\nu}[I_\nu(at^{\frac{1}{2}})-\mathbf{L}_{-\nu}(at^{\frac{1}{2}})]$ $\mathrm{Re}\,\nu > -1$	$(\frac{1}{2}ap)^{-\nu}[p\Gamma(\frac{1}{2}-\nu)]^{-1}$ $\cdot\exp(\frac{1}{4}a^2/p)\Gamma(\frac{1}{2}-\nu,\frac{1}{4}a^2/p)$ $\mathrm{Re}\,p > 0$
.32	$\mathbf{H}_0(t)-Y_0(t)$	$2\pi^{-1}(p^2+1)^{-\frac{1}{2}}$ $\cdot\log\{[1+(1+p^{-2})^{\frac{1}{2}}][p^{-1}+(1+p^{-2})^{\frac{1}{2}}]\}$ $\mathrm{Re}\,p > 1$
.33	$t[J_1(at)\mathbf{H}_0(at)$ $-J_0(at)\mathbf{H}_1(at)]$	$2(\pi p)^{-1}a^2(p^2+a^2)^{-3/2}$ $\mathrm{Re}\,p > \lvert\mathrm{Im}\,a\rvert$
.34	$t[J_\nu(at)\mathbf{H}'_\nu(at)$ $-J'_\nu(at)\mathbf{H}_\nu(at)]$ $\mathrm{Re}\,\nu > -1$	$2(\pi p)^{-1}a^{2\nu}(p^2+a^2)^{-\nu-\frac{1}{2}}$ $\mathrm{Re}\,p > \lvert\mathrm{Im}\,a\rvert$
.35	$t[I_1(at)\mathbf{L}_0(at)$ $-I_0(at)\mathbf{L}_1(at)]$	$2(\pi p)^{-1}a^2(p^2-a^2)^{-3/2}$ $\mathrm{Re}\,p > \lvert\mathrm{Re}\,a\rvert$

	$f(t)$	$g(p)=\int_0^\infty f(t)e^{-pt}dt$
16.36	$t[I_\nu(at)\mathbf{L}'_\nu(at)$ $-I'_\nu(at)\mathbf{L}_\nu(at)]$ $\operatorname{Re}\nu>-1$	$2(\pi p)^{-1}a^{2\nu}(p^2-a^2)^{-\nu-\frac{1}{2}}$ $\operatorname{Re}p>\lvert\operatorname{Re}a\rvert$
16.37	$t^\nu[I_\nu(at)-\mathbf{L}_\nu(at)]$ $\operatorname{Re}\nu>-\frac{1}{2}$	$(\frac{1}{2}p)^{-\frac{1}{2}}(a/p)^\nu\Gamma(1+2\nu)$ $\cdot(a^2-p^2)^{-\frac{1}{2}\nu-\frac{1}{4}}P_{-\nu-\frac{1}{2}}^{-\nu-\frac{1}{2}}(a/p)$ $\operatorname{Re}p>\lvert\operatorname{Re}a\rvert$
16.38	$\operatorname{ber}(at)$	$[\frac{1}{2}(p^4+a^4)^{-\frac{1}{2}}+\frac{1}{2}p^2(p^4+a^4)^{-1}]^{\frac{1}{2}}$ $\operatorname{Re}p>2^{-\frac{1}{2}}[\lvert\operatorname{Re}a\rvert+\lvert\operatorname{Im}a\rvert]$
16.39	$\operatorname{bei}(at)$	$[\frac{1}{2}(p^4+a^4)^{-\frac{1}{2}}-\frac{1}{2}p^2(p^4+a^4)^{-1}]^{\frac{1}{2}}$ $\operatorname{Re}p>2^{-\frac{1}{2}}[\lvert\operatorname{Re}a\rvert+\lvert\operatorname{Im}a\rvert]$
16.40	$\operatorname{ber}_\nu(at)+i\operatorname{bei}_\nu(at)$ $\operatorname{Re}\nu>-1$	$a^\nu e^{i3\pi\nu/4}(p^2-ia)^{-\frac{1}{2}}[p+(p^2-ia^2)^{\frac{1}{2}}]^{-\nu}$ $\operatorname{Re}p>2^{-\frac{1}{2}}\lvert\operatorname{Re}a\rvert$
16.41	$\operatorname{ker}_\nu(at)+i\operatorname{kei}_\nu(at)$ $\lvert\operatorname{Re}\nu\rvert<1$	$\frac{1}{2}\pi\csc(\pi\nu)a^{-\nu}e^{-i\frac{\pi}{4}\nu}(p^2-ia^2)^{-\frac{1}{2}}$ $\cdot\{[p+(p^2-ia)^{\frac{1}{2}}]^\nu-[p-(p^2-ia)^{\frac{1}{2}}]^\nu\}$ $\operatorname{Re}p>2^{-\frac{1}{2}}\lvert\operatorname{Re}a\rvert$

	$f(t)$	$g(p) = \int_0^\infty f(t)e^{-pt}dt$
16.42	$t^{\frac{1}{2}\nu}\mathrm{ber}_\nu(at^{\frac{1}{2}})$ $\mathrm{Re}\ \nu > -1$	$p^{-1}(\frac{1}{2}a/p)^\nu\cos(\frac{1}{4}a^2/p+\frac{3}{4}\pi\nu)$ $\mathrm{Re}\ p > 0$
16.43	$t^{\frac{1}{2}\nu}\mathrm{bei}_\nu(at^{\frac{1}{2}})$ $\mathrm{Re}\ \nu > -1$	$p^{-1}(\frac{1}{2}a/p)^\nu\sin(\frac{1}{4}a^2/p+\frac{3}{4}\pi\nu)$ $\mathrm{Re}\ p > 0$
16.44	$\mathrm{ber}[a(t^2+bt)^{\frac{1}{2}}]$ $+i\ \mathrm{bei}[a(t^2+bt)^{\frac{1}{2}}]$	$(p^2-ia^2)^{-\frac{1}{2}}\exp\{\frac{1}{2}b[p-(p^2-ia^2)^{\frac{1}{2}}]\}$ $\mathrm{Re}(p\pm i^{\frac{1}{2}}a) > 0$
16.45	$(t+\frac{1}{2}b)\{\mathrm{ber}[a(t^2+bt)^{\frac{1}{2}}]$ $+i\ \mathrm{bei}[a(t^2+bt)^{\frac{1}{2}}]\}$	$a(p^2-ia^2)^{-\frac{3}{2}}p[\frac{1}{2}b(p^2-ia^2)^{\frac{1}{2}}+1]$ $\cdot\exp\{\frac{1}{2}b[p-(p^2-ia^2)^{\frac{1}{2}}]\}$ $\mathrm{Re}(p\pm ai^{\frac{1}{2}}) > 0$
16.46	$(t^2+bt)^{\frac{1}{2}}\{\mathrm{ber}_1[a(t^2+bt)^{\frac{1}{2}}]$ $+i\ \mathrm{bei}_1[a(t^2+bt)^{\frac{1}{2}}]\}$	$a(p^2-ia^2)^{-\frac{3}{2}}[\frac{1}{2}b(p^2-ia^2)^{\frac{1}{2}}+1]$ $\cdot e^{i3\pi/4}\exp\{\frac{1}{2}b[p-(p^2-ia^2)^{\frac{1}{2}}]\}$ $\mathrm{Re}(p\pm ai^{\frac{1}{2}}) > 0$
16.47	$(t^2+bt)^{-\frac{1}{2}}\{\mathrm{ber}_1[a(t^2+bt)^{\frac{1}{2}}]$ $+i\ \mathrm{bei}_1[a(t^2+bt)^{\frac{1}{2}}]\}$	$(\frac{1}{2}ab)^{-1}e^{\frac{1}{2}bp-3\pi i/4}$ $\{e^{-bp}-\exp[-b(p^2-ia^2)^{\frac{1}{2}}]\}$ $\mathrm{Re}(p\pm ai^{\frac{1}{2}}) > 0$

	$f(t)$	$g(p) = \int_0^\infty f(t)e^{-pt}dt$
16.48	$t^{\frac{1}{2}\nu}(t+b)^{-\frac{1}{2}\nu}\{\mathrm{ber}_\nu[a(t^2+bt)^{\frac{1}{2}}]$ $+i\ \mathrm{bei}_\nu[a(t^2+bt)^{\frac{1}{2}}]\}$ $\mathrm{Re}\ \nu > -1$	$a^\nu(p^2-ia^2)^{-\frac{1}{2}}[p+(p^2-ia^2)^{\frac{1}{2}}]^{-\nu}$ $\cdot e^{3\pi i\nu/4}\exp\{\frac{1}{2}b[p-(p^2-ia^2)^{\frac{1}{2}}]\}$ $\mathrm{Re}(p\pm ai^{\frac{1}{2}}) > 0$
16.49	$\mathrm{ker}[a(t^2+bt)^{\frac{1}{2}}]$ $+i\ \mathrm{kei}[a(t^2+bt)^{\frac{1}{2}}]$	$(p^2-ia^2)^{-\frac{1}{2}}\exp\{\frac{1}{2}b[p-(p-ia^2)^{\frac{1}{2}}]\}$ $\cdot\log\{a^{-1}i^{-\frac{1}{2}}[p+(p^2-ia^2)^{\frac{1}{2}}]\}$ $\mathrm{Re}(p+ai^{\frac{1}{2}}) > 0$
16.50	$[\mathrm{ber}'_\nu(at^{\frac{1}{2}})]^2+[\mathrm{bei}'_\nu(at^{\frac{1}{2}})]^2$ $\mathrm{Re}\ \nu > 0$	$16a^{-4}p\ I_\nu(\frac{1}{2}a^2/p)]$ $\mathrm{Re}\ p > 0$
16.51	$t^{\frac{1}{2}}[\mathrm{ber}_\nu(at^{\frac{1}{2}})\mathrm{bei}'_\nu(at^{\frac{1}{2}})$ $-\mathrm{bei}_\nu(at^{\frac{1}{2}})\mathrm{ber}'_\nu(at^{\frac{1}{2}})]$ $\mathrm{Re}\ \nu > -2$	$\frac{1}{2}ap^{-2}I_\nu(\frac{1}{2}a^2/p)$ $\mathrm{Re}\ p > 0$
16.52	$[\mathrm{ber}_\nu(at^{\frac{1}{2}})]^2+[\mathrm{bei}_\nu(at^{\frac{1}{2}})]^2$ $\mathrm{Re}\ \nu > -1$	$p^{-1}I_\nu(\frac{1}{2}a^2/p)$ $\mathrm{Re}\ p > 0$
16.53	$2t^{-\frac{1}{2}}[\mathrm{ber}_\nu(at^{\frac{1}{2}})\mathrm{bei}'_\nu\ (at^{\frac{1}{2}})$ $+\mathrm{bei}_\nu(at^{\frac{1}{2}})\mathrm{ber}'_\nu(at^{\frac{1}{2}})]$ $\mathrm{Re}\ \nu > 0$	$2a^{-1}I_\nu(\frac{1}{2}a^2/p)$ $\mathrm{Re}\ p > 0$

	$f(t)$	$g(p) = \int_0^\infty f(t)e^{-pt}dt$
16.54	$\mathrm{ker}[2(at)^{\frac{1}{2}}]$ $\mathrm{Re}\ p > 0$	$-\frac{1}{2}p^{-1}[\cos(a/p)\,\mathrm{Ci}(a/p)$ $+\sin(a/p)\,\mathrm{si}(a/p)]$
16.55	$\mathrm{kei}[2(at)^{\frac{1}{2}}]$ $\mathrm{Re}\ p > 0$	$-\frac{1}{2}p^{-1}[\sin(a/p)\,\mathrm{Ci}(a/p)$ $-\cos(a/p)\,\mathrm{si}(a/p)]$
16.56	$t^{-\frac{1}{2}}\mathrm{ker}_\nu(at^{\frac{1}{2}})$ $-1 < \mathrm{Re}\ \nu < 1$ $\mathrm{Re}\ p > 0$	$\frac{1}{4}\pi(\pi/p)^{\frac{1}{2}}\sec(\frac{1}{2}\pi\nu)$ $\cdot[J_{\frac{1}{2}\nu}(\frac{1}{8}a^2/p)\sin(\frac{1}{8}a^2/p-\frac{1}{4}\pi\nu)$ $-Y_{\frac{1}{2}\nu}(\frac{1}{8}a^2/p)\cos(\frac{1}{8}a^2/p-\frac{1}{4}\pi\nu)]$
16.57	$t^{-\frac{1}{2}}\mathrm{kei}_\nu(at^{\frac{1}{2}})$ $-1 < \mathrm{Re}\ \nu < 1$ $\mathrm{Re}\ p > 0$	$-\frac{1}{4}\pi(\pi/p)^{\frac{1}{2}}\sec(\frac{1}{2}\pi\nu)$ $\cdot[J_{\frac{1}{2}\nu}(\frac{1}{8}a^2/p)\cos(\frac{1}{8}a^2/p-\frac{1}{4}\pi\nu)$ $+Y_{\frac{1}{2}\nu}(\frac{1}{8}a^2/p)\sin(\frac{1}{8}a^2/p-\frac{1}{4}\pi\nu)]$

	$f(t)$	$g(p) = \int_0^\infty f(t)e^{-pt}dt$
16.58	$\frac{d^n}{dt^n}\{[\mathrm{ber}_\nu(at^{\frac{1}{2}})]^2+[\mathrm{bei}_\nu(at^{\frac{1}{2}})]^2\}$ $n = 0,1,2,\cdots,\ \mathrm{Re}\ \nu > n-1$	$p^{n-1}I_\nu(\frac{1}{2}a^2/p)$ $\mathrm{Re}\ p > 0$
16.59	$\mathrm{Ji}_o(at)$	$p^{-1}\log[p/a+(1+p^2/a^2)^{\frac{1}{2}}]$ $\mathrm{Re}\ p > 0$
16.60	$\mathrm{Ji}_\nu(at)$ $\mathrm{Re}\ \nu > 0$	$(\nu p)^{-1}\{1-[(1+p^2/a^2)^{\frac{1}{2}}-p/a]^\nu\}$ $\mathrm{Re}\ p > 0$
16.61	$\mathrm{Yi}_o(at)$	$p^{-1}\log^2[p/a+(1+p^2/a^2)^{\frac{1}{2}}]$ $\mathrm{Re}\ p > 0$
16.62	$\mathrm{Ki}_o(at)$	$p^{-1}\{\frac{1}{2}\log^2[p/a+(p^2/a^2-1)^{\frac{1}{2}}]+\frac{\pi^2}{8}\}$ $\mathrm{Re}\ p > a$
16.63	$\mathrm{Ki}_\nu(at)$ $-1 < \mathrm{Re}\ \nu < 1$	$\frac{1}{2}\pi(p\nu)^{-1}\csc(\pi\nu)a^{-\nu}$ $\cdot\{[p+(p^2-a^2)^{\frac{1}{2}}]^\nu$ $+[p-(p^2-a^2)^{\frac{1}{2}}]^\nu-2a^\nu\cos(\frac{1}{2}\pi\nu)\}$ $\mathrm{Re}\ p > a$

1.17 Whittaker Functions and Special Cases*

	$f(t)$	$g(p) = \int_0^\infty f(t)e^{-pt}dt$
17.1	$C(at)$	$\frac{1}{2}a^{\frac{1}{2}}p^{-1}(p^2+a^2)^{-\frac{1}{2}}[(p^2+a^2)^{\frac{1}{2}}+p]^{\frac{1}{2}}$ $\mathrm{Re}\, p > 0$
17.2	$S(at)$	$\frac{1}{2}a^{\frac{1}{2}}p^{-1}(p^2+a^2)^{-\frac{1}{2}}(p^2+a^2)^{\frac{1}{2}}-p]^{\frac{1}{2}}$ $\mathrm{Re}\, p > 0$
17.3	$C(at^{\frac{1}{2}})$	$\frac{1}{4}\pi^{\frac{1}{2}}ap^{-3/2}\exp(-{}^1\!/_8 a^2/p)I_{-\frac{1}{4}}({}^1\!/_8 a^2/p)$ $\mathrm{Re}\, p > 0$
17.4	$S(at^{\frac{1}{2}})$	$\frac{1}{4}\pi^{\frac{1}{2}}ap^{-3/2}\exp(-{}^1\!/_8 a^2/p)I_{\frac{1}{4}}({}^1\!/_8 a^2/p)$ $\mathrm{Re}\, p > 0$
17.5	$C(at^{\frac{1}{2}})-S(at^{\frac{1}{2}})$	$\frac{1}{2}(2\pi)^{-\frac{1}{2}}p^{-3/2}\exp(-{}^1\!/_8 a^2/p)K_{\frac{1}{4}}({}^1\!/_8 a^2/p)$ $\mathrm{Re}\, p > 0$
17.6	$C(a/t)$	$\frac{1}{2}p^{-1}\{1-e^{-(2ap)^{\frac{1}{2}}}[\cos(2ap)^{\frac{1}{2}} - \sin(2ap)^{\frac{1}{2}}]\}$ $\mathrm{Re}\, p > 0$

The Fresnel, exponential, sine, cosine and error-integrals; incomplete gamma and parabolic cylinder functions.

	$f(t)$	$g(p) = \int_0^\infty f(t)e^{-pt}dt$
17.7	$S(a/t)$	$\frac{1}{2}p^{-1}\{1-e^{-(2ap)^{\frac{1}{2}}}[\cos(2ap)^{\frac{1}{2}}+\sin(2ap)^{\frac{1}{2}}]\}$ Re $p > 0$
17.8	$\cos(at^2)C(at^2)$ $+\sin(at^2)S(at^2)$	$\frac{1}{4}(2\pi/a)^{\frac{1}{2}}[\mathrm{Ci}(\frac{1}{4}p^2/a)\sin(\frac{1}{4}p^2/a)$ $-\mathrm{si}(\frac{1}{4}p^2/a)\cos(\frac{1}{4}p^2/a)]$ Re $p > 0$
17.9	$\cos(at^2)S(at^2)$ $-\sin(at^2)C(at^2)$	$\frac{1}{4}(2\pi/a)^{\frac{1}{2}}[\mathrm{Ci}(\frac{1}{4}p^2/a)\cos(\frac{1}{4}p^2/a)$ $+\mathrm{si}(\frac{1}{4}p^2/a)\sin(\frac{1}{4}p^2/a)]$ Re $p > 0$
17.10	$C(at^2)$	$p^{-1}\{\cos(\frac{1}{4}p^2/a)[\frac{1}{2}-S(\frac{1}{4}p^2/a)]$ $-\sin(\frac{1}{4}p^2/a)[\frac{1}{2}-C(\frac{1}{4}p^2/a)]$ Re $p > 0$
17.11	$S(at^2)$	$p^{-1}\{\cos(\frac{1}{4}p^2/a)[\frac{1}{2}-C(\frac{1}{4}p^2/a)]$ $+\sin(\frac{1}{4}p^2/a)[\frac{1}{2}-S(\frac{1}{4}p^2/a)]\}$ Re $p > 0$
17.12	$\mathrm{Ei}(-at)$	$-p^{-1}\log(1+p/a)$ Re $p > 0$
17.13	$\overline{\mathrm{Ei}}(at)$	$-p^{-1}\log(p/a-1)$ Re $p > a$
17.14	$t^{-\frac{1}{2}}\mathrm{Ei}(-2at)$	$-(\pi/p)^{\frac{1}{2}}\log[1+p/a+(p^2/a^2+2p/a)^{\frac{1}{2}}]$ Re $p > -2a$

	$f(t)$	$g(p) = \int_0^\infty f(t)e^{-pt}dt$
17.15	$t^{-\frac{1}{2}}\mathrm{Ei}(-2at)$	$-(-\pi/p)^{\frac{1}{2}}\arctan\{[a^2(p+a)^{-2}-1]^{\frac{1}{2}}\}$ $-2a < p < 0$
17.16	$t^{-\frac{1}{2}}\overline{\mathrm{Ei}}(2at)$	$-(\pi/p)^{\frac{1}{2}}\log[p/a-1+(p^2/a^2-2p/a)^{\frac{1}{2}}]$ $\mathrm{Re}\, p > 2a$
17.17	$\sin(at)\mathrm{Ei}(-t)$	$(p^2+a^2)^{-1}\{p\arctan[a(p+1)^{-1}]$ $-\frac{1}{2}a\log[(p+1)^2+a^2]\}$ $\mathrm{Re}\, p > 0$
17.18	$\cos(at)\mathrm{Ei}(-t)$	$-(p^2+a^2)^{-1}\{a\arctan[a(p+1)^{-1}]$ $+\frac{1}{2}p\log[(p+1)^2+a^2]\}$ $\mathrm{Re}\, p > 0$
17.19	$\sin t\,\overline{\mathrm{Ei}}(t)$	$(p^2+1)^{-1}\{p\arctan[(p-1)^{-1}]$ $-\frac{1}{2}\log(p^2-2p+2)\}$ $\mathrm{Re}\, p > 1$
17.20	$\cos t\,\overline{\mathrm{Ei}}(t)$	$-(p^2+1)^{-1}\{\arctan[(p-1)^{-1}]$ $+\frac{1}{2}p\log(p^2-2p+2)\}$ $\mathrm{Re}\, p > 1$
17.21	$t^{-3/2}e^{-b/t}\mathrm{Ei}(-at)$ $\mathrm{Re}\, b > 0$	$(\pi/b)^{\frac{1}{2}}\{\exp[2(bp)^{\frac{1}{2}}]\mathrm{Ei}(-z_1)$ $+\exp[-2(bp)^{\frac{1}{2}}]\mathrm{Ei}(-z_2)\}$ $z_{\frac{1}{2}} = 2b^{\frac{1}{2}}[(p+a)^{\frac{1}{2}}\pm p^{\frac{1}{2}}]$ $\mathrm{Re}\, p > -a$

	$f(t)$	$g(p) = \int_0^\infty f(t)e^{-pt}dt$
17.22	$t^{-3/2}e^{-b/t}\overline{Ei}(at)$ $\mathrm{Re}\, b > 0$	$(\pi/b)^{\frac{1}{2}}\{\exp[2(bp)^{\frac{1}{2}}]Ei(-z_1)$ $+ \exp(-2(bp)^{\frac{1}{2}}]\overline{Ei}(z_2)\}$ $z_{\frac{1}{2}} = 2b^{\frac{1}{2}}[p^{\frac{1}{2}} \pm (p-a)^{\frac{1}{2}}]$ $\mathrm{Re}\, p > a$
17.23	$t^{-\frac{1}{2}}[Ei(-2at^{\frac{1}{2}})+\overline{Ei}(2at^{\frac{1}{2}})]$	$(\pi/p)^{\frac{1}{2}}\overline{Ei}(a^2/p)$ $\mathrm{Re}\, p > 0$
17.24	$t^{-\frac{1}{2}}[e^{2at^{\frac{1}{2}}}Ei(-2at^{\frac{1}{2}})$ $+e^{-2at^{\frac{1}{2}}}\overline{Ei}(2at^{\frac{1}{2}})]$	$(\pi/p)^{\frac{1}{2}}e^{a^2/p}Ei(-a^2/p)$ $\mathrm{Re}\, p > 0$
17.25	$t^{-\frac{1}{2}}\{\exp[t^{\frac{1}{2}}(a+b)]$ $\cdot[Ei(-2at^{\frac{1}{2}})+Ei(-2bt^{\frac{1}{2}})]$ $+\exp[-t^{\frac{1}{2}}(a+b)]$ $\cdot[\overline{Ei}(2at^{\frac{1}{2}})+\overline{Ei}(2bt^{\frac{1}{2}})]\}$	$2(\pi/p)^{\frac{1}{2}}\exp[\frac{1}{4}(a+b)^2/p]$ $\cdot Ei(-ab/p)$ $\mathrm{Re}\, p > 0$
17.26	$t^{-\frac{1}{2}}e^{-b/t}Ei(-a/t)$ $\mathrm{Re}\, b \geqq 0$	$(\pi/p)^{\frac{1}{2}}\{\exp[2(bp)^{\frac{1}{2}}]Ei(-z_1)$ $+\exp[-2(bp)^{\frac{1}{2}}]Ei(-z_2)\}$ $z_{\frac{1}{2}} = 2p^{\frac{1}{2}}[(b+a)^{\frac{1}{2}} \pm b^{\frac{1}{2}}]$ $\mathrm{Re}\, p > 0$

	$f(t)$	$g(p) = \int_0^\infty f(t)e^{-pt}dt$
17.27	$t^{-\frac{1}{2}}e^{-b/t}\overline{Ei}(-a/t)$ Re $b \geqq 0$	$(\pi/p)^{\frac{1}{2}}\{\exp[2(bp)^{\frac{1}{2}}]Ei(-z_1)$ $+\exp[-2(bp)^{\frac{1}{2}}]\overline{Ei}(z_2)\}$ $z_{\frac{1}{2}} = 2p^{\frac{1}{2}}[b^{\frac{1}{2}}\pm(b-a)^{\frac{1}{2}}]$ Re $p > a$
17.28	$t^{-\frac{1}{2}}e^{a/t}Ei(-a/t)$	$2(\pi/p)^{\frac{1}{2}}\{\cos[2(ap)^{\frac{1}{2}}]Ci[2(ap)^{\frac{1}{2}}]$ $+\sin[2(ap)^{\frac{1}{2}}]Si[2(ap)^{\frac{1}{2}}]$ $-\frac{1}{2}\pi\sin[(ap)^{\frac{1}{2}}]\}$ Re $p > 0$
17.29	$Ei(-a/t)$	$-2p^{-1}K_o[2(ap)^{\frac{1}{2}}]$ Re $p > 0$
17.30	$Si(at)$	$p^{-1}\text{arccot}(p/a)$ Re $p > 0$
17.31	$Ci(at)$	$-\frac{1}{2}p^{-1}\log(1+p^2/a^2)$ Re $p > 0$
17.32	$si(at)$	$-p^{-1}\arctan(p/a)$ Re $p > 0$
17.33	$\sin(at)Ci(bt)$	$\frac{1}{2}(p^2+a^2)^{-1}b\{p\arctan[2ap(p^2+b^2-a^2)^{-1}]$ $-\frac{1}{2}a\log[b^{-2}(p^2+b^2-a^2)^2+4a^2p^2/b^2]\}$ Re $p > 0$

	$f(t)$	$g(p) = \int_0^\infty f(t)e^{-pt}dt$
17.34	$\cos(at)\,\mathrm{si}(bt)$	$-\frac{1}{2}b(p^2+a^2)^{-1}\{p \arctan[2bp(b^2-a^2-p^2)^{-1}] + \frac{1}{2}a \log[\frac{(b+a)^2+p^2}{(b-a)^2+p^2}]\}$ Re $p > 0$
17.35	$\cos t\ \mathrm{Ci}(t)$	$-\frac{1}{2}(p^2+1)^{-1}\{\arctan(2/p) + \frac{1}{2}p \log[p^2(p^2+4)]\}$ Re $p > 0$
17.36	$\sin t\ \mathrm{si}(t)-\cos t\,\mathrm{Ci}(t)$	$(p^2+1)^{-1}[\frac{1}{2}p \log(p^2+4)+\arctan(2/p)]$ Re $p > 0$
17.37	$\sin t\ \mathrm{Ci}(t)+\cos t\ \mathrm{si}(t)$	$-(p^2+1)^{-1}[\frac{1}{2}\log(p^2+4)-p \arctan(2/p)]$ Re $p > 0$
17.38	$\sin t\ \mathrm{Ci}(t)-\cos t\ \mathrm{si}(t)$	$-(p^2+1)^{-1}\log p$ Re $p > 0$
17.39	$\sin t\ \mathrm{si}(t)+\cos t\ \mathrm{Ci}(t)$	$-(p^2+1)^{-1}p \log p$ Re $p > 0$
17.40	$\sin(at)\mathrm{Si}(at) + \cos(at)\mathrm{Ci}(at)$	$-p(p^2+a^2)^{-1}\log(p/a)$ Re $p > 0$
17.41	$\cos(at)\mathrm{Si}(at) - \sin(at)\mathrm{Ci}(at)$	$a(p^2+a^2)^{-1}\log(p/a)$ Re $p > 0$

	$f(t)$	$g(p) = \int_0^\infty f(t)e^{-pt}dt$
17.42	$t^{-\frac{1}{2}}[\sin(at)\mathrm{Ci}(2at)$ $-\cos(at)\mathrm{Si}(2at)]$	$-(\frac{1}{2}\pi)^{\frac{1}{2}}a(p^2+a^2)^{-\frac{1}{2}}[p+(p^2+a^2)^{\frac{1}{2}}]^{-\frac{1}{2}}$ $\cdot\log[p/a+(1+p^2/a^2)^{\frac{1}{2}}]$ Re $p > 0$
17.43	$t^{-\frac{1}{2}}[\cos(at)\mathrm{Ci}(2at)$ $+\sin(at)\mathrm{Si}(2at)]$	$-(\frac{1}{2}\pi)^{\frac{1}{2}}(p^2+a^2)^{-\frac{1}{2}}[p+(p^2+a^2)^{\frac{1}{2}}]^{\frac{1}{2}}$ $\cdot\log[p/a+(1+p^2/a^2)^{\frac{1}{2}}]$ Re $p > 0$
17.44	$t^{-3/2}e^{-b/t}[\sin(at)\mathrm{Ci}(2at)$ $-\cos(at)\mathrm{Si}(2at)]$ Re $b > 0$	$(\pi/b)^{\frac{1}{2}}\{e^{-z_2}[\sin z_1\mathrm{Ci}(2z_1)$ $-\cos z_1\mathrm{Si}(2z_1)]$ $-\sin z_1 e^{z_2}\mathrm{Ei}(-2z_2)\}$ $z_{\frac{1}{2}} = (2b)^{\frac{1}{2}}[(p^2+a^2)^{\frac{1}{2}}\mp p]$ Re $p > 0$
17.45	$t^{-3/2}e^{-b/t}[\cos(at)\mathrm{Ci}(2at)$ $+\sin(at)\mathrm{Si}(2at)]$ Re $b > 0$	$(\pi/b)^{\frac{1}{2}}\{e^{-z_2}[\cos z_1\mathrm{Ci}(2z_1)$ $+\sin z_1\mathrm{Si}(2z_1)]$ $+\cos z_1\mathrm{Ei}(-2z_2)$ $z_{\frac{1}{2}} = (2b)^{\frac{1}{2}}[(p^2+a^2)^{\frac{1}{2}}\mp p]$ Re $p > 0$
17.46	$t^{-\frac{1}{2}}\mathrm{Ci}(2at^{\frac{1}{2}})$	$\frac{1}{2}(\pi/p)^{\frac{1}{2}}\mathrm{Ei}(-a^2/p)$ Re $p > 0$

	$f(t)$	$g(p) = \int_0^\infty f(t)e^{-pt}dt$
17.47	$t^{-\frac{1}{2}}[\cos(2at^{\frac{1}{2}})\mathrm{Ci}(2at^{\frac{1}{2}})$ $+\sin(2at^{\frac{1}{2}})\mathrm{Si}(2at^{\frac{1}{2}})]$	$\frac{1}{2}(\pi/p)^{\frac{1}{2}}e^{-a^2/p}\overline{\mathrm{Ei}}(a^2/p)$ $\mathrm{Re}\, p > 0$
17.48	$t^{-\frac{1}{2}}[\cos(2at^{\frac{1}{2}})\mathrm{Ci}(2at^{\frac{1}{2}})$ $+\sin(2at^{\frac{1}{2}})\mathrm{si}(2at^{\frac{1}{2}})]$	$(\frac{1}{2}\pi/p)^{\frac{1}{2}}e^{-a^2/p}$ $\cdot[i\pi\mathrm{Erf}(iap^{-\frac{1}{2}})+\overline{\mathrm{Ei}}(a^2/p)]$ $\mathrm{Re}\, p > 0$
17.49	$t^{-\frac{1}{2}}\{[\mathrm{Ci}(2at^{\frac{1}{2}})+\mathrm{Ci}(2bt^{\frac{1}{2}})]\cos[t^{\frac{1}{2}}(a-b)]$ $+[\mathrm{Si}(2at^{\frac{1}{2}})-\mathrm{Si}(2bt^{\frac{1}{2}})]\sin[t^{\frac{1}{2}}(a-b)]\}$	$(\pi/p)^{\frac{1}{2}}\exp[-\frac{1}{4}(a-b)^2/p]$ $\cdot\mathrm{Ei}(-ab/p)$ $\mathrm{Re}\, p > 0$
17.50	$t^{-\frac{1}{2}}\{[\mathrm{Ci}(2at^{\frac{1}{2}})+\mathrm{Ci}(2bt^{\frac{1}{2}})]\cos[t^{\frac{1}{2}}(a+b)]$ $+[\mathrm{Si}(2at^{\frac{1}{2}})+\mathrm{Si}(2bt^{\frac{1}{2}})]\sin[t^{\frac{1}{2}}(a+b)]\}$	$(\pi/p)^{\frac{1}{2}}\exp[-\frac{1}{4}(a+b)^2/p]$ $\cdot\overline{\mathrm{Ei}}(ab/p)$ $\mathrm{Re}\, p > 0$
17.51	$t^{-\frac{1}{2}}e^{-b/t}[\sin(a/t)\mathrm{Ci}(2a/t)$ $-\cos(a/t)\mathrm{Si}(2a/t)]$ $\mathrm{Re}\, b \geqq 0$	$(\pi/p)^{\frac{1}{2}}\{e^{-z_2}[\sin z_1\mathrm{Ci}(2z_2)$ $-\cos z_1\mathrm{Si}(2z_1)]$ $-\sin z_1 e^{z_2}\mathrm{Ei}(-2z_2)\}$ $z_{\frac{1}{2}} = (2p)^{\frac{1}{2}}[(b^2+a^2)^{\frac{1}{2}}\mp b]$ $\mathrm{Re}\, p > 0$

	$f(t)$	$g(p) = \int_0^\infty f(t) e^{-pt} dt$
7.52	$t^{-\frac{1}{2}} e^{-b/t} [\cos(a/t) \mathrm{Ci}(2a/t)$ $+\sin(a/t) \mathrm{Si}(2a/t)]$ $\mathrm{Re}\, b \geqq 0$	$(\pi/p)^{\frac{1}{2}} \{ e^{-z_2} [\cos z_1 \mathrm{Ci}(2z_1)$ $+\sin z_1 \mathrm{Si}(2z_1)]$ $+\cos z_1 e^{z_2} \mathrm{Ei}(-2z_2) \}$ $z_{\frac{1}{2}} = (2p)^{\frac{1}{2}} [(b^2+a^2)^{\frac{1}{2}} \mp b]$ $\mathrm{Re}\, p > 0$
7.53	$\mathrm{si}(a/t)$	$2p^{-1} \mathrm{kei}[2(ap)^{\frac{1}{2}}]$ $\mathrm{Re}\, p > 0$
7.54	$\mathrm{Ci}(a/t)$	$-2p^{-1} \mathrm{ker}[2(ap)^{\frac{1}{2}}]$ $\mathrm{Re}\, p > 0$
7.55	$\mathrm{Erf}(at)$	$p^{-1} e^{\frac{1}{4}p^2/a^2} \mathrm{Erfc}(\frac{1}{2}p/a)$ $\mathrm{Re}\, p > 0$
7.56	$\mathrm{Erfc}(at)$	$p^{-1} [1-e^{\frac{1}{4}p^2/a^2} \mathrm{Erfc}(\frac{1}{2}p/a)]$ $\mathrm{Re}\, p > 0$
7.57	$\exp(-a^2t^2) \mathrm{Erf}(iat)$	$(2ai)^{-1} \pi^{\frac{1}{2}} e^{\frac{1}{4}p^2/a^2} \mathrm{Ei}(-\frac{1}{4}p^2/a^2)$ $\mathrm{Re}\, p > 0$
7.58	$\exp(-a^2t^2) \mathrm{Erfc}(iat)$	$(2a)^{-1} \pi^{\frac{1}{2}} e^{\frac{1}{4}p^2/a^2}$ $\cdot [\mathrm{Erfc}(\frac{1}{2}p/a) + i\pi^{-1} \mathrm{Ei}(-\frac{1}{4}p^2/a^2)]$ $\mathrm{Re}\, p > 0$

	$f(t)$	$g(p) = \int_0^\infty f(t)e^{-pt}dt$
17.59	$\mathrm{Erf}(at+b)$	$p^{-1}[\exp(\frac{1}{4}p^2/a^2+pb/a)\mathrm{Erfc}(b+\frac{1}{2}p/a) + \mathrm{Erf}b]$ Re $p > 0$
17.60	$\mathrm{Erfc}(at+b)$	$p^{-1}[\mathrm{Erfc}b-\exp(\frac{1}{4}p^2/a^2+pb/a) \cdot\mathrm{Erfc}(b+\frac{1}{2}p/a)]$ Re $p > 0$
17.61	$e^{a^2t^2}\mathrm{Erfc}(at)$	$-\frac{1}{2}\pi^{-\frac{1}{2}}a^{-1}e^{-\frac{1}{4}p^2/a^2} \cdot[i\pi\mathrm{Erf}(\frac{1}{2}ip/a)+\overline{\mathrm{Ei}}(\frac{1}{4}p^2/a^2)]$ Re $p > 0$
17.62	$\mathrm{Erf}(at^{\frac{1}{2}})$	$ap^{-1}(a^2+p)^{-\frac{1}{2}}$ Re $p > 0$
17.63	$\mathrm{Erfc}(at^{\frac{1}{2}})$	$p^{-1}(a^2+p)^{-\frac{1}{2}}[(a^2+p)^{\frac{1}{2}}-a]$ Re $p > 0$
17.64	$t^{-1}e^{at}\mathrm{Erf}[(at)^{\frac{1}{2}}]$	$\log[(p^{\frac{1}{2}}+a^{\frac{1}{2}})/(p^{\frac{1}{2}}-a^{\frac{1}{2}})]$ Re $p > \|\mathrm{Re}\ a\|$
17.65	$t^{\mu}e^{\frac{1}{2}at}\mathrm{Erfc}[(at^{\frac{1}{2}})]$ Re $\mu > -1$	$2^{2\mu+2}\pi^{-\frac{1}{2}}(\mu+1)^{-1}\Gamma(\frac{3}{2}+\mu) \cdot(4p^2+a)^{-\mu-1}{}_2F_1(\frac{1}{2},1+\mu;2+\mu;\frac{4p-a}{4p+a})$ Re $p > \frac{1}{4}\mathrm{Re}\ a$

	$f(t)$	$g(p) = \int_0^\infty f(t)e^{-pt}dt$
17.66	$\mathrm{Erf}(at^{-\frac{1}{2}})$	$p^{-1}[1-\exp(-2ap^{\frac{1}{2}})]$ Re $p > 0$
17.67	$\mathrm{Erfc}(at^{-\frac{1}{2}})$	$p^{-1}\exp(-2ap^{\frac{1}{2}})$ Re $p > 0$
17.68	$\mathrm{Erfc}(at^{\frac{1}{2}}+bt^{-\frac{1}{2}})$	$(p+a^2)^{-\frac{1}{2}}[a+(p+a^2)^{\frac{1}{2}}]^{-1}$ $\cdot\exp\{-2b[a+(p+a^2)^{\frac{1}{2}}]\}$ Re $p > 0$
17.69	$t^{-\frac{1}{2}}e^{a^2/t}\mathrm{Erfc}(at^{-\frac{1}{2}})$	$2(\pi p)^{-\frac{1}{2}}[\mathrm{Ci}(2ap^{\frac{1}{2}})\sin(2ap^{\frac{1}{2}})$ $-\mathrm{si}(2ap^{\frac{1}{2}})\cos(2ap^{\frac{1}{2}})]$ Re $p > 0$
17.70	$t^{\nu}e^{a^2/t}\mathrm{Erfc}(at^{-\frac{1}{2}})$ Re $\nu > -\frac{3}{2}$	$-\pi\sec(\pi\nu)a^{\nu+1}p^{-\frac{1}{2}\nu-\frac{1}{2}}$ $\cdot[\mathbf{H}_{-\nu-1}(2ap^{\frac{1}{2}})-Y_{-\nu-1}(2ap^{\frac{1}{2}})]$ Re $p > 0$
17.71	$t^{\nu}e^{-a^2/t}\mathrm{Erf}(iat^{-\frac{1}{2}})$	$i\pi\sec(\pi\nu)a^{\nu+1}p^{-\frac{1}{2}\nu-\frac{1}{2}}$ $\cdot[\mathbf{L}_{-\nu-1}(2ap^{\frac{1}{2}})-I_{\nu+1}(2ap^{\frac{1}{2}})]$ Re $p > 0$
7.72	$\exp(-\frac{1}{4}at^2)$ $\cdot[D_{2\nu}(-at)-D_{2\nu}(at)]$	$(2\pi)^{\frac{1}{2}}a^{-2\nu-1}p^{2\nu}[\Gamma(-\nu)]^{-1}$ $\cdot\exp(\frac{1}{2}p^2/a^2)\Gamma(-\nu,\frac{1}{2}p^2/a^2)$ Re $p > 0$

	$f(t)$	$g(p) = \int_0^\infty f(t)e^{-pt}dt$
17.73	$t^{-3/4}D_\nu[(2at)^{\frac{1}{2}}]$	$2^{1+\frac{1}{2}\nu}\pi^{\frac{1}{2}}a^{\frac{1}{4}}[(2p+a)/(2p-a)]^{-\frac{1}{2}(\frac{1}{4}+\frac{1}{2}\nu)}$ $\cdot(4p^2-a^2)^{-\frac{1}{4}}e^{-i\frac{\pi}{4}}q^{\frac{1}{4}}_{-\frac{1}{2}\nu-3/4}\{[2p/(4p^2-a^2)^{-1}]^{\frac{1}{2}}\}$ $\operatorname{Re} p > \frac{1}{2}\lvert\operatorname{Re} a\rvert$
17.74	$t^{-\frac{1}{2}-\frac{1}{2}\nu}D_\nu(at^{\frac{1}{2}})$ $\operatorname{Re}\nu < 1$	$2^{\frac{1}{2}\nu}\pi^{\frac{1}{2}}(p+\frac{1}{4}a^2)^{-\frac{1}{2}}[(p+\frac{1}{4}a^2)^{\frac{1}{2}}+2^{-\frac{1}{2}}a]^{\nu}$ $\operatorname{Re} p > -\frac{1}{4}\operatorname{Re} a^2$
17.75	$t^{-3/2-\frac{1}{2}\nu}D_\nu(at^{\frac{1}{2}})$ $\operatorname{Re}\nu < -1$	$-2^{1+\frac{1}{2}\nu}\pi^{\frac{1}{2}}(\nu+1)^{-1}[(p+\frac{1}{4}a^2)^{\frac{1}{2}}+2^{-\frac{1}{2}}a]^{\nu+1}$ $\operatorname{Re} p > -\frac{1}{4}\operatorname{Re} a^2$
17.76	$t^{-\frac{1}{2}}[D_\nu(-bt^{\frac{1}{2}})+D_\nu(bt^{\frac{1}{2}})]$	$\pi 2^{1+\frac{1}{2}\nu}[\Gamma(\frac{1}{2}-\frac{1}{2}\nu)]^{-1}(p+\frac{1}{4}b^2)^{-\frac{1}{2}}$ $\cdot[(p-\frac{1}{4}b^2)/(p+\frac{1}{4}b^2)]^{\frac{1}{2}\nu}$ $\operatorname{Re} p > -\frac{1}{4}\operatorname{Re} b^2$
17.77	$D_\nu(-bt^{\frac{1}{2}})-D_\nu(bt^{\frac{1}{2}})$	$\pi 2^{\frac{1}{2}+\frac{1}{2}\nu}[\Gamma(-\frac{1}{2}\nu)]^{-1}b(p-\frac{1}{4}b^2)^{\frac{1}{2}\nu-\frac{1}{2}}$ $\cdot(p+\frac{1}{4}b^2)^{-\frac{1}{2}\nu-1}$ $\operatorname{Re} p > -\frac{1}{4}\operatorname{Re} b^2$

	$f(t)$	$g(p) = \int_0^\infty f(t)e^{-pt}dt$
17.78	$t^{\mu}D_{\nu}(at^{\frac{1}{2}})$ $\mathrm{Re}\ \mu > -1$	$\pi^{\frac{1}{2}}2^{\frac{1}{2}\nu-\mu}a^{-2\mu-2}[\Gamma(\tfrac{3}{2}+\mu-\tfrac{1}{2}\nu)]^{-1}$ $\cdot\Gamma(2+2\mu)\,{}_2F_1(\tfrac{3}{2}+\mu,1+\mu;\tfrac{3}{2}+\mu-\tfrac{1}{2}\nu;\tfrac{1}{2}-2p/a^2)$ $= \pi^{\frac{1}{2}}2^{\frac{1}{2}\nu+1}[\Gamma(\tfrac{3}{2}+\mu-\tfrac{1}{2}\nu)]^{-1}\Gamma(2+2\mu)$ $\cdot(4p+a^2)^{-\mu-1}\,{}_2F_1(-\tfrac{1}{2}\nu,1+\mu;\tfrac{3}{2}+\mu-\tfrac{1}{2}\nu;\frac{4p-a^2}{4p+a^2})$ $=\pi^{\frac{1}{2}}2^{\frac{3}{2}+\mu}\Gamma(2+2\mu)(4p+a^2)^{-\frac{1}{2}(\mu+\frac{1}{2}\nu+\frac{3}{2})}$ $\cdot(4p-a^2)^{-\frac{1}{2}(\mu-\frac{1}{2}\nu+\frac{1}{2})}P^{\frac{1}{2}\nu-\mu-\frac{1}{2}}_{\frac{1}{2}\nu+\mu+\frac{1}{2}}[(\tfrac{1}{2}+2p/a^2)^{-\frac{1}{2}}]$ $\mathrm{Re}\ p > \tfrac{1}{4}\mathrm{Re}(a^2)$
17.79	$t^{\nu-1}\exp(\tfrac{1}{4}a^2t)$ $\cdot D_{2\nu+2n-1}(at^{\frac{1}{2}})$ $n=0,1,2,\cdots \mathrm{Re}\ \nu>0$	$2^{\frac{1}{2}-2n-\nu}(n!)^{-1}\pi^{\frac{1}{2}}\Gamma(2n+2\nu)p^{-n-\nu}$ $\cdot(a^2-2p)^n\,{}_2F_1[n+\nu,\tfrac{1}{2}-\nu;n+1;1-\tfrac{1}{2}a^2/p]$ $\mathrm{Re}\ p > 0$
17.80	$t^{-\frac{1}{2}-\frac{1}{2}\nu}\exp(\tfrac{1}{4}a^2t)D_{\nu}(at^{\frac{1}{2}})$ $\mathrm{Re}\ \nu < 1$	$(\pi/p)^{\frac{1}{2}}(a+2^{\frac{1}{2}}p^{\frac{1}{2}})^{\nu}$ $\quad\mathrm{Re}\ p > 0$
17.81	$t^{-\frac{1}{2}\nu-\frac{1}{2}}\exp(-\tfrac{1}{4}a^2t)D_{\nu}(at^{\frac{1}{2}})$ $\mathrm{Re}\ \nu < 1$	$(2a)^{\nu}\pi^{\frac{1}{2}}(p+\tfrac{1}{2}a^2)^{-\frac{1}{2}}[(2p+a^2)^{\frac{1}{2}}-a]^{-\nu}$ $\mathrm{Re}\ p > -\tfrac{1}{2}\mathrm{Re}\ a^2$

	$f(t)$	$g(p) = \int_0^\infty f(t)e^{-pt}dt$
17.82	$t^{-\frac{1}{2}\nu-3/2}\exp(\frac{1}{4}a^2t)D_\nu(at^{\frac{1}{2}})$ $\mathrm{Re}\ \nu < -1$	$(2\pi)^{\frac{1}{2}}(\nu+1)^{-1}(a+2^{\frac{1}{2}}p^{\frac{1}{2}})^{\nu+1}$ $\mathrm{Re}\ p > 0$
17.83	$t^{-\frac{1}{2}\nu-3/2}\exp(-\frac{1}{4}a^2t)D_\nu(at^{\frac{1}{2}})$ $\mathrm{Re}\ \nu < -1$	$-\pi^{\frac{1}{2}}(\nu+1)^{-1}2^{\nu+3/2}p^{\nu+1}[(2p+a^2)^{\frac{1}{2}}-a]^{-\nu-1}$ $\mathrm{Re}\ p > -\frac{1}{2}\mathrm{Re}\ a^2$
17.84	$t^{\nu-1}\exp(\frac{1}{4}a^2t)D_\mu(at^{\frac{1}{2}})$ $\mathrm{Re}\ \nu>0,\ \mathrm{Re}(\nu-\frac{1}{2}\mu)>-\frac{1}{2}$	$(2\pi)^{\frac{1}{2}}\Gamma(2\nu)2^{-\nu}p^{-\frac{1}{4}-\frac{1}{2}\nu-\frac{1}{4}\mu}$ $\cdot(p-\frac{1}{2}a^2)^{\frac{1}{4}(1+\mu-2\nu)}P^{\frac{1}{2}+\frac{1}{2}\mu-\nu}_{\frac{1}{2}(\nu+\mu+1)}[a(2p)^{-\frac{1}{2}}]$ $\mathrm{Re}\ p > 0$
17.85	$t^{-\frac{1}{2}-\frac{1}{2}\nu}\exp(-\frac{1}{4}a^2/t)D_\nu(a/t)$	$2^{\frac{1}{2}\nu}\pi^{\frac{1}{2}}p^{\frac{1}{2}\nu-\frac{1}{2}}\exp[-a(2p)^{\frac{1}{2}}]$ $\mathrm{Re}\ p > 0$
17.86	$t^{\frac{1}{2}\nu}\exp(\frac{1}{4}a^2/t)D_\nu(a/t)$ $\mathrm{Re}\ \nu > -1$	$2^{\frac{1}{4}-\frac{1}{2}\nu}a^{\frac{1}{4}}p^{-\frac{1}{2}\nu-3/4}$ $\cdot S_{\frac{1}{2}+\nu,\frac{1}{2}}[a(2p)^{\frac{1}{2}}]$ $\mathrm{Re}\ p > 0$
17.87	$(e^t-1)^{-\frac{1}{2}-\nu}\exp[\frac{a^2}{4}e^{-t}(1-e^{-t})^{-1}]$ $\cdot D_{2\nu}[a(1-e^{-t})^{-\frac{1}{2}}]$	$2^{\nu+p+\frac{1}{2}}\Gamma(p+\frac{1}{2}+\nu)D_{-2p-1}(a)$ $\mathrm{Re}(p+\nu) > -\frac{1}{2}$

	$f(t)$	$g(p) = \int_0^\infty f(t)e^{-pt}dt$
17.88	$(1-e^{-t})^{-\frac{1}{2}\nu-\frac{1}{2}}$ $\cdot\exp[-\frac{a^2}{4}(e^t-1)^{-1}]$ $\cdot D_\nu[a(e^t-1)^{-1}]$	$\pi^{\frac{1}{2}}2^{\frac{1}{2}\nu-p+3/4}\Gamma(\frac{1}{2}+2p)e^{a^2/4}$ $\cdot[\Gamma(3/4+p-\frac{1}{2}\nu)]^{-1}D_{-2p-\frac{1}{2}}(a)$ Re $p > -\frac{1}{4}$
17.89	$\gamma(\nu,at)$ Re $\nu > 0$	$p^{-1}\Gamma(\nu)(1+p/a)^{-\nu}$ Re $p > 0$
17.90	$\Gamma(\nu,at)$	$p^{-1}\Gamma(\nu)[1-(1+p/a)^{-\nu}]$ Re $p > 0$
17.91	$\Gamma(\nu,a/t)$	$2p^{-1}(ap)^{\frac{1}{2}\nu}K_\nu[2(ap)^{\frac{1}{2}}]$ Re $p > 0$
17.92	$\gamma(\nu,at^2)$ Re $\nu > 0$	$2^{-\nu-1}\Gamma(2\nu)p^{-1}\exp(1/8p^2/b)D_{-2\nu}[(2b)^{-\frac{1}{2}}p]$ Re $p > 0$
17.93	$t^{-3/4}e^{-\frac{1}{2}at}M_{\mu,-\frac{1}{4}}(at)$	$(\pi^2a/p)^{\frac{1}{4}}p^\mu(p+a)^{-\mu-\frac{1}{4}}$ Re $p >$ Max[0,-Re a]
17.94	$t^{-\frac{1}{4}}e^{-\frac{1}{2}at}M_{\mu,\frac{1}{4}}(at)$	$\frac{1}{2}\pi^{\frac{1}{2}}(a/p)^{3/4}p^\mu(p+a)^{-\mu-3/4}$ Re $p >$ Max[0,-Re a]

	$f(t)$	$g(p) = \int_0^\infty f(t) e^{-pt} dt$
17.95	$t^{-5/4} e^{\frac{1}{2}at} M_{-\frac{1}{4}, \frac{1}{2}n+\frac{1}{4}}(at)$	$2\Gamma(n+3/2) a^{\frac{1}{4}} [\Gamma(1+\frac{1}{2}n)]^{-1} q_n[(p/a)^{\frac{1}{2}}]$ $n = 1,2,3,\cdots$ Re $p > \|\text{Re } a\|$
17.96	$t^{\nu-\frac{1}{2}} M_{\mu,\nu}(at)$ Re $\nu > -\frac{1}{2}$	$a^{\nu+\frac{1}{2}} \Gamma(2\nu+1)(p+\frac{1}{2}a)^{-\mu-\nu-\frac{1}{2}}$ $\cdot (p-\frac{1}{2}a)^{\mu-\nu-\frac{1}{2}}$ Re $p > \frac{1}{2}\|\text{Re } a\|$
17.97	$t^{-1} M_{\mu,\nu}(at)$ Re $\nu > -\frac{1}{2}$	$2\Gamma(1+2\nu)[\Gamma(\frac{1}{2}+\nu+\mu)]^{-1}$ $\cdot [(2p+a)/(2p-a)]^{-\frac{1}{2}\mu} e^{-i\pi\mu} q^{\mu}_{\nu-\frac{1}{2}}(2p/a)$ $= (2\pi a)^{\frac{1}{2}} \Gamma(1+2\nu)(4p^2-a^2)^{-\frac{1}{4}}$ $\cdot [(2p+a)/(2p-a)]^{-\frac{1}{2}\mu}$ $\cdot p^{-\nu}_{\mu-\frac{1}{2}}[2p(4p^2-a^2)^{-\frac{1}{2}}]$ Re $p > \frac{1}{2}\|\text{Re } a\|$
17.98	$t^{-1} W_{\mu,\nu}(at)$ $-\frac{1}{2} < \text{Re } \nu < \frac{1}{2}$	$\pi \sec(\pi\nu)[(2p+a)/(2p-a)]^{-\frac{1}{2}\mu}$ $\cdot p^{\mu}_{\nu-\frac{1}{2}}(2p/a)$ $= (2\pi a)^{\frac{1}{2}} \sec(\pi\nu)[(2p+a)/(2p-a)]^{-\frac{1}{2}\mu}$ $\cdot (4p^2-a^2)^{-\frac{1}{4}} e^{i\pi\nu} q^{-\nu}_{-\mu-\frac{1}{2}}[2p(4p^2-a^2)^{-\frac{1}{2}}]$ Re $p > \frac{1}{2}\|\text{Re } a\|$

	$f(t)$	$g(p) = \int_0^\infty f(t)e^{-pt}dt$
17.99	$t^{-2\nu-\mu-1}M_{\mu,\nu}(at)$ $\mathrm{Re}(\nu+\mu) < \frac{1}{2}$	$a^{\frac{1}{2}}\Gamma(1+2\nu)\Gamma(\frac{1}{2}-\nu-\mu)(p-\frac{1}{2}a)^{\nu+\mu-\frac{1}{2}}$ $(p+\frac{1}{2}a)^{\nu}P^{-2\nu}_{\nu+\mu-\frac{1}{2}}[(2p+3a)/(2p-a)]$ $\mathrm{Re}\ p > \frac{1}{2}\lvert\mathrm{Re}\ a\rvert$
17.100	$t^{-\mu-\frac{1}{2}}M_{\mu,\nu}(at)$	$2^{2\nu-\mu+1}a^{\frac{1}{2}}\Gamma(\nu-\mu+1)\Gamma(1+2\nu)$ $\cdot(4p^2-a^2)^{-\frac{1}{2}}(2p-a)^{\mu}P^{-2\nu}_{2\mu-1}\{[(2p+a)/(2p-a)]^{\frac{1}{2}}\}$ $=2^{5/4+\mu}a^{\frac{1}{4}}\Gamma(1+2\nu)[\Gamma(\frac{1}{2}+\nu-\mu)]^{-1}$ $\cdot(2p+a)^{-\frac{1}{2}}(2p-a)^{\mu-\frac{1}{4}}e^{-i\pi(\frac{1}{2}-2\mu)}$ $\cdot\ q^{\frac{1}{2}-2\mu}_{2\nu-\frac{1}{2}}[(\frac{1}{2}+p/a)^{\frac{1}{2}}]$ $\mathrm{Re}\ p > \frac{1}{2}\lvert\mathrm{Re}\ a\rvert$
17.101	$t^{-\mu-\frac{1}{2}}W_{\mu,\nu}(at)$ $\mathrm{Re}(\pm\nu-\mu) > -1$	$a^{\frac{1}{4}}2^{5/4-3\mu}\Gamma(1+\nu-\mu)\Gamma(1-\nu-\mu)$ $\cdot(2p-a)^{\mu-\frac{1}{4}}(2p+a)^{-\frac{1}{2}}$ $\cdot P^{2\mu-\frac{1}{2}}_{2\nu-\frac{1}{2}}[(\frac{1}{2}+p/a)^{\frac{1}{2}}]$ $=2^{2\nu-\mu+2}a^{\frac{1}{2}}(2p+a)^{-\frac{1}{2}}(2p-a)^{\mu-\frac{1}{2}}$ $\cdot\Gamma(1+\nu-\mu)[\Gamma(\frac{1}{2}-\nu-\mu)]^{-1}$ $\cdot e^{i2\pi\nu}q^{-2\nu}_{-2\mu}\{[(2p+a)/(2p-a)]^{\frac{1}{2}}\}$ $\mathrm{Re}\ p > \frac{1}{2}\lvert\mathrm{Re}\ a\rvert$

	$f(t)$	$g(p) = \int_0^\infty f(t) e^{-pt} dt$
17.102	$t^{-\mu-3/2} M_{\mu,\nu}(at)$ $\mathrm{Re}(\nu-\mu) > 0$	$2^{2\nu-\mu} a^{1/2} \Gamma(\nu-\mu)\Gamma(1+2\nu)$ $(2p-a)^{\mu} P_{2\mu}^{-2\nu}\{[(2p+a)/(2p-a)]^{1/2}\}$ $= 2^{5/4+\mu} a^{1/4} \Gamma(1+2\nu)[\Gamma(\tfrac{1}{2}+\nu-\mu)]^{-1}$ $\cdot (2p-a)^{\mu+1/4} e^{i\pi(2\mu+1/2)} Q_{2\nu-1/2}^{-2\mu-1/2}[(\tfrac{1}{2}+p/a)^{1/2}]$ $\mathrm{Re}\, p > \tfrac{1}{2}\lvert \mathrm{Re}\, a \rvert$
17.103	$t^{-\mu-3/2} W_{\mu,\nu}(at)$ $\mathrm{Re}(\pm\nu-\mu) > 0$	$a^{1/4} 2^{3\mu-3/4} \Gamma(\mu-\nu)\Gamma(-\nu-\mu)$ $\cdot (2p-a)^{1/4+\mu} P_{2\nu-1/2}^{2\mu+1/2}[(\tfrac{1}{2}+p/a)^{1/2}]$ $= 2^{2\nu-\mu+1} a^{1/2} \Gamma(\mu-\nu)[\Gamma(\tfrac{1}{2}-\mu-\nu)]^{-1}$ $\cdot (2p-a)^{\mu} e^{i2\pi\nu} Q_{-2\mu-1}^{-2\nu}\{[(2p+a)/(2p-a)]^{1/2}\}$ $\mathrm{Re}\, p > \tfrac{1}{2}\lvert \mathrm{Re}\, a \rvert$
17.104	$t^{\lambda} M_{\mu,\nu}(at)$ $\mathrm{Re}(\lambda+\nu) > 3/2$	$a^{\nu+1/2} \Gamma(\lambda+\nu+3/2)(p-\tfrac{1}{2}a)^{-\lambda-\nu-3/2}$ $\cdot {}_2F_1[\lambda+\nu+3/2, \nu+\mu+\tfrac{1}{2}; 2\nu+1; (\tfrac{1}{2}-p/a)^{-1}]$ $= a^{\nu+1/2} \Gamma(\lambda+\nu+3/2)(p+\tfrac{1}{2}a)^{-\lambda-\nu-3/2}$ $\cdot {}_2F_1[\lambda+\nu+3/2, \nu-\mu+\tfrac{1}{2}; 2\nu+1; (\tfrac{1}{2}+p/a)^{-1}]$ $\mathrm{Re}\, p > \tfrac{1}{2}\lvert \mathrm{Re}\, a \rvert$

	$f(t)$	$g(p) = \int_0^\infty f(t)e^{-pt}dt$
17.105	$t^{\lambda}W_{\mu,\nu}(at)$ $\mathrm{Re}(\lambda\pm\nu) - > 3/2$	$a^{-\lambda-1}[\Gamma(2+\lambda-\mu)]^{-1}\Gamma(3/2+\lambda+\nu)\Gamma(3/2+\lambda-\nu)$ $\cdot {}_2F_1[\lambda+\nu+3/2, \lambda-\nu+3/2; \lambda-\mu+2; \tfrac{1}{2}-p/a]$ $= a^{\frac{1}{2}+\nu}[\Gamma(2+\lambda-\mu)]^{-1}\Gamma(3/2+\lambda+\nu)$ $\cdot\Gamma(3/2+\lambda-\nu)(p+\tfrac{1}{2}a)^{-\lambda-\nu-3/2}$ $\cdot {}_2F_1[3/2+\lambda+\nu, \tfrac{1}{2}+\nu-\mu; \lambda+2-\mu; (2p-a)/(2p+a)]$ $\mathrm{Re}\, p > \tfrac{1}{2}\lvert\mathrm{Re}\, a\rvert$
17.106	$t^{-1}\exp(-\tfrac{1}{2}a/t)$ $\cdot W_{\frac{1}{2},\nu}(a/t)$	$2\pi^{-\frac{1}{2}}(2ap)^{\frac{1}{2}}K_{\nu+\frac{1}{2}}[(ap)^{\frac{1}{2}}]K_{\nu-\frac{1}{2}}[(ap)^{\frac{1}{2}}]$ $\mathrm{Re}\, p > 0$
17.107	$t^{-1}\exp(\tfrac{1}{2}a/t)$ $\cdot W_{-\frac{1}{2},\mu}(a/t)$	$(4\mu)^{-1}(ap\pi^3)^{\frac{1}{2}}\{H^{(1)}_{\mu+\frac{1}{2}}[(ap)^{\frac{1}{2}}]H^{(2)}_{\mu-\frac{1}{2}}[(ap)^{\frac{1}{2}}]$ $+ H^{(1)}_{\mu-\frac{1}{2}}[(ap)^{\frac{1}{2}}]H^{(2)}_{\mu+\frac{1}{2}}[(ap)^{\frac{1}{2}}]\}$ $\mathrm{Re}\, p > 0$
17.108	$t^{\nu}\exp(\tfrac{1}{2}a/t)$ $\cdot W_{\nu,\nu}(a/t)$	$(a\pi/p)^{\frac{1}{2}}\Gamma(\tfrac{1}{2}+2\nu)p^{-\nu}$ $\cdot[\mathbf{H}_{2\nu}[2(ap)^{\frac{1}{2}}]-Y_{2\nu}[2(ap)^{\frac{1}{2}}]$ $\mathrm{Re}\, p > 0$

	$f(t)$	$g(p) = \int_0^\infty f(t)e^{-pt}dt$
17.109	$t^{3\nu-\frac{1}{2}}\exp(\frac{1}{2}a/t)W_{\nu,\nu}(a/t)$ $\mathrm{Re}\,\nu > -\frac{1}{4}$	$\frac{1}{2}\Gamma(\frac{1}{2}+2\nu)a^{\nu+\frac{1}{2}}p^{-2\nu}$ $\cdot H_{2\nu}^{(1)}[(ap)^{\frac{1}{2}}]H_{2\nu}^{(2)}[(ap)^{\frac{1}{2}}]$ $\mathrm{Re}\,p > 0$
17.110	$t^{-3\nu-\frac{1}{2}}\exp(-\frac{1}{2}a/t)W_{\nu,\nu}(a/t)$	$2\pi^{-\frac{1}{2}}a^{\frac{1}{2}-\nu}p^{-2\nu}\{K_{2\nu}[(ap)^{\frac{1}{2}}]\}^2$ $\mathrm{Re}\,p > 0$
17.111	$t^{\mu}\exp(\frac{1}{2}a/t)W_{\mu,\nu}(a/t)$ $\mathrm{Re}(\mu\pm\nu) > -\frac{1}{2}$	$2^{1-2\mu}a^{\frac{1}{2}}p^{-\mu-\frac{1}{2}}S_{2\mu,2\nu}[2(ap)^{\frac{1}{2}}]$ $\mathrm{Re}\,p > 0$
17.112	$t^{-\mu}\exp(-\frac{1}{2}a/t)W_{\mu,\nu}(a/t)$	$2a^{\frac{1}{2}}p^{\mu-\frac{1}{2}}K_{2\nu}[2(ap)^{\frac{1}{2}}]$ $\mathrm{Re}\,p > 0$
17.113	$t^{2\nu}\exp(-\frac{1}{2}at^2)M_{\mu,\nu}(at^2)$ $\mathrm{Re}\,\nu > -\frac{1}{2}$	$2^{-1-\mu-3\nu}a^{-\frac{1}{2}\nu-\frac{1}{2}\mu}\Gamma(2+4\nu)$ $\cdot p^{\mu-\nu-1}\exp(\frac{1}{8}p^2/a)W_{\alpha,\beta}(\frac{1}{4}p^2/a)$ $2\alpha = -3\nu-\mu-1;\ 2\beta = \nu-\mu+1$ $\mathrm{Re}\,p > 0$

	$f(t)$	$g(p) = \int_0^\infty f(t)e^{-pt}dt$
17.114	$t^{2\nu-1}\exp(-\frac{1}{2}at^2)M_{\mu,\nu}(at^2)$ $\mathrm{Re}\ \nu > -\frac{1}{4}$	$2^{-\mu-3\nu}a^{\frac{1}{2}-\frac{1}{2}\nu-\frac{1}{2}\mu}\Gamma(1+4\nu)$ $p^{\mu-\nu-1}\exp(\frac{1}{8}p^2/a)W_{\alpha,\beta}(\frac{1}{4}p^2/a)$ $2\alpha = -3\nu-\mu;\ 2\beta = \nu-\mu;$ $\mathrm{Re}\ p > 0$
17.115	$(e^t-1)^{\nu-\frac{1}{2}}\exp(-\frac{1}{2}ae^t)$ $\cdot M_{\mu,\nu}[a(e^t-1)]$ $\mathrm{Re}\ \nu > -\frac{1}{2}$	$[\Gamma(p+1)]^{-1}\Gamma(1+2\nu)\Gamma(\frac{1}{2}+\mu-\nu+p)$ $\cdot W_{-\mu-\frac{1}{2}p,\nu-\frac{1}{2}p}(a)$ $\mathrm{Re}\ p > -\frac{1}{2}+\mathrm{Re}(\nu-\mu)$
17.116	$e^t(e^t-1)^{-\mu-1}\exp[-\frac{1}{2}a(e^t-1)^{-1}]$ $\cdot W_{\mu,\nu}[a(e^t-1)^{-1}]$	$a^{-1}\Gamma(p+\mu)W_{-p,\nu}(a)$ $\mathrm{Re}\ p > -\mathrm{Re}\ \mu$
17.117	$(1-e^{-t})^{-\mu}\exp[-\frac{1}{2}a(e^t-1)^{-1}]$ $\cdot W_{\mu,\nu}[a(e^t-1)^{-1}]$	$\Gamma(\frac{1}{2}+\nu+p)\Gamma(\frac{1}{2}-\nu+p)[\Gamma(1-\mu+p)]^{-1}$ $\cdot e^{\frac{1}{2}a}W_{-p,\nu}(a)$ $\mathrm{Re}(\frac{1}{2}+p\pm\nu) > 0$

1.18 Elliptic Functions

	$f(t)$	$g(p) = \int_0^\infty f(t)e^{-pt}dt$
18.1	$\theta_1(z\|at)$	$(ap)^{-\frac{1}{2}}\sinh[2z(p/a)^{\frac{1}{2}}]\operatorname{sech}[(p/a)^{\frac{1}{2}}]$ $-\frac{1}{2} \leqq z \leqq \frac{1}{2}, \quad \operatorname{Re} p > 0$
18.2	$\theta_2(z\|at)$	$(ap)^{-\frac{1}{2}}\sinh[(1-2z)(p/a)^{\frac{1}{2}}]\operatorname{sech}[(p/a)^{\frac{1}{2}}]$ $0 \leqq z \leqq 1, \quad \operatorname{Re} p > 0$
18.3	$\theta_3(z\|at)$	$(ap)^{-\frac{1}{2}}\cosh[(1-2z)(p/a)^{\frac{1}{2}}]\operatorname{csch}[(p/a)^{\frac{1}{2}}]$ $0 \leqq z \leqq 1, \quad \operatorname{Re} p > 0$
18.4	$\theta_4(z\|at)$	$(ap)^{-\frac{1}{2}}\cosh[2z(p/a)^{\frac{1}{2}}]\operatorname{csch}[(p/a)^{\frac{1}{2}}]$ $-\frac{1}{2} \leqq z \leqq \frac{1}{2}, \quad \operatorname{Re} p > 0$
18.5	$\theta_2(n\|at)$ $n = 0,1,2,\cdots$	$(-1)^n(ap)^{-\frac{1}{2}}\tanh[(p/a)^{\frac{1}{2}}]$ $\operatorname{Re} p > 0$
18.6	$\theta_3(n\|at)$ $n = 0,1,2,\cdots$	$(ap)^{-\frac{1}{2}}\coth[(p/a)^{\frac{1}{2}}]$ $\operatorname{Re} p > 0$
18.7	$\theta_4(n\|at)$ $n = 0,1,2,\cdots$	$(ap)^{-\frac{1}{2}}\operatorname{csch}[(p/a)^{\frac{1}{2}}] \quad \operatorname{Re} p > 0$

	$f(t)$	$g(p) = \int_0^\infty f(t)e^{-pt}dt$
18.8	$\frac{\partial}{\partial z}\,\theta_1(z\|at)$	$2a^{-1}\cosh[2z(p/a)^{\frac{1}{2}}]\operatorname{sech}[(p/a)^{\frac{1}{2}}]$ $-\frac{1}{2} < z < \frac{1}{2}, \quad \mathrm{Re}\, p > 0$
18.9	$\frac{\partial}{\partial z}\,\theta_2(z\|at)$	$-2a^{-1}\cosh[(1-2z)(p/a)^{\frac{1}{2}}]\operatorname{sech}[(p/a)^{\frac{1}{2}}]$ $0 < z < 1, \quad \mathrm{Re}\, p > 0$
18.10	$\frac{\partial}{\partial z}\,\theta_3(z\|at)$	$-2a^{-1}\sinh[(1-2z)(p/a)^{\frac{1}{2}}]\operatorname{csch}[(p/a)^{\frac{1}{2}}]$ $0 < z < 1, \quad \mathrm{Re}\, p > 0$
18.11	$\frac{\partial}{\partial z}\,\theta_4(z\|at)$	$2a^{-1}\sinh[2z(p/a)^{\frac{1}{2}}]\operatorname{csch}[(p/a)^{\frac{1}{2}}]$ $-\frac{1}{2} < z < \frac{1}{2}, \quad \mathrm{Re}\, p > 0$
18.12	$\hat{\theta}_1(z\|at)$	$(ap)^{-\frac{1}{2}}\cosh[2z(p/a)^{\frac{1}{2}}]\operatorname{sech}[(p/a)^{\frac{1}{2}}]$ $-\frac{1}{2} \leqq z \leqq \frac{1}{2} \quad \mathrm{Re}\, p > 0$
18.13	$\hat{\theta}_2(z\|at)$	$(ap)^{-\frac{1}{2}}\cosh[(1-2z)(p/a)^{\frac{1}{2}}]\operatorname{sech}[(p/a)^{\frac{1}{2}}]$ $0 \leqq z \leqq 1, \quad \mathrm{Re}\, p > 0$
18.14	$\hat{\theta}_3(z\|at)$	$(ap)^{-\frac{1}{2}}\sinh[(1-2z)(p/a)^{\frac{1}{2}}]\operatorname{csch}[(p/a)^{\frac{1}{2}}]$ $0 \leqq z \leqq 1, \quad \mathrm{Re}\, p > 0$

	$f(t)$	$g(p) = \int_0^\infty f(t)e^{-pt}dt$
18.15	$\hat{\theta}_4(z\|at)$	$-(ap)^{-\frac{1}{2}}\sinh[2z(p/a)^{\frac{1}{2}}]\operatorname{csch}[(p/a)^{\frac{1}{2}}]$ $-\frac{1}{2} \leqq z \leqq \frac{1}{2}$, Re $p > 0$
18.16	$\frac{\partial}{\partial z}\hat{\theta}_1(z\|at)$	$2a^{-1}\sinh[2z(p/a)^{\frac{1}{2}}]\operatorname{sech}[(p/a)^{\frac{1}{2}}]$ $-\frac{1}{2} < z < \frac{1}{2}$, Re $p > 0$
18.17	$\frac{\partial}{\partial z}\hat{\theta}_2(z\|at)$	$-2a^{-1}\sinh[(1-2z)(p/a)^{\frac{1}{2}}]\operatorname{sech}[(p/a)^{\frac{1}{2}}]$ $0 < z < 1$, Re $p > 0$
18.18	$\frac{\partial}{\partial z}\hat{\theta}_3(z\|at)$	$-2a^{-1}\cosh[(1-2z)(p/a)^{\frac{1}{2}}]\operatorname{csch}[(p/a)^{\frac{1}{2}}]$ $0 < z < 1$, Re $p > 0$
18.19	$\frac{\partial}{\partial z}\hat{\theta}_4(z\|at)$	$-2a^{-1}\cosh[2z(p/a)^{\frac{1}{2}}]\operatorname{csch}[(p/a)^{\frac{1}{2}}]$ $-\frac{1}{2} < z < \frac{1}{2}$, Re $p > 0$

1.19 Gauss' Hypergeometric Function

	$f(t)$	$g(p) = \int_0^\infty f(t)e^{-pt}dt$
19.1	$t^{\nu-1}{}_2F_1(1,\frac{1}{2};\nu;-t/a)$ $\mathrm{Re}\,\nu > 0$	$(\pi a)^{\frac{1}{2}}\Gamma(\nu)p^{-\nu+\frac{1}{2}}e^{ap}\mathrm{Erfc}[(ap)^{\frac{1}{2}}]$ $\mathrm{Re}\,p > 0$
19.2	$t^{a-1}{}_2F_1(\frac{1}{2}+\nu,\frac{1}{2}-\nu;a;-\frac{1}{2}t)$ $\mathrm{Re}\,a > 0$	$\pi^{-\frac{1}{2}}\Gamma(a)(2p)^{\frac{1}{2}-a}K_\nu(p)$ $\mathrm{Re}\,p > 0$
19.3	$t^{c-1}{}_2F_1(a,b;c;-ht)$ $\mathrm{Re}\,c > 0$	$\Gamma(c)p^{-c}(p/h)^{\frac{1}{2}(a+b-1)}\exp(\frac{1}{2}p/h)$ $\cdot W_{\frac{1}{2}(1-a-b),\frac{1}{2}a-\frac{1}{2}b}(\frac{1}{2}p/h),$ $\mathrm{Re}\,p > 0$
19.4	$0 \quad t < 1$ $(t^2-1)^{2\alpha-\frac{1}{2}}$ $\cdot{}_2F_1(\alpha-\frac{1}{2}\nu,\alpha+\frac{1}{2}\nu;2\alpha+\frac{1}{2};1-t^2)$ $t > 1$ $\mathrm{Re}\,\alpha > -\frac{1}{4}$	$\pi^{-\frac{1}{2}}2^{2\alpha}p^{-2\alpha}\Gamma(\frac{1}{2}+2\alpha)K_\nu(p)$ $\mathrm{Re}\,p > 0$
19.5	$[(a+t)(b+t)]^{-\nu}$ $\cdot{}_2F_1(\nu,\nu;1;z)$ $z = [(a+t)(b+t)]^{-1}$ $\cdot t(a+b+t)$ $\lvert\arg(a,b)\rvert < \pi$	$\pi^{-1}(ab)^{\frac{1}{2}-\nu}\exp[\frac{1}{2}(a+b)p]$ $\cdot K_{\nu-\frac{1}{2}}(\frac{1}{2}ap)K_{\nu-\frac{1}{2}}(\frac{1}{2}bp)$ $\mathrm{Re}\,p > 0$ $\lvert\arg(ap,bp)\rvert < \pi$

	$f(t)$	$g(p) = \int_0^\infty f(t)e^{-pt}dt$
19.6	$t^{-\frac{1}{2}}(1+a/t)^{\mu}(1+b/t)^{\nu}$ $\cdot {}_2F_1(-\mu,-\nu;\frac{1}{2}-\mu-\nu;z)$ $z = [(a+t)(b+t)]^{-1}$ $\cdot t(a+b+t)$ $\mathrm{Re}(\mu+\nu)<1, \lvert\arg(a,b)<\pi$	$2^{-\mu-\nu}\Gamma(\frac{1}{2}-\mu-\nu)p^{-\frac{1}{2}}$ $\cdot\exp[\frac{1}{2}p(a+b)]$ $\cdot D_{2\mu}[(2ap)^{\frac{1}{2}}]D_{2\nu}[(2bp)^{\frac{1}{2}}]$ $\mathrm{Re}\, p > 0, \lvert\arg(ap,bp) < \pi$
19.7	$t^{c-1}(a+t)^{-\mu}(b+t)^{-\nu}$ $\cdot {}_2F_1(\mu,\nu;c;z)$ $z = [(a+t)(b+t)]^{-1}$ $\cdot t(a+b+t)$ $\lvert\arg(a,b)\rvert < \pi, \mathrm{Re}\, c > 0$	$\Gamma(c)(ab)^{\frac{1}{2}(c-\mu-\nu-1)}p^{-1}$ $\cdot\exp[\frac{1}{2}p(a+b)]$ $\cdot W_{\alpha,\beta}(ap)W_{\gamma,\beta}(bp)$ $\alpha = \frac{1}{2}(1+\nu-c-\mu), \beta=\frac{1}{2}(\mu+\nu-c)$ $\gamma = \frac{1}{2}(1-\nu-c+\mu),$ $\lvert\arg(ap,bp)\rvert < \pi, \mathrm{Re}\, p > 0$
19.8	$(1-e^{-t})^{\nu}{}_2F_1(-n,\nu+b+n;b;e^{-t})$ $n = 0,1,2,\cdots$	$[B(p,b-p)]^{-1}$ $\cdot B(p,\nu+n+1)B(p,b+n-p)$
19.9	$(1-e^{-t})^{\nu}$ $\cdot {}_2F_1(a,b;c;he^{-t})$ $\mathrm{Re}\,\nu > -1$	$B(p,\nu+1)\,{}_3F_2\left[{a,b,p; \atop c,p+\nu+1;}h\right]$ $\mathrm{Re}\, p > 0, \lvert\arg(1-h)\rvert < \pi$

	$f(t)$	$g(p) = \int_0^\infty f(t)e^{-pt}dt$
9.10	$(1-e^{-t})^{c-1}$ $\cdot {}_2F_1(a,b;c;1-e^{-t})$ Re $c > 0$	$\Gamma(p)\Gamma(c-a-b+p)\Gamma(c)$ $\cdot [\Gamma(c-a+p)\Gamma(c-b+p)]^{-1}$ $\lvert\arg(1-h)\rvert<\pi$, Re $p>\mathrm{Max}[0,\mathrm{Re}(a+b-c)]$
9.11	$(1-e^{-t})^{c-1}$ $\cdot {}_2F_1(a,b;c;h-he^{-t})$ Re $c > 0$	$B(p,c)\,{}_2F_1(a,b;p+c;h)$ Re $p > 0$, $\lvert\arg(1-h)\rvert < \pi$
9.12	$(1-e^{-t})^{\lambda-1}$ $\cdot {}_2F_1(a,b;c;h-he^{-t})$ Re $\lambda > 0$ $\lvert\arg(1-h)\rvert < \pi$	$B(p,\lambda)\,{}_3F_2(a,b,\lambda;c,p+\lambda;h)$ Re $p > 0$

.20 Miscellaneous Functions

0.1	$\nu(t)$	$(p \log p)^{-1}$ Re $p > 1$
0.2	$\nu(t,a)$ Re $a \geqq 0$	$p^{-a-1}(\log p)^{-1}$ Re $p > 1$

	$f(t)$	$g(p) = \int_0^\infty f(t)e^{-pt}dt$
20.3	$\mu(t,a)$ $\mathrm{Re}\ a > -1$	$\Gamma(a+1)p^{-1}(\log p)^{-a-1}$ $\mathrm{Re}\ p > 1$
20.4	$t^n\nu(t)$ $n = 1,2,3,\cdots$	$p^{-n-1}\sum_{k=0}^{n} a_k k!(\log p)^{-k-1}$ $\mathrm{Re}\ p > 1$ The a_k given by $(s+1)(s+2)\cdots(s+n)$ $= \sum_{k=0}^{n} a_k s^k$
20.5	$t^n\nu(t,a)$ $n = 1,2,3\cdots$ $\mathrm{Re}\ a \geqq 0$	$p^{-n-1-a}\sum_{k=0}^{n} b_k k!(\log p)^{-k-1}$ $\mathrm{Re}\ p > 1$ The b_k given by $(s+a+1)(s+a+2)\cdots(s+a+n)$ $= \sum_{k=0}^{n} b_k s^k$
20.6	$t^n\mu(t,a)$ $\mathrm{Re}\ a > -1$ $n = 1,2,3\cdots$	$p^{-n-1}\sum_{k=0}^{n} a_k\Gamma(k+a+1)\log p)^{-k-a-1}$ $\mathrm{Re}\ p > 1$ The a_k given as before

	$f(t)$	$g(p) = \int_0^\infty f(t)e^{-pt}dt$	
20.7	$(1-e^{-t})^{-1}\nu(t)$	$\int_0^\infty \zeta(x+1,p)dx$	Re $p > 1$
20.8	$t^{-\frac{1}{2}}\nu(2t^{\frac{1}{2}})$	$2\pi^{\frac{1}{2}}p^{-\frac{1}{2}}\nu(p^{-1})$	Re $p > 0$
20.9	$\nu(e^{-t})$	$\int_0^\infty [(p+x)\Gamma(1+x)]^{-1}dx$	Re $p > 0$
20.10	$\nu(1-e^{-t})$	$\Gamma(p)\nu(1,p)$	Re $p > 0$
20.11	$\nu(2t^{\frac{1}{2}},2a)$	$\frac{1}{2}\pi^{\frac{1}{2}}p^{-3/2}\nu(p^{-1},a-\frac{1}{2})$	Re $p > 0$
20.12	$t^{-\frac{1}{2}}\nu(2t^{\frac{1}{2}},2a)$	$2(\pi/p)^{\frac{1}{2}}\nu(p^{-1},a)$	Re $p > 0$
20.13	$t^{-\frac{1}{2}}\mu(2t^{\frac{1}{2}},a)$	$2^{a+1}(\pi/p)^{\frac{1}{2}}\mu(p^{-1},a)$	Re $p > 0$

1.21 Generalized Hypergeometric Functions

	$f(t)$	$g(p) = \int_0^\infty f(t)e^{-pt}dt$
21.1	${}_0F_1(\ ;n+1)$ $n = 0,1,2,\cdots$	$p^{-1}{}_1F_1(1;n+1;p^{-1})$ $\mathrm{Re}\ p > 0$
21.2	$t^{\gamma-1}{}_1F_1(\alpha;\gamma;\lambda t)$ $\mathrm{Re}(\gamma,\lambda) > 0$	$\Gamma(\gamma)p^{\alpha-\gamma}(p-\lambda)^{-\alpha}$ $\mathrm{Re}\ p > 0$
21.3	${}_0F_2(\ ;\tfrac{1}{2},1;-t)$	$p^{-1}\cos(2p^{-\frac{1}{2}})$ $\quad \mathrm{Re}\ p > 0$
21.4	$t^{\nu}{}_0F_2(\ ;\nu+1;\tfrac{3}{2};-t/a)$ $\mathrm{Re}\ \nu > -1$	$\tfrac{1}{2}a^{\frac{1}{2}}\Gamma(1+\nu)p^{-\nu-\frac{1}{2}}\sin[2(ap)^{\frac{1}{2}}]$ $\mathrm{Re}\ p > 0$
21.5	$t^{\nu}{}_0F_2(\ ;\nu+1,\tfrac{1}{2};-t/a)$ $\mathrm{Re}\ \nu > -1$	$\Gamma(1+\nu)p^{-\nu-1}\cos[2(ap)^{\frac{1}{2}}]$ $\mathrm{Re}\ p > 0$
21.6	$t^{\mu-1}{}_0F_2(\ ;\mu,\nu;-at)$ $\mathrm{Re}\ \mu > 0$	$a^{\frac{1}{2}-\frac{1}{2}\nu}\Gamma(\nu)\Gamma(\mu)p^{\frac{1}{2}\nu-\frac{1}{2}-\mu}$ $\cdot J_{\nu-1}[2(a/p)^{\frac{1}{2}}]$ $\quad \mathrm{Re}\ p > 0$
21.7	$t^{2\nu}{}_0F_2(\ ;2\nu+1,\nu+1;-t^2)$ $\mathrm{Re}\ \nu > -1$	$\Gamma(\nu+1)\Gamma(2\nu+1)p^{-2}\exp(-2p^{-2})$ $\cdot I_{\nu}(2p^{-2})$ $\quad \mathrm{Re}\ p > 0$

	$f(t)$	$g(p) = \int_0^\infty f(t)e^{-pt}dt$
21.8	$t^{\mu-1}{}_1F_2(1;\mu,\nu;at)$ $\mathrm{Re}\,\nu > 1,\ \mathrm{Re}(\nu-\mu) > 1$	$(\nu-1)\Gamma(\mu)a^{1-\nu}p^{\nu-\mu-1}e^{a/p}$ $\cdot\gamma(\nu-1,a/p)$
21.9	$t^{\nu-1}{}_1F_2(-n;\mu,\nu;at)$ $\mathrm{Re}\,\nu>0,\ n=0,1,2,\cdots$	$n![(\mu)_n]^{-1}\Gamma(\nu)p^{-\nu}L_n^{\mu-1}(a/p)$ $\mathrm{Re}\,p > 0$
21.10	$t^{\mu-1}{}_1F_2(\tfrac{1}{2}\nu;\nu,\mu;at)$ $\mathrm{Re}\,\mu > 0$	$2^{\nu-1}a^{\frac{1}{2}-\frac{1}{2}\nu}\Gamma(\tfrac{1}{2}+\tfrac{1}{2}\nu)\Gamma(\mu)p^{\frac{1}{2}\nu-\mu-\frac{1}{2}}$ $\cdot\exp(\tfrac{1}{2}a/p)I_{\nu-\frac{1}{2}}(\tfrac{1}{2}a/p)$ $\mathrm{Re}\,p > 0$
21.11	$t^{2\mu-1}{}_1F_2(\nu;\mu,\tfrac{1}{2}+\mu;-a^2t^2)$ $\mathrm{Re}\,\mu > 0$	$\Gamma(2\mu)p^{2\nu-2\mu}(p^2+4a^2)^{-\nu}$ $\mathrm{Re}\,p > 2\lvert\mathrm{Im}\,a\rvert$
21.12	$t^{\nu-1}{}_2F_2(-n,n;\nu,\tfrac{1}{2};t)$ $\mathrm{Re}\,\nu > 0,\ n=0,1,2,\cdots$	$\Gamma(\nu)p^{-\nu}\cos[2n\arcsin(p^{-\frac{1}{2}})]$ $\mathrm{Re}\,p > 0$
21.13	$t^{\nu-1}{}_2F_2(-n,n+1;\nu,\tfrac{3}{2};t)$ $\mathrm{Re}\,\nu > 0,\ n=0,1,2,\cdots$	$(2n+1)^{-1}\Gamma(\nu)p^{-\nu}$ $\cdot\sin[(2n+1)\arcsin(p^{-\frac{1}{2}})]$ $\mathrm{Re}\,p > 0$

	$f(t)$	$g(p) = \int_0^\infty f(t)e^{-pt}dt$
21.14	$t^{\nu-1}{}_2F_2(-n,n+1;1,\nu;t)$ $\mathrm{Re}\ \nu > 0$	$\Gamma(\nu)p^{-\nu}P_n(1-2/p)$ $\mathrm{Re}\ p > 0$
21.15	$t^{\nu-1}{}_2F_2(-n,\nu+n;\nu,\mu;t)$ $\mathrm{Re}\ \nu > 0,\ n=0,1,2,\cdots$	$\Gamma(\nu)p^{-\nu}{}_2F_1(-n,\nu+n;\mu;p^{-1})$ $\mathrm{Re}\ p > 0$
21.16	$t^{\nu-1}{}_2F_2(-n,n+2\mu;\tfrac{1}{2}+\mu;\nu;t)$ $\mathrm{Re}\ \nu > 0,\ n=0,1,2,\cdots$	$nB(n,2\mu)\Gamma(\nu)p^{-\nu}$ $\cdot C_n^{\mu}(1-2/p)$ $\mathrm{Re}\ p > 0$
21.17	$t^{\mu-1}{}_2F_2(\nu,\tfrac{1}{2}+\nu;\tfrac{1}{2}+2\nu,\mu;at)$ $\mathrm{Re}\ \mu > 0$	$\pi^{-\frac{1}{2}}2^{2\nu}a^{-\nu}[\Gamma(2\nu)]^{-1}\Gamma(\mu)$ $\cdot\Gamma(\tfrac{1}{2}+2\nu)p^{\nu-\mu}Q_{2\nu-1}[(p/a)^{\frac{1}{2}}]$ $\mathrm{Re}\ p > 0$
21.18	$t^{\mu+\nu-1}e^{-\frac{1}{2}t^2}$ $\cdot{}_2F_2(\mu,\nu;\tfrac{1}{2}\mu+\tfrac{1}{2}\nu,$ $\tfrac{1}{2}\mu+\tfrac{1}{2}\nu+\tfrac{1}{2};\tfrac{1}{4}t^2)$	$\Gamma(\mu+\nu)e^{\frac{1}{4}p^2}D_{-\mu}(p)D_{-\nu}(p)$ $\mathrm{Re}(\mu+\nu) > 0$
21.19	$t^{\mu-1}{}_0F_3(\ ;\nu,\mu,\tfrac{1}{2}+\mu;-a^2t^2)$ $\mathrm{Re}\ \mu > 0$	$\Gamma(\nu)\Gamma(2\mu)(2a)^{-1\nu}p^{\nu-2\mu-1}$ $\cdot J_{\nu-1}(4a/p)$ $\mathrm{Re}\ p > 0$

	$f(t)$	$g(p) = \int_0^\infty f(t)e^{-pt}dt$
21.20	$t^{2\mu-1}{}_0F_3(\,;\nu,\mu,\tfrac{1}{2}+\mu;a^2t^2)$ $\mathrm{Re}\ \mu > 0$	$\Gamma(\nu)\Gamma(2\mu)(2a)^{1-\nu}p^{\nu-2\mu-1}$ $\cdot I_{\nu-1}(4a/p) \qquad \mathrm{Re}\ p > 0$
21.21	$t^{3\mu-1}{}_1F_3(\nu;\mu,$ $\mu+1/3,\mu+2/3;-a^3t^3)$ $\mathrm{Re}\ \mu > 0$	$\Gamma(3\mu)a^{3\nu-3\mu}(p^3+27a^3)^{-\nu}$ $\mathrm{Re}\ p > 3\lvert \mathrm{Im} a\rvert$
21.22	$t^{2\mu}{}_1F_4(1;3/2,3/2+\nu,$ $\mu+\tfrac{1}{2},\mu+1;a^2t^2)$ $\mathrm{Re}\ \mu > -\tfrac{1}{2}$	$\tfrac{1}{2}\pi^{\frac{1}{2}}\Gamma(3/2+\nu)\Gamma(1+2\mu)(2a)^{-\nu-1}$ $\cdot p^{\nu-2\mu}\mathbf{L}_\nu(4a/p) \qquad \mathrm{Re}\ p > 0$
21.23	$t^{2\mu}{}_1F_4(1;3/2,3/2+\nu,$ $\tfrac{1}{2}+\mu,\mu+1;-a^2t^2)$ $\mathrm{Re}\ \mu > -\tfrac{1}{2}$	$\tfrac{1}{2}\pi^{\frac{1}{2}}\Gamma(3/2+\nu)\Gamma(1+2\mu)(2a)^{-\nu-1}$ $\cdot p^{\nu-2\mu}\mathbf{H}_\nu(4a/p) \qquad \mathrm{Re}\ p > 0$
21.24	$t^{2\alpha-1}{}_3F_2(1,\tfrac{1}{2}-\mu+\nu,$ $\tfrac{1}{2}-\mu-\nu;\alpha,\tfrac{1}{2}+\alpha;-k^2t^2)$ $\mathrm{Re}\ \alpha > 0,\ \mathrm{Re}\ k > 0$	$\Gamma(2\alpha)k^{2\mu-1}p^{1-2\alpha-2\mu}S_{2\mu,2\nu}(p/k)$ $\mathrm{Re}\ p > 0$

	$f(t)$	$g(p) = \int_0^\infty f(t)e^{-pt}dt$
21.25	$t^{2\alpha-1}\,{}_4F_3(\tfrac{1}{2}-\mu+\nu,\tfrac{1}{2}-\mu-\nu,$ $\tfrac{1}{2}-\mu,1-\mu;1-2\mu,\alpha,\tfrac{1}{2}+\alpha;$ $-k^2t^2)$ $\operatorname{Re} k > 0$, $\operatorname{Re} \alpha > 0$	$\Gamma(2\alpha)k^{2\mu}p^{-2\alpha-2\mu}$ $\cdot W_{\mu,\nu}(ip/k)W_{\mu,\nu}(-ip/k)$ $\operatorname{Re} p > 0$
21.26	$t^{\gamma}\,{}_rF_s[\alpha_1,\alpha_2,\cdots,\alpha_r;$ $\beta_1,\beta_2,\cdots,\beta_s;(kt)^n]$ $n=1,2,3,\cdots,r+n\leqq s+1$ $\operatorname{Re} \gamma > -1$	$\Gamma(1+\gamma)p^{-\gamma-1}\,{}_{r+n}F_s[\alpha_1,\alpha_2,\cdots,\alpha_r,$ $\frac{\gamma+1}{n},\cdots,\frac{\gamma+n}{n};\beta_1,\beta_2,\cdots,\beta_s;$ $(nk/p)^n]$ $\operatorname{Re} p > 0$ if $r+n \leqq s$
21.27	$(1-e^{-t})^{\nu}\,{}_rF_s(\alpha_1,\alpha_2,\cdots,\alpha_r;$ $\beta_1,\beta_2,\cdots,\beta_s;ke^{-t})$ $\operatorname{Re} \nu > -1$, $r \leqq s$	$B(\nu+1,p)\,{}_{r+1}F_{s+1}(\alpha_1,\alpha_2,\cdots,\alpha_r,p;$ $\beta_0,\beta_1,\cdots,\beta_s,p+\nu+1;k)$ valid for $r = s+1$ if $\lvert k\rvert < 1$, $\operatorname{Re} p > 0$
21.28	$(1-e^{-t})^{\nu}\,{}_rF_s[\alpha_1,\alpha_2,\cdots,\alpha_r;$ $\beta_1,\beta_2,\cdots,\beta_s;k(1-e^{-t})]$ $\operatorname{Re} \nu > -1$, $r \leqq s$	$B(\nu+1,p)\,{}_{r+1}F_{s+1}(\alpha_1,\alpha_2,\cdots,\alpha_r,\nu+1;$ $\beta_1,\beta_2,\cdots,\beta_s,p+\nu+1;k)$ valid for $r = s+1$ if $\lvert k\rvert < 1$, $\operatorname{Re} p > 0$

	$f(t)$	$g(p) = \int_0^\infty f(t)e^{-pt}dt$
21.29	$t^{\nu}\Phi_2(a,b,c;xt,y)$ $\mathrm{Re}\ \nu > -1$	$p^{-\nu-1}\Gamma(1+\nu)\Psi_3(a,b,\nu+1,c;\ \frac{x}{p},y)$ $\mathrm{Re}\ p > \mathrm{Max}(0,\ \mathrm{Re}\ x)$
21.30	$t^{\nu}\Phi_2(a,b,c;xt,yt)$ $\mathrm{Re}\ \nu > -1$	$p^{-\nu-1}\Gamma(1+\nu)F_1(\nu+1,a,b,c;\ \frac{x}{p},\ \frac{y}{p})$ $\mathrm{Re}\ p > \mathrm{Max}(0,\ \mathrm{Re}\ x,\ \mathrm{Re}\ y)$
21.31	$t^{\nu}\Phi_3(b,c;xt,y)$ $\mathrm{Re}\ \nu > -1$	$p^{-\nu-1}\Gamma(1+\nu)\Psi_4(\nu+1,b,c;\ \frac{x}{p},\ y)$ $\mathrm{Re}\ p > \mathrm{Max}(0,\ \mathrm{Re}\ x)$
21.32	$t^{\nu}\Phi_3(b,c;x,yt)$ $\mathrm{Re}\ \nu > -1$	$p^{-\nu-1}\Gamma(\nu+1)\Phi_2(b,\nu+1,c;x,\ \frac{y}{p})$ $\mathrm{Re}\ p > \mathrm{Max}(0,\ \mathrm{Re}\ y)$
21.33	$t^{\nu-1}\Phi_3(b,\nu;xt,yt)$ $\mathrm{Re}\ \nu > 0$	$p^{-\nu}(1 - \frac{x}{p})^{-b}\Gamma(\nu)e^{y/p}$ $\mathrm{Re}\ p > \mathrm{Max}(0,\ \mathrm{Re}\ x)$
21.34	$t^{\nu}\Phi_3(b,c;xt,yt)$ $\mathrm{Re}\ \nu > -1$	$p^{-\nu-1}\Gamma(1+\nu)\Phi_1(\nu+1,b,c;\ \frac{x}{p},\ \frac{y}{p})$ $\mathrm{Re}\ p > \mathrm{Max}(0,\ \mathrm{Re}\ x)$
21.35	$t^{\nu}\Psi_3(a,b,c,d;x,yt)$ $\mathrm{Re}\ \nu > -1$	$p^{-\nu-1}\Gamma(1+\nu)F_3(a,b,c,\nu+1;d;x,\ \frac{x}{p})$ $\mathrm{Re}\ p > \mathrm{Max}(0,\ \mathrm{Re}\ y)$

	$f(t)$	$g(p) = \int_0^\infty f(t)e^{-pt}dt$
21.36	$t^\nu Y_4(a,b,c;x,yt)$ $\text{Re } \nu > -1$	$p^{-\nu-1}\Gamma(1+\nu)Y_3(a,\nu+1,b,c;x,\frac{y}{p})$ $\text{Re } p > \text{Max}(0, \text{Re } y)$
21.37	$t^\nu Y_1(a,b,c,d;x,yt)$ $\text{Re } \nu > -1$	$p^{-\nu-1}\Gamma(1+\nu)F_2(a,b,\nu+1,c,d;x,\frac{y}{p})$ $\text{Re } p > \text{Max}(0, \text{Re } y)$
21.38	$t^\nu Y_2(a,b,c;xt,y)$ $\text{Re } \nu > -1$	$p^{-\nu-1}\Gamma(1+\nu)Y_1(a,\nu+1,b,c;\frac{x}{p},y)$ $\text{Re } p > \text{Max}(0, \text{Re } x)$
21.39	$t^\nu Y_2(a,b,c;xt,yt)$ $\text{Re } \nu > -1$	$p^{-\nu-1}\Gamma(1+\nu)F_4(\nu+1,a,b,c;\frac{x}{p},\frac{y}{p})$ $\text{Re } p > \text{Max}(0, \text{Re } x, \text{Re } y)$
21.40	$t^\nu \Phi_3(b,c;x,yt^2)$ $\text{Re } \nu > \frac{1}{2}$	$p^{-\nu-1}\Gamma(1+\nu)F_1(\frac{1}{2}+\frac{1}{2}\nu,b,1+\frac{1}{2}\nu,c;$ $4yp^{-2},x)$ $\text{Re } p > 2\lvert\text{Re } y^{\frac{1}{2}}\rvert$
21.41	$t^\nu Y_4(a,b,c;x,yt^2)$ $\text{Re } \nu > \frac{1}{2}$	$p^{-\nu-1}\Gamma(1+\nu)$ $\cdot F_3(a,\frac{1}{2}+\frac{1}{2}\nu,b,1+\frac{1}{2}\nu,c;x,4yp^{-2})$ $\text{Re } p > 2\lvert\text{Re } y^{\frac{1}{2}}\rvert$

Part II. Inverse Laplace Transforms

2.1 General Formulas

	$g(p) = \int_0^\infty f(t)e^{-pt}dt$	$f(t)$
1.1	$g(p)$	$f(t)$
1.2	$g(ap) \qquad a > 0$	$a^{-1}f(t/a)$
1.3	$g(ap-b) \qquad a > 0$	$a^{-1}e^{bt/a}f(t/a)$
1.4	$g(p+a)-g(p)$	$(e^{-at}-1)f(t)$
1.5	$g(p-a)+g(p+a)$	$2f(t)\cosh(at)$
1.6	$g(p-a)-g(p+a)$	$2f(t)\sinh(at)$
1.7	$g(p-ia)+g(p+ia)$	$2f(t)\cos(at)$
1.8	$g(p-ia)-g(p+ia)$	$2if(t)\sin(at)$
1.9	$\Delta_p^n \; g(p)$ $n = 1,2,3,\cdots$	$(e^{-at}-1)^n f(t)$
1.10	$pg(p)$	$f'(t)+f(0)$

	$g(p) = \int_0^\infty f(t)e^{-pt}dt$	$f(t)$
1.11	$p^n g(p)$ $n = 1,2,3,\cdots$	$f^{(n)}(t) + \sum_{k=0}^{n-1} f^{(k)}(0)p^{n-k-1}$
1.12	$p^{-n} g(p)$ $n = 1,2,3,\cdots$	$\int_0^t \cdots \int_0^t f(u)(du)^n$
1.13	$(p-1)(p-2)\cdots(p-n)$ $n = 1,2,3,\cdots$	$(e^t \frac{d}{dt})^n f(t)$ if $f^{(k)}(0)=0$ for $k=0,1,\cdots n-1$
1.14	$e^{-bp/a} g(p/a)$ $a,b > 0$	$0 \qquad t < b/a$ $af(at-b) \qquad t > b/a$
1.15	$(1-e^{-ap})^{-1}(1+e^{-ap})g(p)$ $a > 0$	$f(t)+2 \sum_{n=1}^{[t/a]} f(t-an)$
1.16	$(b+e^{ap})^{-\nu} g(p)$ $a,\nu > 0$	$\sum_{n=0}^{[t/a-\nu]} \binom{-\nu}{n} b^n f(t-a\nu-n)$

	$g(p) = \int_0^\infty f(t)e^{-pt}dt$	$f(t)$
1.17	$(1+be^{-ap})^{\nu}$	$\sum_{n=0}^{[t/a]} \binom{\nu}{n} b^n f(t-an)$
1.18	$g_1(p)g_2(p)$	$\int_0^t f_1(u)f_2(t-u)du$
1.19	$p^{-\frac{1}{2}}g(p^{-1})$	$(\pi t)^{-\frac{1}{2}} \int_0^\infty \cos[2(ut)^{\frac{1}{2}}]f(u)du$
1.20	$p^{-1}g(p^{-1})$	$\int_0^\infty f(u)J_0[2(ut)^{\frac{1}{2}}]du$
1.21	$p^{-\frac{3}{2}}g(p^{-1})$	$\pi^{-\frac{1}{2}} \int_0^\infty u^{-\frac{1}{2}}\sin[2(ut)^{\frac{1}{2}}]f(u)du$
1.22	$p^{-2\nu-1}g(p^{-1})$ $\mathrm{Re}\,\nu > -\frac{1}{2}$	$t^{\nu} \int_0^\infty u^{-\nu}J_{2\nu}[2(ut)^{\frac{1}{2}}]f(u)du$
1.23	$p^{-\frac{1}{2}}g(-p^{-1})$	$(\pi t)^{-\frac{1}{2}} \int_0^\infty \cosh[2(ut)^{\frac{1}{2}}]f(u)du$
1.24	$p^{-\frac{3}{2}}g(-p^{-1})$	$\pi^{-\frac{1}{2}} \int_0^\infty u^{-\frac{1}{2}}\sinh[2(tu)^{\frac{1}{2}}]f(u)du$

	$g(p) = \int_0^\infty f(t)e^{-pt}dt$	$f(t)$
1.25	$p^{-1}g(p+p^{-1})$	$\int_0^t f(u)J_0\{2[u(t-u)]^{\frac{1}{2}}\}du$
1.26	$p^{-2\nu-1}g(p+a/p)$	$\int_0^t f(u)[(t-u)/au]^{\nu}$ $\cdot J_{2\nu}[2(aut-au^2)^{\frac{1}{2}}]du$
1.27	$g(p^{\frac{1}{2}})$	$\frac{1}{2}\pi^{-\frac{1}{2}}t^{-\frac{3}{2}}\int_0^\infty ue^{-\frac{1}{4}u^2/t}f(u)du$
1.28	$p^{-\frac{1}{2}}g(p^{\frac{1}{2}})$	$(\pi t)^{\frac{1}{2}}\int_0^\infty e^{-\frac{1}{4}u^2/t}f(u)du$
1.29	$p^{\frac{1}{2}n-\frac{1}{2}}g(p^{\frac{1}{2}})$ $n = 0,1,2,\cdots$	$(\pi t)^{-\frac{1}{2}}(2t)^{-\frac{1}{2}n}\int_0^\infty e^{-\frac{1}{4}u^2/t}He_n[(2t)^{-\frac{1}{2}}u]f(u)du$
1.30	$p^{\nu}g(p^{\frac{1}{2}})$	$(2/\pi)^{\frac{1}{2}}(2t)^{-\nu-1}\int_0^\infty \exp(-\frac{1}{8}u^2/t)$ $\cdot D_{2\nu+1}[(2t)^{-\frac{1}{2}}u]f(u)du$
1.31	$g(p+p^{\frac{1}{2}})$	$\frac{1}{2}\pi^{-\frac{1}{2}}\int_0^t u(t-u)^{-\frac{3}{2}}\exp[-\frac{1}{4}u^2(t-u)^{-1}]f(u)du$

	$g(p) = \int_0^\infty f(t)e^{-pt}dt$	$f(t)$
1.32	$p^{-\frac{1}{2}}g(p+p^{\frac{1}{2}})$	$\pi^{-\frac{1}{2}}\int_0^t (t-u)^{-\frac{1}{2}}\exp[-\frac{1}{4}u^2(t-u)^{-1}]f(u)du$
1.33	$g[(p^2+a^2)^{\frac{1}{2}}]$	$f(t)-a\int_0^t f(u)u(t^2-u^2)^{-\frac{1}{2}}J_1[a(t^2-u^2)^{\frac{1}{2}}]du$
1.34	$(p^2+a^2)^{-\frac{1}{2}}g[(p^2+a^2)^{\frac{1}{2}}]$	$\int_0^t f(u)J_0[a(t^2-u^2)^{\frac{1}{2}}]du$
1.35	$p(p^2+a^2)^{-\frac{1}{2}}$ $\cdot g[(p^2+a^2)^{\frac{1}{2}}]$	$f(t)-at\int_0^t f(u)(t^2-u^2)^{-\frac{1}{2}}J_1[a(t^2-u^2)^{\frac{1}{2}}]du$
1.36	$(p^2+a^2)^{-\frac{1}{2}}[(p^2+a^2)^{\frac{1}{2}}-p]^{\nu}$ $\cdot g[(p^2+a^2)^{\frac{1}{2}}]$ $R\ \nu>-1$	$a^{\nu}\int_0^t [(t-u)/(t+u)]^{\frac{1}{2}\nu}J_{\nu}[a(t^2-u^2)^{\frac{1}{2}}]f(u)du$
1.37	$g[(p^2-a^2)^{\frac{1}{2}}]$	$f(t)+a\int_0^t f(u)u(t^2-u^2)^{-\frac{1}{2}}I_1[a(t^2-u^2)^{\frac{1}{2}}]du$
1.38	$(p^2-a^2)^{-\frac{1}{2}}g[(p^2-a^2)^{\frac{1}{2}}]$	$\int_0^t f(u)I_0[a(t^2-u^2)^{\frac{1}{2}}]du$

	$g(p) = \int_0^\infty f(t)e^{-pt}dt$	$f(t)$
1.39	$p(p^2-a^2)^{-\frac{1}{2}}$ $\cdot g[(p^2-a^2)^{\frac{1}{2}}]$	$f(t)+at\int_0^t f(u)(t^2-u^2)^{-\frac{1}{2}}I_1[a(t^2-u^2)^{\frac{1}{2}}]du$
1.40	$(p^2-a^2)^{-\frac{1}{2}}[p-(p^2-a^2)^{\frac{1}{2}}]^\nu$ $\cdot g[(p^2-a^2)^{\frac{1}{2}}]$	$a^\nu \int_0^t [(t-u)/(t+u)]^{\frac{1}{2}\nu}I_\nu[a(t^2-u^2)^{\frac{1}{2}}]f(u)du$
1.41	$(p^2+a^2)^{-\frac{1}{2}}[(p^2+a^2)^{\frac{1}{2}}-p]^\nu$ $\cdot g[(p^2+a^2)^{\frac{1}{2}}-p]^\nu$ Re $\nu > -1$	$a^\nu t^{\frac{1}{2}\nu}\int_0^\infty (t+2u)^{-\frac{1}{2}\nu}J_\nu[a(t^2+2ut)^{\frac{1}{2}}]f(u)du$
1.42	$(p^2+a^2)^{-\frac{1}{2}}[(p^2+a^2)^{\frac{1}{2}}-p]^\nu$ $\cdot g[p-(p^2+a^2)^{\frac{1}{2}}]$ Re $\nu > -1$	$a^\nu t^{\frac{1}{2}\nu}\int_0^\infty (t-2u)^{-\frac{1}{2}\nu}J_\nu[a(t^2-2ut)^{\frac{1}{2}}]f(u)du$
1.43	$(p^2-a^2)^{-\frac{1}{2}}[p-(p^2-a^2)^{\frac{1}{2}}]^\nu$ $\cdot g[(p^2-a^2)^{\frac{1}{2}}-p]$ Re $\nu > -1$	$a^\nu t^{\frac{1}{2}\nu}\int_0^\infty (t+2u)^{-\frac{1}{2}\nu}I_\nu[a(t^2+2ut)^{\frac{1}{2}}]f(u)du$

	$g(p) = \int_0^\infty f(t)e^{-pt}dt$	$f(t)$
1.44	$(p^2-a^2)^{-\frac{1}{2}}[p-(p^2-a^2)^{\frac{1}{2}}]^{\nu}$ $\cdot g[p-(p^2-a^2)^{\frac{1}{2}}]$ $\mathrm{Re}\,\nu > -1$	$a^{\nu}t^{\frac{1}{2}\nu}\int_0^\infty (t-2u)^{-\frac{1}{2}\nu}I_{\nu}[a(t^2-2ut)^{\frac{1}{2}}]f(u)du$
1.45	$g(\log p^a)$	$\int_0^\infty [\Gamma(au)]^{-1}t^{au-1}f(u)du$
1.46	$p^{-1}g(\log p)$	$\int_0^\infty [\Gamma(1+u)]^{-1}t^{u}f(u)du$
1.47	$g^{(n)}(p)$ $n = 0,1,2,\cdots$	$(-1)^n t^n f(t)$
1.48	$p^n g^{(m)}(p)$ $m \geqq n, m,n = 0,1,2,\cdots$	$(-1)^m \frac{d^n}{dt^n}[t^m f(t)]$
1.49	$\frac{\partial}{\partial a} g(s,a)$	$\frac{\partial}{\partial a} f(t,a)$
1.50	$\int_p^\infty g(u)du$	$t^{-1}f(t)$

	$g(p) = \int_0^\infty f(t)e^{-pt}dt$	$f(t)$
1.51	$p^{-1} \int_p^\infty g(u)du$	$\int_0^t z^{-1}f(z)dz$
1.52	$p^{-1} \int_0^p g(u)du$	$\int_t^\infty z^{-1}f(z)dz$
1.53	$\int_p^\infty \cdots \int_p^\infty g(u)(du)^n$	$t^{-n}f(t)$
1.54	$(1-e^{-ap})^{-1} \int_0^a e^{-pu}f(u)du$	$f(t+a)=f(t)$
1.55	$(1+e^{-ap})^{-1} \int_0^a e^{-pu}f(u)du$	$f(t+a) = -f(t)$
1.56	$\int_0^\infty (e^{pu}-1)^{-1}f(u)du$	$\sum_{m=1}^{n} n^{-1}f(t/n)$
1.57	$\int_0^\infty e^{-\frac{1}{4}p^2/u^2}g(u^2)du$	$\pi^{\frac{1}{2}}f(t^2)$
1.58	$\int_0^\infty u^{-\frac{1}{2}}e^{-\frac{1}{4}p^2/u^2}g(u)du$	$2\pi^{\frac{1}{2}}f(t^2)$

	$g(p) = \int_0^\infty f(t)e^{-pt}dt$	$f(t)$
1.59	$p\int_0^\infty u^{-3/2}e^{-\frac{1}{4}p^2/u}g(u)du$	$4\pi^{\frac{1}{2}}t\ f(t^2)$
1.60	$\int_0^\infty J_0[2(pu)^{\frac{1}{2}}]g(u)du$	$t^{-1}f(t^{-1})$
1.61	$p^{-\frac{1}{2}}\int_0^\infty u^{\frac{1}{2}}J_1[2(up)^{\frac{1}{2}}]g(u)du$	$f(t^{-1})$
1.62	$\int_0^\infty J_p(au)g(u)du$ $a > 0$	$f(a\ \sinh t)$
1.63	$\int_0^\infty u^{n-2}e^{-\frac{1}{4}p^2u^2}He_n(2^{-\frac{1}{2}}up)$ $\cdot g(u^{-2})du$	$\pi^{\frac{1}{2}}2^{\frac{1}{2}n}t^n f(t^2)$ $n = 0,1,2,\cdots$
1.64	$\int_0^\infty t^{\nu-2}e^{-\frac{1}{4}p^2u^2}D_\nu(pu)$ $\cdot g(\frac{1}{2}u^{-2})du$	$2^{\frac{1}{2}}\pi^{\frac{1}{2}}t^\nu f(t^2)$

2.2 Rational Functions

	$g(p) = \int_0^\infty f(t)e^{-pt}dt$	$f(t)$
2.1	p^{-1}	1
2.2	$(p+a)^{-1}$	e^{-at}
2.3	$p^{-n} \quad n = 1,2,3,\cdots$	$[(n-1)!]^{-1}t^{n-1}$
2.4	$(p+a)^{-n} \quad n = 1,2,3,\cdots$	$[(n-1!]^{-1}t^{n-1}e^{-at}$
2.5	$p^{-1}(p+a)^{-1}$	$a^{-1}(1-e^{-at})$
2.6	$p^{-2}(p+a)^{-1}$	$a^{-2}(at-1+e^{-at})$
2.7	$p^{-3}(p+a)^{-1}$	$a^{-3}(1-e^{-at})-a^{-2}t+\frac{1}{2}a^{-1}t^2$
2.8	$p(p+a)^{-2}$	$e^{-at}(1-at)$
2.9	$p^{-1}(p+a)^{-2}$	$a^{-2}[1-(1+at)e^{-at}]$
2.10	$p^{-2}(p+a)^{-2}$	$a^{-2}t(1+e^{-at})-2a^{-3}(1-e^{-at})$
2.11	$p(p+a)^{-3}$	$te^{-at}(1-\frac{1}{2}at)$

	$g(p) = \int_0^\infty f(t)e^{-pt}dt$	$f(t)$
2.12	$p^2(p+a)^{-3}$	$e^{-at}(1-2at+\frac{1}{2}a^2t^2)$
2.13	$p^{-1}(p+a)^{-3}$	$a^{-3}(1-e^{-at})-a^{-1}ate^{-at}(a^{-1}+\frac{1}{2}t)$
2.14	$p^{-2}(p+a)^{-3}$	$2a^{-3}t+\frac{1}{2}a^{-2}t^2(1+e^{-at})-3a^{-4}(1-e^{-at})$
2.15	$[(p+a)(p+b)]^{-1}$	$(b-a)^{-1}[e^{-at}-e^{-bt}]$
2.16	$p[(p+a)(p+b)]^{-1}$	$(b-a)^{-1}[be^{-bt}-ae^{-at}]$
2.17	$p^{-1}[(p+a)(p+b)]^{-1}$	$(ab)^{-1}[1+(a-b)^{-1}(be^{-at}-ae^{-bt})]$
2.18	$p^{-2}[(p+a)(p+b)]^{-1}$	$(ab)^{-2}(abt-a-b)$ $+(a-b)^{-1}(b^{-2}e^{-bt}-a^{-2}e^{-at})$
2.19	$[(p+a)(p+b)(p+c)]^{-1}$	$[(b-a)(c-a)]^{-1}e^{-at}+[(a-b)(c-b)]^{-1}e^{-bt}$ $+[(a-c)(b-c)]^{-1}e^{-ct}$
2.20	$p[(p+a)(p+b)$ $\cdot(p+c)]^{-1}$	$a[(a-b)(c-a)]^{-1}e^{-at}$ $+b[(b-a)(c-b)]^{-1}e^{-bt}$ $+c[(c-a)(b-c)]^{-1}e^{-ct}$

	$g(p) = \int_0^\infty f(t)e^{-pt}dt$	$f(t)$
2.21	$p^2[(p+a)(p+b) \cdot (p+c)]^{-1}$	$a^2[(a-b)(a-c)]^{-1}e^{-at} + b^2[(b-a)(b-c)]^{-1}e^{-bt} + c^2[(c-a)(c-b)]^{-1}e^{-ct}$
2.22	$p^{-1}[(p+a)(p+b) \cdot (p+c)]^{-1}$	$(abc)^{-1} - [a(a-b)(a-c)]^{-1}e^{-at} - [b(b-a)(b-c)]^{-1}e^{-bt} - [c(c-a)(c-b)]^{-1}e^{-ct}$
2.23	$[(p+a)(p+b) \cdot (p+c)(p+d)]^{-1}$	$[(b-a)(c-a)(d-a)]^{-1}e^{-at} + [(a-b)(c-b)(d-b)]^{-1}e^{-bt} + [(c-a)(c-b)(d-c)]^{-1}e^{-ct}$
2.24	$(p+a)^{-2}(p+b)^{-1}$	$(b-a)^{-2}\{e^{-bt}+[(b-a)t-1]e^{-at}\}$
2.25	$p(p+a)^{-2}(p+b)^{-1}$	$(a-b)^{-2}\{a(a-b)te^{-at}-b[e^{-bt}-e^{-at}]\}$
2.26	$p^2(p+a)^{-2}(p+b)^{-1}$	$[(a-b)^{-2}a(a-2b)-a^2t(a-b)^{-1}]e^{-at} + b^2(a-b)^{-2}e^{-bt}$
2.27	$p^{-1}(p+a)^{-2}(p+b)^{-1}$	$(b-a)^{-2}\{b^{-1}(1-e^{-bt})-a^{-1}(1-e^{-at}) + (b-a)[a^{-2}(1-e^{-at})-a^{-1}te^{-at}]\}$

	$g(p) = \int_0^\infty f(t)e^{-pt}dt$	$f(t)$
2.28	$(p+a)^{-1}(p+b)^{-3}$	$(a-b)^{-3}\{e^{-bt}[1-(a-b)t+\frac{1}{2}(a-b)^2t^2]-e^{-at}\}$
2.29	$p(p+a)^{-1}(p+b)^{-3}$	$(a-b)^{-3}\{a(e^{-at}-e^{-bt})$ $+e^{-bt}[a(a-b)-\frac{1}{2}b(a-b)^{-1}t^2]\}$
2.30	$[(p+a)(p+b)]^{-2}$	$(a-b)^{-2}t(e^{-at}+e^{-bt})$ $+2(a-b)^{-3}(e^{-at}-e^{-bt})$
2.31	$(p^2+a^2)^{-1}$	$a^{-1}\sin(at)$
2.32	$(p^2-a^2)^{-1}$	$a^{-1}\sinh(at)$
2.33	$p(p^2+a^2)^{-1}$	$\cos(at)$
2.34	$p(p^2-a^2)^{-1}$	$\cosh(at)$
.35	$p^{-1}(p^2+a^2)^{-1}$	$2a^{-2}\sin^2(\frac{1}{2}at)$
.36	$p^{-1}(p^2-a^2)^{-1}$	$2a^{-2}\sinh^2(\frac{1}{2}at)$
.37	$p^{-2}(p^2+a^2)^{-1}$	$a^{-3}[at-\sin(at)]$

	$g(p) = \int_0^\infty f(t)e^{-pt}dt$	$f(t)$
2.38	$p^{-2}(p^2-a^2)^{-1}$	$a^{-3}[\sinh(at)-at]$
2.39	$(p^2+a^2)^{-2}$	$\frac{1}{2}a^{-2}[a^{-1}\sin(at)-t\cos(at)]$
2.40	$(p^2-a^2)^{-2}$	$\frac{1}{2}a^{-2}[t\cosh(at)-a^{-1}\sinh(at)]$
2.41	$p(p^2+a^2)^{-2}$	$\frac{1}{2}a^{-1}t\sin(at)$
2.42	$p(p^2-a^2)^{-2}$	$\frac{1}{2}a^{-1}t\sinh(at)$
2.43	$p^{-1}(p^2+ap+b)^{-1}$	$b^{-1}+(r_1-r_2)^{-1}(r_1^{-1}e^{r_1t}-r_2^{-1}e^{r_2t})$
	r_1 and r_2 are the roots of $p^2+ap+b = 0$	
2.44	$(p^2+ap+b)^{-1}$ r_1, r_2 as before	$(r_1-r_2)^{-1}(e^{r_1t}-e^{r_2t})$
2.45	$(p^3-a^3)^{-1}$	$\frac{1}{3}a^{-2}[e^{at}-e^{-\frac{1}{2}at}\cos(3^{\frac{1}{2}}at/2)$ $-3^{\frac{1}{2}}e^{-\frac{1}{2}at}\sin(3^{\frac{1}{2}}at/2)]$

	$g(p) = \int_0^\infty f(t)e^{-pt}dt$	$f(t)$
2.46	$(p^3+a^3)^{-1}$	$\frac{1}{3}a^{-2}[e^{-at}-e^{\frac{1}{2}at}\cos(3^{\frac{1}{2}}at/2)$ $+3^{\frac{1}{2}}e^{\frac{1}{2}at}\sin(3^{\frac{1}{2}}at/2)]$
2.47	$p(p^3+a^3)^{-1}$	$\frac{1}{3}a^{-2}[ae^{\frac{1}{2}at}\cos(3^{\frac{1}{2}}at/2)-ae^{-at}$ $+3^{\frac{1}{2}}a\ e^{\frac{1}{2}at}\sin(3^{\frac{1}{2}}at/2)]$
2.48	$p(p^3-a^3)^{-1}$	$\frac{1}{3}a^{-2}[ae^{at}-ae^{-\frac{1}{2}at}\cos(3^{\frac{1}{2}}at/2)$ $+3^{\frac{1}{2}}a\ e^{-\frac{1}{2}at}\sin(3^{\frac{1}{2}}at/2)]$
2.49	$p^2(p^3+a^3)^{-1}$	$\frac{1}{3}e^{-at}+\frac{2}{3}e^{\frac{1}{2}at}\cos(3^{\frac{1}{2}}at/2)$
2.50	$p^2(p^3-a^3)^{-1}$	$\frac{1}{3}e^{at}+\frac{2}{3}e^{-\frac{1}{2}at}\cos(3^{\frac{1}{2}}at/2)$
2.51	$(p^4+a^4)^{-1}$	$2^{-\frac{1}{2}}a^{-3}[\sin(2^{-\frac{1}{2}}at)\cosh(2^{-\frac{1}{2}}at)$ $-\cos(2^{-\frac{1}{2}}at)\sinh(2^{-\frac{1}{2}}at)]$
2.52	$(p^4-a^4)^{-1}$	$\frac{1}{2}\ a^{-3}[\sinh(at)-\sin(at)]$
2.53	$p(p^4+a^4)^{-1}$	$a^{-2}\sin(2^{-\frac{1}{2}}at)\sinh(2^{-\frac{1}{2}}at)$

	$g(p) = \int_0^\infty f(t)e^{-pt}dt$	$f(t)$
2.54	$p(p^4-a^4)^{-1}$	$\frac{1}{2}a^{-2}[\cosh(at)-\cos(at)]$
2.55	$p^2(p^4+a^4)^{-1}$	$2^{-\frac{1}{2}}a^{-1}[\cos(2^{-\frac{1}{2}}at)\sinh(2^{-\frac{1}{2}}at)$ $+\sin(2^{-\frac{1}{2}}at)\cosh(2^{-\frac{1}{2}}at)]$
2.56	$p^2(p^4-a^4)^{-1}$	$\frac{1}{2}a^{-1}[\sinh(at)+\sin(at)]$
2.57	$p^3(p^4+a^4)^{-1}$	$\cos(2^{-\frac{1}{2}}at)\cosh(2^{-\frac{1}{2}}at)$
2.58	$p^3(p^4-a^4)^{-1}$	$\frac{1}{2}[\cos(at)+\cosh(at)]$
2.59	$(p^n+a^n)^{-1}$ $n = 1,2,3,\cdots$	$n^{-1}a^{1-n}\sum_{k=1}^{n} z\, e^{-azt}$ $z = \exp[i\pi(2k-1)/n]$
2.60	$(p^{2n}+a^{2n})^{-1}$ $n = 1,2,3,\cdots$	$-n^{-1}a^{1-2n}\sum_{k=1}^{n} z\sinh(azt)$ $z = \exp[i\pi(k-\frac{1}{2})/n]$
2.61	$(p^n-a^n)^{-1}$ $n = 1,2,3,\cdots$	$n^{-1}a^{1-n}\sum_{k=1}^{n} z\, e^{azt}$ $z = e^{i2\pi k/n}$

	$g(p) = \int_0^\infty f(t)e^{-pt}dt$	$f(t)$
2.62	$(p^{2n}-a^{2n})^{-1}$ $n = 1,2,3,\cdots$	$n^{-1}a^{1-2n} \sum_{k=1}^{n} z \sinh(azt)$ $z = e^{i\pi k/n}$
2.63	$[(p+a)(p^2+b^2)]^{-1}$	$(a^2+b^2)^{-1}[e^{-at}-\cos(bt)+ab^{-1}\sin(bt)]$
2.64	$[(p+a)(p^2-b^2)]^{-1}$	$(a^2-b^2)^{-1}[e^{-at}-\cosh(bt)+ab^{-1}\sinh(bt)]$
2.65	$[(p^2+a^2)(p^2+b^2)]^{-1}$	$(ab)^{-1}(a^2-b^2)^{-1}[a \sin(bt)-b \sin(at)]$
2.66	$[(p^2+a^2)(p^2-b^2)]^{-1}$	$(ab)^{-1}(a^2+b^2)^{-1}[a \sinh(bt)-b \sin(at)]$
2.67	$[(p^2-a^2)(p^2-b^2)]^{-1}$	$(ab)^{-1}(a^2-b^2)^{-1}[b \sinh(at)-a \sinh(bt)]$
2.68	$p[(p^2+a^2)(p^2+b^2)]^{-1}$	$(a^2-b^2)^{-1}[\cos(bt)-\cos(at)]$
2.69	$p[(p^2+a^2)(p^2-b^2)]^{-1}$	$(a^2+b^2)^{-1}[\cosh(bt)-\cos(at)]$
2.70	$p^2[(p^2+a^2)(p^2+b^2)]^{-1}$	$(a^2-b^2)^{-1}[a \sin(at)-b \sin(bt)]$
2.71	$p^2[(p^2+a^2)(p^2-b^2)]^{-1}$	$(a^2+b^2)^{-1}[b \sinh(bt)+a \sin(at)]$

	$g(p) = \int_0^\infty f(t)e^{-pt}dt$	$f(t)$
2.72	$p^{-1}(a+p)^{-n}$ $n = 1,2,3 \cdots$	$a^{-n}\{1-e^{-at}[1+\frac{at}{1!} + \cdots + \frac{(at)^{n-1}}{(n-1)!}]\}$
2.73	$p^{-n-1}(p-1)^{n}$ $n = 0,1,2,\cdots$	$L_n(t)$
2.74	$p^{-1}(1+ap)_n$ $n = 0,1,2,\cdots$	$(n!)^{-1}(1-e^{-t/a})^{n}$
2.75	$(p+\frac{1}{2})^{-n-1}(p-\frac{1}{2})^{n}$ $n = 0,1,2,\cdots$	$e^{-\frac{1}{2}t}L_n(t)$
2.76	$[P(p)]^{-1}$ $P(p)=(p+a_1)(p+a_2)$ $\cdots(p+a_n)$	$\sum_{m=1}^{n} e^{-a_m t}[P_m(-a_m)]^{-1}$ $P_m(p) = P(p)(p+a_m)^{-1}$
2.77	$[P(p)]^{-1}$ $P(p)=(p-a_1)(p-a_2)$ $\cdots(p-a_n)$ $a_i \neq a_k$ for $i \neq k$	$\sum_{m=1}^{n} e^{a_m t}[P_m(a_m)]^{-1}$ $P_m(p) = P(p)(p-a_m)^{-1}$

	$g(p) = \int_0^\infty f(t)e^{-pt}dt$	$f(t)$
2.78	$p^k[P(p)]^{-1}$ $P(p) = (p-a_1)(p-a_2)\cdots(p-a_n)$ $k \leqq n-1,\ a_i \neq a_k$ for $i \neq k$	$\sum_{m=1}^{n} a_m^k e^{a_m t} [P_n(a_m)]^{-1}$ $P_m(p) = P(p)(p-a_m)^{-1}$
2.79	$\{(p^2+a^2)(p^2+3^2a^2)\cdots[p^2+(2n+1)^2a^2]\}^{-1}$	$[(2n+1)!]^{-1}a^{-2n-1}\sin^{2n+1}(at)$ $n = 0,1,2,\cdots$
2.80	$[p(p^2+2^2a^2)(p^2+4^2a^2)\cdots(p^2+4n^2a^2)]^{-1}$	$[(2n)!]^{-1}a^{-2n}\sin^{2n}(at)$ $n = 1,2,3,\cdots$
2.81	$\{(p^2-a^2)(p^2-3a^2)\cdots[p^2-(2n+1)^2a^2]\}^{-1}$	$[(2n+1)!]^{-1}a^{-2n-1}\sinh^{2n+1}(at)$ $n = 0,1,2,\cdots$
2.82	$\{p(p^2-2^2a^2)(p^2-4a^2)\cdots[p^2-(2n)^2a^2]\}^{-1}$	$[(2n)!]^{-1}a^{-2n}\sinh^{2n}(at)$ $n = 1,2,3,\cdots$
2.83	$p(p^2+2^2a^2)(p^2+4^2a^2)\cdots[p^2+(2n)^2a^2]$ $\cdot\{(p^2+a^2)(p^2+3^2a^2)\cdots[p^2+(2n+1)^2a^2]\}^{-1}$	$P_{2n+1}[\cos(at)]$ $n = 1,2,3,\cdots$

	$g(p) = \int_0^\infty f(t)e^{-pt}dt$	$f(t)$
2.84	$(p^2+a^2)(p^2+3^2a^2)$ $\cdots[p^2+(2n-1)^2a^2]$ $\cdot\{p(p^2+2^2a^2)\cdots[p^2+(2n)^2a^2]\}^{-1}$	$P_{2n}[\cos(at)]$ $n = 1,2,3,\cdots$
2.85	$p(p^2-2^2a^2)(p^2-4^2a^2)$ $\cdots[p^2-(2n)^2a^2]$ $\cdot\{(p^2-a^2)(p^2-3^2a^2)$ $\cdots[p^2-(2n+1)^2a^2]\}^{-1}$	$P_{2n+1}[\cosh(at)]$ $n = 1,2,3,\cdots$
2.86	$(p^2-a^2)(p^2-3^2a^2)$ $\cdots[p^2-(2n-1)^2a^2]$ $\cdot\{p(p^2-2^2a^2)$ $\cdots[p^2-(2n)^2a^2]\}^{-1}$	$P_{2n}[\cosh(at)]$ $n = 1,2,3,\cdots$
2.87	$(p-1)(p-2)$ $\cdots(p-n+1)$ $\cdot\{(p+n)(p+n-2)$ $\cdots(p-n+2)\}^{-1}$	$P_n(e^{-t})$ $n = 2,3,4,\cdots$

2.3 Irrational Algebraic Functions

	$g(p) = \int_0^\infty f(t)e^{-pt}dt$	$f(t)$
3.1	$p^{-\frac{1}{2}}$	$(\pi t)^{-\frac{1}{2}}$
3.2	$p^{-3/2}$	$2(t/\pi)^{\frac{1}{2}}$
3.3	$p^{-n-\frac{1}{2}}$ $n = 0,1,2,\cdots$	$\pi^{-\frac{1}{2}}2^{2n}n![(2n)!]^{-1}t^{n-\frac{1}{2}}$
3.4	$(p+a)^{-\frac{1}{2}}$	$(\pi t)^{-\frac{1}{2}}e^{-at}$
3.5	$(p+a)^{-n-\frac{1}{2}}$ $n = 0,1,2,\cdots$	$\pi^{-\frac{1}{2}}2^{2n}n![(2n)!]^{-1}e^{-at}t^{n-\frac{1}{2}}$
3.6	$(p+a)^{\frac{1}{2}}-(p+b)^{\frac{1}{2}}$	$\frac{1}{2}t^{-1}(\pi t)^{-\frac{1}{2}}(e^{-bt}-e^{-at})$
3.7	$(p-a)^{\frac{1}{2}}-(p-b)^{\frac{1}{2}}$	$\frac{1}{2}t^{-1}(\pi t)^{-\frac{1}{2}}(e^{bt}-e^{at})$
3.8	$p^{-1}(p+a)^{\frac{1}{2}}$	$(\pi t)^{-\frac{1}{2}}e^{-at}+a^{\frac{1}{2}}\mathrm{Erf}[(at)^{\frac{1}{2}}]$
3.9	$p^{-3/2}(p+a)^{\frac{1}{2}}$	$e^{-\frac{1}{2}at}[(1+at)I_0(\frac{1}{2}at)+atI_1(\frac{1}{2}at)]$
3.10	$p^{-1}(p+a)^{-\frac{1}{2}}$	$a^{-\frac{1}{2}}\mathrm{Erf}[(at)^{\frac{1}{2}}]$

	$g(p) = \int_0^\infty f(t)e^{-pt}dt$	$f(t)$
3.11	$p^{-1}(p+b)^{\frac{1}{2}}(p+a)^{-\frac{1}{2}}$	$\exp[-\frac{1}{2}(a+b)t]$ $+b\int_0^t \exp[-\frac{1}{2}(a+b)u]I_0(\frac{1}{2}au-\frac{1}{2}bu)du$
3.12	$p^{-\frac{1}{2}}(p-a)^{-1}$	$a^{-\frac{1}{2}}e^{at}\mathrm{Erf}[(at)^{\frac{1}{2}}]$
3.13	$p^{-3/2}(p-a)^{-1}$	$a^{-3/2}e^{at}\mathrm{Erf}[(at)^{\frac{1}{2}}]-2a^{-1}(t/\pi)^{\frac{1}{2}}$
3.14	$p^{-3/2}(p+a)$	$(\pi t)^{-\frac{1}{2}}(1+2at)$
3.15	$(p+b)^{-1}(p+a)^{\frac{1}{2}}$	$(\pi t)^{-\frac{1}{2}}e^{-at}+(a-b)^{\frac{1}{2}}\mathrm{Erf}[(a-b)^{\frac{1}{2}}t^{\frac{1}{2}}]$
3.16	$(p+a)^{-1}(p+b)^{-\frac{1}{2}}$	$(b-a)^{-\frac{1}{2}}e^{-at}\mathrm{Erf}[(b-a)^{\frac{1}{2}}t^{\frac{1}{2}}]$
3.17	$[(p+a)/(p-a)]^{\frac{1}{2}}-1$	$a[I_0(at)+I_1(at)]$
3.18	$[(p+a)(p+b)]^{-\frac{1}{2}}$	$\exp[-\frac{1}{2}(a+b)t]I_0(\frac{1}{2}at-\frac{1}{2}bt)$
3.19	$(p+a)^{-\frac{1}{2}}(p+b)^{-3/2}$	$t\exp[-\frac{1}{2}(a+b)t][I_0(\frac{1}{2}at-\frac{1}{2}bt)$ $+I_1(\frac{1}{2}at-\frac{1}{2}bt)]$

	$g(p) = \int_0^\infty f(t)e^{-pt}dt$	$f(t)$
3.20	$(p+a)^{\frac{1}{2}}(p+b)^{-3/2}$	$\exp[-\frac{1}{2}(a+b)t]\{[1+(a-b)t]I_0(\frac{1}{2}at-\frac{1}{2}bt)$ $+(a-b)tI_1(\frac{1}{2}at-\frac{1}{2}bt)\}$
3.21	$p^{-\frac{1}{2}}(p^{\frac{1}{2}}+a)^{-1}$	$e^{a^2t}\mathrm{Erfc}(at^{\frac{1}{2}})$
3.22	$p^{-1}(p^{\frac{1}{2}}+a)^{-1}$	$a^{-1}[1-a^2t\mathrm{Erfc}(at^{\frac{1}{2}})]$
3.23	$p^{-3/2}(p^{\frac{1}{2}}+a)$	$a^{-1}2(t/\pi)^{\frac{1}{2}}+a^{-2}e^{a^2t}\mathrm{Erfc}(at^{\frac{1}{2}})-a^{-2}$
3.24	$(p^{\frac{1}{2}}+a)^{-1}$	$(\pi t)^{-\frac{1}{2}}-ae^{a^2t}\mathrm{Erfc}(at^{\frac{1}{2}})$
3.25	$(p^{\frac{1}{2}}+a)^{-2}$	$(-2a(t/\pi)^{\frac{1}{2}}+(1-2a^2t)e^{a^2t}$ $\cdot[\mathrm{Erf}(at^{\frac{1}{2}})-1]$
3.26	$p^{-1}(p^{\frac{1}{2}}+a)^{-2}$	$a^{-2}+(2t-a^{-2})e^{a^2t}\mathrm{Erfc}(at^{\frac{1}{2}})$ $-2\ a^{-1}(t/\pi)^{\frac{1}{2}}$
3.27	$p^{-\frac{1}{2}}(p^{\frac{1}{2}}+a)^{-2}$	$2(t/\pi)^{-\frac{1}{2}}-2at\ e^{a^2t}\mathrm{Erfc}(at^{\frac{1}{2}})$
3.28	$(p^{\frac{1}{2}}+a)^{-3}$	$2(1+a^2t)(t/\pi)^{\frac{1}{2}}-at(3+2a^2t)$ $\cdot\mathrm{Erfc}(at^{\frac{1}{2}})$

	$g(p) = \int_0^\infty f(t)e^{-pt}dt$	$f(t)$
3.29	$p^{-1}(p^{\frac{1}{2}}+a)^{-3}$	$a^{-3}+2a^{-2}(t/\pi)^{\frac{1}{2}}(a^2t-1)$ $-(2at^2-a^{-1}t+a^{-3})e^{a^2t}\mathrm{Erfc}(at^{\frac{1}{2}})$
3.30	$p^{\frac{1}{2}}(p^{\frac{1}{2}}+a)^{-3}$	$-2a(t/\pi)^{\frac{1}{2}}(a^2t+2)+(1+2a^4t^2+5a^2t)$ $\cdot e^{a^2t}\mathrm{Erfc}(at^{\frac{1}{2}})$
3.31	$p^{-\frac{1}{2}}(p^{\frac{1}{2}}+a)^{-3}$	$-2at(t/\pi)^{\frac{1}{2}}+(1+2at^2)$ $\cdot e^{a^2t}\mathrm{Erfc}(at^{\frac{1}{2}})$
3.32	$(p^{\frac{1}{2}}+a)^{-4}$	$-\frac{2}{3}(at)^3(\pi t)^{-\frac{1}{2}}(5+2a^2t)+e^{a^2t}$ $\cdot t(\frac{4}{3}a^4t^2+4a^2t+1)\mathrm{Erfc}(at^{\frac{1}{2}})$
3.33	$p^{-1}[(p^{\frac{1}{2}}-a)/(p^{\frac{1}{2}}+a)]$	$2e^{a^2t}\mathrm{Erfc}(at^{\frac{1}{2}})-1$
3.34	$p^{-1}[(p^{\frac{1}{2}}-a)/(p^{\frac{1}{2}}+a)]^2$	$1-8a(t/\pi)^{\frac{1}{2}}$ $+8a^2te^{a^2t}\mathrm{Erfc}(at^{\frac{1}{2}})$
3.35	$p^{-1}[(p^{\frac{1}{2}}-a)/(p^{\frac{1}{2}}+a)]^3$	$-1-8a(t/\pi)^{\frac{1}{2}}(1+2a^2t)+2e^{a^2t}$ $\cdot(1+8a^2t+8a^4t^2)\mathrm{Erfc}(at^{\frac{1}{2}})$

	$g(p) = \int_0^\infty f(t)e^{-pt}dt$	$f(t)$
3.36	$p^{-\frac{1}{2}}[(p^{\frac{1}{2}}+a)(p-b^2)]^{-1}$	$(a^2-b^2)^{-1}[e^{a^2t}\mathrm{Erfc}(at^{\frac{1}{2}}) + ab^{-1}e^{b^2t}\mathrm{Erf}(bt^{\frac{1}{2}})]$
3.37	$p^{\frac{1}{2}}[(p^{\frac{1}{2}}+a)(p-b^2)]^{-1}$	$(a^2-b^2)^{-1}[a^2e^{a^2t}\mathrm{Erfc}(at^{\frac{1}{2}}) + ab\, e^{b^2t}\mathrm{Erfc}(bt^{\frac{1}{2}})-b^2e^{b^2t}]$
3.38	$(p^2+a^2)^{-\frac{1}{2}}$	$J_0(at)$
3.39	$(p^2-a^2)^{-\frac{1}{2}}$	$I_0(at)$
3.40	$(p^2+a^2)^{-n-\frac{1}{2}}$ $n = 1,2,3,\cdots$	$[1\cdot3\cdot5\cdots(2n-1)a^n]^{-1}t^nJ_n(at)$
3.41	$(p^2-a^2)^{-n-\frac{1}{2}}$ $n = 1,2,3,\cdots$	$[1\cdot3\cdot5\cdots(2n-1)a^n]^{-1}t^nI_n(at)$
3.42	$p^{-1}(p^2+a^2)^{-3/2}$	$\frac{1}{2}\pi ta^{-2}[J_1(at)\mathbf{H}_0(at)-J_0(at)\mathbf{H}_1(at)]$
3.43	$p^{-1}(p^2-a^2)^{-3/2}$	$\frac{1}{2}\pi ta^{-2}[I_1(at)\mathbf{L}_0(at)-I_0(at)\mathbf{L}_1(at)]$

	$g(p) = \int_0^\infty f(t)e^{-pt}dt$	$f(t)$
3.44	$(p^2+ap+b)^{-\frac{1}{2}}$	$e^{-\frac{1}{2}at}J_0[t(b-\frac{1}{4}a^2)^{\frac{1}{2}}]$
3.45	$[(p^2+a^2)^{\frac{1}{2}}-p]^{\frac{1}{2}}$	$(2\pi)^{-\frac{1}{2}}t^{-3/2}\sin(at)$
3.46	$p-(p^2-a^2)^{\frac{1}{2}}$	$at^{-1}I_1(at)$
3.47	$(p^2+a^2)^{-\frac{1}{2}}$ $\cdot[p+(p^2+a^2)^{\frac{1}{2}}]^{\frac{1}{2}}$	$(\frac{1}{2}\pi t)^{-\frac{1}{2}}\cos(at)$
3.48	$(p^2+a^2)^{-\frac{1}{2}}$ $\cdot[(p^2+a^2)^{\frac{1}{2}}-p]^{\frac{1}{2}}$	$(\frac{1}{2}\pi t)^{-\frac{1}{2}}\sin(at)$
3.49	$(p^2-a^2)^{-\frac{1}{2}}$ $\cdot[(p^2-a^2)^{\frac{1}{2}}+p]^{\frac{1}{2}}$	$(\frac{1}{2}\pi t)^{-\frac{1}{2}}\cosh(at)$
3.50	$(p^2-a^2)^{-\frac{1}{2}}$ $\cdot[p-(p^2-a^2)^{\frac{1}{2}}]^{\frac{1}{2}}$	$(\frac{1}{2}\pi t)^{-\frac{1}{2}}\sinh(at)$
3.51	$(p^2+a^2)^{-\frac{1}{2}}$ $\cdot[(p^2+a^2)^{\frac{1}{2}}-p]^{n+\frac{1}{2}}$ $n = 0,1,2,\cdots$	$a^{n+\frac{1}{2}}J_{n+\frac{1}{2}}(at)$

	$g(p) = \int_0^\infty f(t)e^{-pt}dt$	$f(t)$
3.52	$(p^2-a^2)^{-\frac{1}{2}}$ $\cdot[p-(p^2-a^2)^{\frac{1}{2}}]^{n+\frac{1}{2}}$ $n = 0,1,2,\cdots$	$a^{n+\frac{1}{2}}I_{n+\frac{1}{2}}(at)$
3.53	$p^{-1}(p^2+a^2)^{-\frac{1}{2}}$ $\cdot[(p^2+a^2)^{\frac{1}{2}}+p]^{\frac{1}{2}}$	$2a^{-\frac{1}{2}}C(at)$
3.54	$p^{-1}(p^2+a^2)^{-\frac{1}{2}}$ $\cdot[(p^2+a^2)^{\frac{1}{2}}-p]^{\frac{1}{2}}$	$2a^{-\frac{1}{2}}S(at)$
3.55	$p^{-3/2}(p^2+4)^{-\frac{1}{2}}$ $\cdot[(p^2+4)^{\frac{1}{2}}+p]^{\frac{1}{2}}$	$2^{\frac{1}{2}}\int_0^t J_0(u)\cos u\,du$
3.56	$p^{-3/2}(p^2+4)^{-\frac{1}{2}}$ $\cdot[(p^2+4)^{\frac{1}{2}}-p]^{\frac{1}{2}}$	$-2^{\frac{1}{2}}\int_0^t J_0(u)\sin u\,du$
3.57	$(p^3+a^3)^{-1/3}$	${}_0F_2(\;;1,\frac{2}{3};-\frac{1}{27}a^3t^3)$
3.58	$(p^3+a^3)^{-\frac{1}{2}}$	$\frac{2}{3}[\Gamma(\frac{7}{6})\Gamma(\frac{5}{6})]^{-1}(\pi t)^{\frac{1}{2}}\,{}_0F_2(\;;\frac{7}{6},\frac{5}{6};-\frac{1}{27}a^3t^3)$

	$g(p) = \int_0^\infty f(t)e^{-pt}dt$	$f(t)$
3.59	$[(p^4+a^4)^{-\frac{1}{2}}$ $+p^2(p^4+a^4)^{-1}]^{\frac{1}{2}}$	$2^{\frac{1}{2}}\mathrm{ber}(at)$
3.60	$[(p^4+a^4)^{-\frac{1}{2}}$ $-p^2(p^4+a^4)^{-1}]^{\frac{1}{2}}$	$2^{\frac{1}{2}}\mathrm{bei}(at)$
3.61	$[(p^2+a^2)^{\frac{1}{2}}-p]^n$ $n = 1,2,3,\cdots$	$na^n t^{-1} J_n(at)$
3.62	$[p-(p^2-a^2)^{\frac{1}{2}}]^n$ $n = 1,2,3,\cdots$	$na^n t^{-1} I_n(at)$
3.63	$p^{-n-\frac{1}{2}}(a-p)^n$ $n = 0,1,2,\cdots$	$2^n n![(2n)!]^{-1}(\pi t)^{-\frac{1}{2}}He_{2n}[(2at)^{\frac{1}{2}}]$
3.64	$p^{-n-3/2}(a-p)^n$ $n = 0,1,2,\cdots$	$2^{n+\frac{1}{2}}n![(2n+1)!]^{-1}(\pi a)^{-\frac{1}{2}}He_{2n+1}[(2at)^{\frac{1}{2}}]$
3.65	$p^{-\frac{1}{2}}(p-a)^n$ $\cdot(p+a)^{-n-1}$ $n = 0,1,2,\cdots$	$\frac{1}{2}ia^{-\frac{1}{2}}\pi^{-1}n!\{D^2_{-n-1}[i(2at)^{\frac{1}{2}}]$ $-D^2_{-n-1}[-i(2at)^{\frac{1}{2}}]\}$

	$g(p) = \int_0^\infty f(t)e^{-pt}dt$	$f(t)$
3.66	$(p-a)^n(p+a)^{-n-\frac{1}{2}}$ $n = 0,1,2,\cdots$	$(-1)^n 2^n n![(2n)!]^{-1}(\pi t)^{-\frac{1}{2}}$ $e^{-at}He_{2n}[2(at)^{\frac{1}{2}}]$
3.67	$(p-a)^n(p+a)^{-n-3/2}$ $n = 0,1,2,\cdots$	$(-1)^n 2^{n+1}(n+1)![(2n+2)!]^{-1}(\pi a)^{-\frac{1}{2}}$ $\cdot e^{-at}He_{2n+1}[2(at)^{\frac{1}{2}}]$
3.68	$(p+a)^{-n-\frac{1}{2}}$ $n = 1,2,3,\cdots$	$[\frac{1}{2}\cdot\frac{3}{2}\cdots(n-\frac{1}{2})]^{-1}\pi^{-\frac{1}{2}}t^{n-\frac{1}{2}}e^{-at}$
3.69	$(p-a)^n(p-a)^{-n-\frac{1}{2}}$ $n = 0,1,2,\cdots$	$(-2)^n n![(2n)!]^{-1}(\pi t)^{-\frac{1}{2}}e^{bt}$ $\cdot He_{2n}[(2t)^{\frac{1}{2}}(a-b)^{\frac{1}{2}}]$
3.70	$(p-a)^n(p-b)^{-n-3/2}$ $n = 0,1,2,\cdots$	$(-2)^n n![(2n+1)!]^{-1}[\frac{1}{2}\pi(a-b)]^{-\frac{1}{2}}$ $\cdot e^{bt}He_{2n+1}[(2t)^{\frac{1}{2}}(a-b)^{\frac{1}{2}}]$
3.71	$(p-a)^n(p-b)^{-m-n-\frac{1}{2}}$ $n = 0,1,2,\cdots$ $m = 1,2,3,\cdots$	$(-2)^n(a-b)^{-m}(\pi t)^{-\frac{1}{2}}e^{bt}$ $\cdot \sum_{k=1}^{m}\binom{m}{k}2^k(n+k)![(2n+2k)!]^{-1}$ $\cdot He_{2n+2k}[(2t)^{\frac{1}{2}}(a-b)^{\frac{1}{2}}]$

	$g(p) = \int_0^\infty f(t)e^{-pt}dt$	$f(t)$
3.72	$(p-a)^n(p-b)^{-m-n-3/2}$	$(-1)^n 2^{n+1/2}(a-b)^{-m}\pi^{-1/2}e^{bt}$ $\cdot \sum_{k=1}^{m} \binom{m}{k} 2^k (n+k)!\,[(2n+2k+1)!]^{-1}$ $\cdot He_{2n+2k+1}[(2t)^{1/2}(a-b)^{1/2}]$

2.4 Powers of Arbitrary Order

	$g(p) = \int_0^\infty f(t)e^{-pt}dt$	$f(t)$
4.1	$p^{-\nu}$	$[\Gamma(\nu)]^{-1}t^{\nu-1}$ Re $\nu > 0$
4.2	$(p\pm a)^{-\nu}$	$[\Gamma(\nu)]^{-1}t^{\nu-1}e^{\pm at}$ Re $\nu > 0$
4.3	$(p\pm ia)^{-\nu}$	$(\Gamma(\nu)]^{-1}t^{\nu-1}e^{\pm iat}$ Re $\nu > 0$
4.4	$\sum_{n=0}^{\infty} z^n(p+n)^{-\nu}$	$[\Gamma(\nu)]^{-1}t^{\nu-1}(1-ze^{-t})^{-1}$ Re $\nu > 0$
4.5	$(p-b)^{-1}(p+a)^{-\nu}$ Re $\nu > 0$	$[\Gamma(\nu)]^{-1}(a+b)^{-\nu}$ $\cdot e^{bt}\gamma[\nu,(a+b)t]$
4.6	$p^n(p+a)^{-n-\nu}$ Re $\nu > 0$	$n![\Gamma(n+\nu)]^{-1}e^{-at}t^{\nu-1}L_n^{\nu-1}(at)$ $n = 0,1,2,\cdots$
4.7	$p^{-n-\nu}(p-a)^n$ Re $\nu > 0$	$n![\Gamma(n+\nu)]^{-1}t^{\nu-1}L_n^{\nu-1}(at)$ $n = 0,1,2,\cdots$
4.8	$(p-b)^{-n-\nu}(p-a)^n$ Re $\nu > 0$	$n![\Gamma(n+\nu)]^{-1}t^{\nu-1}e^{bt}$ $\cdot L_n^{\nu-1}[(a-b)t]$; $n = 0,1,2,\cdots$

	$g(p) = \int_0^\infty f(t)e^{-pt}dt$	$f(t)$
4.9	$(p-a)^n(p-b)^{-\nu}$ $\mathrm{Re}\,\nu > n$	$n![\Gamma(\nu)]^{-1}t^{\nu-n-1}e^{bt}$ $\cdot L_n^{\nu-1-n}[(b-a)t], \quad n = 0,1,2,\cdots$
4.10	$[(p+a)(p+b)]^{-\nu}$ $\mathrm{Re}\,\nu > 0$	$\pi^{\frac{1}{2}}[\Gamma(\nu)]^{-1}(a-b)^{\frac{1}{2}-\nu}t^{\nu-\frac{1}{2}}$ $\cdot \exp[-\frac{1}{2}(a+b)t]I_{\nu-\frac{1}{2}}[\frac{1}{2}(a-b)t]$
4.11	$(p-a)^{\nu}(p+a)^{-\nu-\frac{1}{2}}$	$2^{-1-\nu}\pi^{-1}\Gamma(\frac{1}{2}-\nu)t^{-\frac{1}{2}}$ $\cdot\{D_{2\nu}[2(at)^{\frac{1}{2}}] + D_{2\nu}[-2(at)^{\frac{1}{2}}]\}$
4.12	$(p-a)^{\nu}(p-b)^{-\nu-\frac{1}{2}}$	$(2\pi)^{-1}2^{-\nu}\Gamma(\frac{1}{2}-\nu)t^{-\frac{1}{2}}\exp[\frac{1}{2}(a+b)t]$ $\cdot\{D_{2\nu}[(2at-2bt)^{\frac{1}{2}}]+D_{2\nu}[-(2at-2bt)^{\frac{1}{2}}]\}$
4.13	$(p-a)^{\nu}(p+a)^{-\nu-3/2}$	$2^{-\nu-2}\pi^{-1}a^{-\frac{1}{2}}\Gamma(-\frac{1}{2}-\nu)$ $\cdot\{D_{2\nu+1}[-2(at)^{\frac{1}{2}}]-D_{2\nu+1}[2(at)^{\frac{1}{2}}]\}$
4.14	$(p-a)^{\nu}(p-b)^{-\nu-3/2}$	$2^{-\nu-3/2}\pi^{-1}(a-b)^{-\frac{1}{2}}\Gamma(-\frac{1}{2}-\nu)\exp[\frac{1}{2}(a+b)t]$ $\cdot\{D_{2\nu-1}[-(2at-2bt)^{\frac{1}{2}}]-D_{2\nu+1}[(2at-2bt)^{\frac{1}{2}}]\}$

	$g(p) = \int_0^\infty f(t)e^{-pt}dt$	$f(t)$
4.15	$(p-a)^{\nu}(p+a)^{-\mu}$ $\mathrm{Re}(\mu-\nu) > 0$	$(2a)^{\frac{1}{2}\nu-\frac{1}{2}\mu}[\Gamma(\mu-\nu)]^{-1}t^{\frac{1}{2}\mu-\frac{1}{2}\nu-1}$ $\cdot M_{\frac{1}{2}(\mu+\nu),\frac{1}{2}(\mu-\nu-1)}(2at)$
4.16	$(a-p)^{\mu}(a+p)^{-\nu}$ $\mathrm{Re}(\nu-\mu) > 0$	$(2a)^{\frac{1}{2}\mu-\frac{1}{2}\nu}[\Gamma(\nu)]^{-1}t^{\frac{1}{2}\nu-\frac{1}{2}\mu-1}$ $\cdot W_{\frac{1}{2}(\mu+\nu),\frac{1}{2}(\nu-\mu-1)}(2at)$
4.17	$(p-a)^{\mu}(p-b)^{-\nu}$ $\mathrm{Re}(\nu-\mu) > 0$	$(a-b)^{\frac{1}{2}\mu-\frac{1}{2}\nu}[\Gamma(\nu-\mu)]^{-1}t^{\frac{1}{2}\nu-\frac{1}{2}\mu-1}$ $\cdot\exp[\frac{1}{2}(a+b)t]M_{\frac{1}{2}(\mu+\nu),\frac{1}{2}(\nu-\mu-1)}[(a-b)t]$
4.18	$p^{\mu}(p-b)^{-\nu}$ $\mathrm{Re}(\nu-\mu) > 0$	$[\Gamma(\nu-\mu)]^{-1}t^{\nu-\mu-1}\,{}_1F_1(a;a-c;bt)$
4.19	$p^{-\gamma}(1-x/p)^{-\alpha}(1-y/p)^{-\beta}$	$[\Gamma(\gamma)]^{-1}t^{\gamma-1}\Phi_2(a,b,c;xt,yt)$
4.20	$[(p+a)^{\frac{1}{2}}-(p+b)^{\frac{1}{2}}]^{\nu}$ $\mathrm{Re}\ \nu > 0$	$\frac{1}{2}t^{-1}\nu(a-b)^{\frac{1}{2}\nu}\exp[-\frac{1}{2}(a+b)t]$ $\cdot I_{\frac{1}{2}\nu}[\frac{1}{2}(a-b)t]$

	$g(p) = \int_0^\infty f(t)e^{-pt}dt$	$f(t)$
4.21	$[(p+a)(p+b)]^{-\frac{1}{2}} \cdot [(p+a)^{\frac{1}{2}}-(p+b)^{\frac{1}{2}}]^{\nu}$	$(a-b)^{\frac{1}{2}\nu}\exp[-\frac{1}{2}(a+b)t] \cdot I_{\frac{1}{2}\nu}[\frac{1}{2}(a-b)t]$ Re $\nu > -2$
4.22	$[(p+a)^{\frac{1}{2}}+b^{\frac{1}{2}}]^{\nu}$ Re $\nu < 0$	$-(2/\pi)^{\frac{1}{2}}\nu(2t)^{-1-\frac{1}{2}\nu}\exp(\frac{1}{2}bt-at) \cdot D_{\nu-1}[(2bt)^{\frac{1}{2}}]$
4.23	$(p+a)^{-\frac{1}{2}} \cdot [(p+a)^{\frac{1}{2}}+b^{\frac{1}{2}}]^{\nu}$ Re $\nu < 1$	$(2/\pi)^{\frac{1}{2}}(2t)^{-\frac{1}{2}-\frac{1}{2}\nu}\exp(\frac{1}{2}bt-at) \cdot D_{\nu}[(2bt)^{\frac{1}{2}}]$
4.24	$p^{-\nu}(p+a)^{-\frac{1}{2}} \cdot [(p+a)^{\frac{1}{2}}-a^{\frac{1}{2}}]^{\nu}$	$(2/\pi)^{\frac{1}{2}}(2t)^{\frac{1}{2}\nu-\frac{1}{2}}e^{-\frac{1}{2}at}D_{-\nu}[(2at)^{\frac{1}{2}}]$ Re $\nu > -1$
4.25	$p^{-\nu}[(p+a)^{\frac{1}{2}}-a^{\frac{1}{2}}]^{\nu}$ Re $\nu < 0$	$(2/\pi)^{\frac{1}{2}}\nu(2t)^{\frac{1}{2}\nu-1}e^{-\frac{1}{2}at}D_{-\nu-1}[(2at)^{\frac{1}{2}}]$
4.26	$(a^{\frac{1}{2}}+p^{\frac{1}{2}})^{\nu}$ Re $\nu < 0$	$-(2\pi)^{-\frac{1}{2}}\nu 2^{-\frac{1}{2}\nu}t^{-1-\frac{1}{2}\nu}e^{\frac{1}{2}at} \cdot D_{\nu-1}[(2at)^{\frac{1}{2}}]$
4.27	$(p^2+a^2)^{-\nu}$ Re $\nu > 0$	$\pi^{\frac{1}{2}}(2a)^{\frac{1}{2}-\nu}[\Gamma(\nu)]^{-1}t^{\nu-\frac{1}{2}}J_{\nu-\frac{1}{2}}(at)$

	$g(p) = \int_0^\infty f(t)e^{-pt}dt$	$f(t)$
4.28	$(p^2-a^2)^{-\nu}$ $\mathrm{Re}\,\nu > 0$	$\pi^{\frac{1}{2}}(2a)^{\frac{1}{2}-\nu}[\Gamma(\nu)]^{-1}t^{\nu-\frac{1}{2}}I_{\nu-\frac{1}{2}}(at)$
4.29	$p^{-1}(p^2+a^2)^{-3/2}$	$\frac{1}{2}\pi t a^{-2}[J_1(at)\mathbf{H}_0(at)-J_0(at)\mathbf{H}_1(at)]$
4.30	$p^{-1}(p^2-a^2)^{-3/2}$	$\frac{1}{2}\pi a^{-2}t[I_1(at)\mathbf{L}_0(at)-I_0(at)\mathbf{L}_1(at)]$
4.31	$p^{-1}(p^2+a^2)^{-\nu}$ $\mathrm{Re}\,\nu > -\frac{1}{2}$	$\frac{1}{2}\pi a^{1-2\nu}t[J_{\nu-\frac{1}{2}}(at)\mathbf{H}'_{\nu-\frac{1}{2}}(at)$ $-J'_{\nu-\frac{1}{2}}(at)\mathbf{H}_{\nu-\frac{1}{2}}(at)]$
4.32	$p^{-1}(p^2-a^2)^{-\nu}$ $\mathrm{Re}\,\nu > -\frac{1}{2}$	$\frac{1}{2}\pi a^{1-2\nu}t[I_{\nu-\frac{1}{2}}(at)\mathbf{L}'_{\nu-\frac{1}{2}}(at)$ $-I'_{\nu-\frac{1}{2}}(at)\mathbf{L}_{\nu-\frac{1}{2}}(at)]$
4.33	$p^{-2\mu}(p^2+a^2)^{-\nu}$ $\mathrm{Re}(\mu+\nu) > 0$	$[\Gamma(2\mu+2\nu)]^{-1}t^{2\mu+2\nu-1}$ $\cdot {}_1F_2(\nu;\mu+\nu,\frac{1}{2}+\mu+\nu;-\frac{1}{4}a^2t^2)$
4.34	$p(p^2+a^2)^{-\nu}$ $\mathrm{Re}\,\nu > \frac{1}{2}$	$\pi^{\frac{1}{2}}[\Gamma(\nu)]^{-1}a^{3/2-\nu}(\frac{1}{2}t)^{\nu-\frac{1}{2}}J_{\nu-3/2}(at)$

	$g(p) = \int_0^\infty f(t)e^{-pt}dt$	$f(t)$
4.35	$p(p^2-a^2)^{-\nu}$ $\mathrm{Re}\ \nu > \frac{1}{2}$	$\pi^{\frac{1}{2}}[\Gamma(\nu)]^{-1}a^{\frac{3}{2}-\nu}(\frac{1}{2}t)^{\nu-\frac{1}{2}}I_{\nu-\frac{3}{2}}(at)$
4.36	$[(p+a)^{\frac{1}{2}}-a^{\frac{1}{2}}]^{\nu}$ $\mathrm{Re}\ \nu > 0$	$\frac{1}{2}\nu a^{\frac{1}{2}\nu}t^{-1}e^{-\frac{1}{2}at}I_{\frac{1}{2}\nu}(\frac{1}{2}at)$
4.37	$p^{\frac{1}{2}}(p+a)^{-\frac{1}{2}}$ $\cdot[(p+a)^{\frac{1}{2}}-p^{\frac{1}{2}}]^{\nu}$	$\frac{1}{2}a^{1+\frac{1}{2}\nu}e^{-\frac{1}{2}at}[I'_{\frac{1}{2}\nu}(\frac{1}{2}at)-I_{\frac{1}{2}\nu}(\frac{1}{2}at)]$ $\mathrm{Re}\ \nu > 0$
4.38	$p^{-\frac{1}{2}}(p+a)^{-\frac{1}{2}}$ $\cdot[(p+a)^{\frac{1}{2}}-p^{\frac{1}{2}}]^{\nu}$	$a^{\frac{1}{2}\nu}e^{-\frac{1}{2}at}I_{\nu}(\frac{1}{2}at)$ $\mathrm{Re}\ \nu > -1$
4.39	$p^{-\frac{1}{2}}(p-a)^{-\frac{1}{2}}$ $\cdot[p^{\frac{1}{2}}-(p-a)^{\frac{1}{2}}]^{\nu}$	$a^{\frac{1}{2}\nu}e^{\frac{1}{2}at}I_{\nu}(\frac{1}{2}at)$ $\mathrm{Re}\ \nu > -1$
4.40	$[(p^2+a^2)^{\frac{1}{2}}-p]^{\nu}$	$\nu a^{\nu}t^{-1}J_{\nu}(at)$ $\mathrm{Re}\ \nu > 0$
4.41	$[p-(p^2-a^2)^{\frac{1}{2}}]^{\nu}$	$\nu t^{-1}a^{\nu}I_{\nu}(at)$ $\mathrm{Re}\ \nu > 0$
4.42	$(p^2-a^2)^{-\nu}$ $\cdot[(p+a)^{\nu}+(p-a)^{\nu}]$	$\frac{1}{2}[\Gamma(\nu)]^{-1}t^{\nu-1}\cosh(at)$ $\mathrm{Re}\ \nu > 0$

	$g(p) = \int_0^\infty f(t)e^{-pt}dt$	$f(t)$
4.43	$p[(p^2+a^2)^{\frac{1}{2}}-p]^\nu$ $\mathrm{Re}\,\nu > 1$	$\nu a^{\nu+1}t^{-1}J_{\nu-1}(at)$ $-\nu(\nu+1)a^\nu t^{-2}J_\nu(at)$
4.44	$p[p-(p^2-a^2)^{\frac{1}{2}}]^\nu$ $\mathrm{Re}\,\nu > 1$	$\nu a^{\nu+1}t^{-1}I_{\nu-1}(at)$ $-\nu(\nu+1)a^\nu t^{-2}I_\nu(at)$
4.45	$p^{-1}\{[p+(p^2-a^2)^{\frac{1}{2}}]^\nu$ $+[p-(p^2-a^2)^{\frac{1}{2}}]^\nu\}$	$2a^\nu[\pi^{-1}\nu\sin(\pi\nu)Ki_\nu(at)+\cos(\frac{1}{2}\pi\nu)]$ $-1 < \mathrm{Re}\,\nu < 1$
4.46	$p^{-1}\{[(p^2+a^2)^{\frac{1}{2}}+p]^\nu$ $+\cos(\pi\nu)[(p^2+a^2)^{\frac{1}{2}}-p]^\nu\}$	$a^\nu[1+\cos(\pi\nu)-\nu\sin[\pi\nu)Yi_\nu(at)]$ $-1 < \mathrm{Re}\,\nu < 1$
4.47	$(p^2+a^2)^{-\frac{1}{2}}$ $\cdot[(p^2+a^2)^{\frac{1}{2}}-p]^\nu$	$a^\nu J_\nu(at)$ $\quad \mathrm{Re}\,\nu > -1$
4.48	$(p^2-a^2)^{-\frac{1}{2}}$ $\cdot[p-(p^2-a^2)^{\frac{1}{2}}]^\nu$	$a^\nu I_\nu(at)$ $\quad \mathrm{Re}\,\nu > -1$

	$g(p) = \int_0^\infty f(t)e^{-pt}dt$	$f(t)$
4.49	$(p^2-a^2)^{-\frac{1}{2}}$ $\cdot\{p+(p^2-a^2)^{\frac{1}{2}}]^{\nu}$ $-[p-(p^2-a^2)^{\frac{1}{2}}]\}$	$2\pi^{-1}a^{\nu}\sin(\pi\nu)K_{\nu}(at)$ $-1 < \mathrm{Re}\,\nu < 1$
4.50	$(p^2-ia^2)^{-\frac{1}{2}}$ $\cdot[p-(p^2-ia^2)^{\frac{1}{2}}]^{\nu}$	$(ia)^{\nu}e^{-i3\pi\nu/4}[\mathrm{ber}_{\nu}(at)+i\,\mathrm{bei}_{\nu}(at)]$ $\mathrm{Re}\,\nu > -1$
4.51	$p(p^2+a^2)^{-\frac{1}{2}}$ $\cdot[(p^2+a^2)^{\frac{1}{2}}-p]^{\nu}$	$\frac{1}{2}a^{\nu+1}[J_{\nu-1}(at)-J_{\nu+1}(at)]$ $\mathrm{Re}\,\nu > 0$
4.52	$p(p^2-a^2)^{-\frac{1}{2}}$ $\cdot[p-(p^2-a^2)^{\frac{1}{2}}]^{\nu}$	$\frac{1}{2}a^{\nu+1}[I_{\nu-1}(at)+I_{\nu+1}(at)]$

2.5 Exponential Functions

	$g(p) = \int_0^\infty f(t)e^{-pt}dt$	$f(t)$
5.1	e^{-ap}	$\delta(t-a)$
5.2	$p^{-1}e^{-ap}$	$H(t-a)$
5.3	$p^{-1/2}e^{-ap}$	$0 \quad t < a$ $\pi^{-1/2}(t-a)^{-1/2} \quad t > a$
5.4	$p^{-3/2}e^{-ap}$	$0 \quad t < a$ $2\pi^{-1/2}(t-a)^{1/2} \quad t > a$
5.5	$p^{-\nu}e^{-ap}$ $\mathrm{Re}\,\nu > 0$	$0 \quad t < a$ $[\Gamma(\nu)]^{-1}(t-a)^{\nu-1} \quad t > a$
5.6	$p^{-1}(e^{-ap}-e^{-bp})$ $a < b$	$0 \quad t < a$ $1 \quad a < t < b$ $0 \quad t > a$
5.7	$p^{-1}(e^{-ap}-e^{-bp})^2$ $a < b$	$0 \quad t < 2a$ $1 \quad 2a < t < a+b$ $-1 \quad a+b < t < 2b$ $0 \quad t > 2b$

	$g(p) = \int_0^\infty f(t)e^{-pt}dt$	$f(t)$	
5.8	$(p+b)^{-1}e^{-ap}$	0	$t < a$
		$e^{-b(t-a)}$	$t > a$
5.9	$[p(p+b)]^{-1}e^{-ap}$	0	$t < a$
		$b^{-1}[1-e^{-b(t-a)}]$	$t > a$
5.10	$p^{-2}(e^{-ap}-e^{-bp})$	0	$t < a$
	$a < b$	$t - a$	$a < t < b$
		$b - a$	$t > b$
5.11	$p^{-2}(e^{-ap}-e^{-bp})^2$	0	$t < 2a$
	$a < b$	$t - 2a$	$2a < t < a+b$
		$2b - t$	$a+b < t < 2b$
		0	$t > 2b$
5.12	$p^{-3}(e^{-ap}-e^{-bp})$	0	$t < a$
	$a < b$	$\frac{1}{2}(t-a)^2$	$a < t < b$
		$t(b-a)+\frac{1}{2}(a^2-b^2)$	$t > b$

	$g(p) = \int_0^\infty f(t)e^{-pt}dt$	$f(t)$	
5.13	$p^{-3}(e^{-ap}-e^{-bp})^2$	0	$t < 2a$
	$a < b$	$\frac{1}{2}(t-2a)^2$	$2a < t < a+b$
		$(b-a)^2-\frac{1}{2}(t-2b)^2$	$a+b < t < 2b$
		$(b-a)^2$	$t > 2b$
5.14	$p^{-3}(e^{-ap}-e^{-bp})^3$	0	$t < 3a$
		$\frac{1}{2}(t-3a)^2$	$3a < t < 2a+b$
		$\frac{3}{4}(b-a)^2-[t-\frac{3}{2}(a+b)]^2$	$2a+b<t<a+2b$
		$\frac{1}{2}(3b-t)^2$	$a+2b < t < 3b$
		0	$t > 3b$
5.15	$(p+c)(p^2+b^2)^{-1}$	0	$t < a$
	$\cdot e^{-ap}$	$\cos[b(t-a)]+c/b \sin[b(t-a)]$	$t > a$
5.16	$(p+c)(p^2-b^2)^{-1}$	0	$t < a$
	$\cdot e^{-ap}$	$\cosh[b(t-a)]+c/b \sinh[b(t-a)]$	$t>a$
5.17	$(p^2+1)^{-1}(1+e^{-\pi p})$	$\sin t$	$t < \pi$
		0	$t > \pi$

	$g(p) = \int_0^\infty f(t)e^{-pt}dt$	$f(t)$	
5.18	$p(p^2+1)^{-1}(1+e^{-\pi p})$	$\cos t$ 0	$t < \pi$ $t > \pi$
5.19	$(p^2+a^2)^{-1}e^{-bp}$	0 $a^{-1}\sin[a(t-b)]$	$t < b$ $t > b$
5.20	$p(p^2+a^2)^{-1}e^{-bp}$	0 $\cos[a(t-b)]$	$t < b$ $t > b$
5.21	$p^{-1}(p^2+a^2)^{-1}e^{-bp}$	0 $2a^{-2}\sin^2[\frac{1}{2}a(t-b)]$	$t < b$ $t > b$
5.22	$p^{-2}(p^2+a^2)^{-1}$ $\cdot e^{-bp}$	0 $a^{-3}\{a(t-b)-\sin[a(t-b)]\}$	$t < b$ $t > b$
5.23	$(p^2-a^2)^{-1}e^{-bp}$	0 $a^{-1}\sinh[a(t-b)]$	$t < b$ $t > b$
5.24	$p(p^2-a^2)^{-1}e^{-bp}$	0 $\cosh[a(t-b)]$	$t < b$ $t > b$

	$g(p) = \int_0^\infty f(t) e^{-pt} dt$	$f(t)$	
5.25	$p^{-1}(p^2-a^2)^{-1} e^{-bp}$	$2a^{-2} \sinh^2[\frac{1}{2}a(t-b)]$	$t > b$
5.26	$p^{-2}(p^2-a^2)^{-1}$ $\cdot e^{-bp}$	0	$t < b$
		$a^{-3}\{\sinh[a(t-b)]-a(t-b)]\}$	$t < b$
5.27	$(p^2+a^2)^{-2}$ $\cdot e^{-bp}$	0	$t < b$
		$\frac{1}{2}a^{-3}\sin[a(t-b)]$ $-\frac{1}{2}a^{-2}(t-b)\cos[a(t-b)]$	$t > b$
.28	$p(p^2+a^2)^{-2}$ $\cdot e^{-bp}$	0	$t < b$
		$\frac{1}{2}a^{-1}(t-b)\sin[a(t-b)]$	$t > b$
.29	$(p^2-a^2)^{-2}$ $\cdot e^{-bp}$	0	$t < b$
		$\frac{1}{2}a^{-2}(t-b)\cosh[a(t-b)]$ $-\frac{1}{2}a^{-3}\sinh[a(t-b)]$	$t > b$
.30	$p(p^2-a^2)^{-2}$ $\cdot e^{-bp}$	0	$t < b$
		$\frac{1}{2}a^{-1}(t-b)\sinh[a(t-b)]$	$t > b$

	$g(p) = \int_0^\infty f(t)e^{-pt}dt$	$f(t)$
5.31	$e^{-\frac{1}{2}\pi p}$ $[p(p^2+2^2)(p^2+4^2)\cdots(p^2+4n^2)]^{-1}$ $n = 1,2,3,\cdots$	$0 \quad t < \frac{1}{2}\pi$ $[(2n)!]^{-1}\cos^{2n}t \quad t > \frac{1}{2}\pi$
5.32	$e^{-\frac{1}{2}\pi p}$ $\{(p^2+1^2)(p^2+3^2)\cdots[p^2+(2n+1)^2]\}^{-1}$	$0 \quad t < \frac{1}{2}\pi$ $-[(2n+1)!]^{-1}\cos^{2n+1}t \quad t > \frac{1}{2}\pi$
5.33	$p^{-1}(1+e^{ap})^{-1}$	$0 \quad 2na<t<(2n+1)a$ $1 \quad (2n+1)a<t<(2n+2)a$ $n = 0,1,2,\cdots$
5.34	$p^{-1}(1+e^{-ap})^{-1}$	$1 \quad 2na<t<(2n+1)a$ $0 \quad (2n+1)a<t<(2n+2)a$ $n = 0,1,2,\cdots$
5.35	$p^{-1}(e^{ap}-1)^{-1}$	$n \quad na<t<(n+1)a$ $n = 0,1,2,\cdots$
5.36	$p^{-1}(1-e^{-ap})^{-1}$	$n+1 \quad na<t<(n+1)a$ $n = 0,1,2,\cdots$

	$g(p) = \int_0^\infty f(t)e^{-pt}dt$	$f(t)$	
5.37	$p^{-1}(1-e^{-ap})^{-1}$	1	$na<t<na+b$
	$\cdot(1-e^{-bp})$	0	$na+b<t<(n+1)a$
	$0<b<a$		$n = 0,1,2,\cdots$
5.38	$p^{-1}(1+a^{-ap})^{-1}$	1	$2na<t<(2n+1)a$
	$\cdot(1-e^{-ap})$	-1	$(2n+1)a<t<(2n+2)a$
			$n = 0,1,2,\cdots$
5.39	$p^{-1}(1+e^{ap})^{-1}$	0	$0<t<a$
	$\cdot(1-e^{-ap})$	1	$(2n+1)a<t<(2n+2)a$
		-1	$(2n+2)a<t<(2n+3)a$
			$n = 0,1,2,\cdots$
5.40	$p^{-1}(e^{ap}-e^{-ap})^{-1}$	0	$2na<t<(2n+1)a$
	$\cdot(1-e^{-ap})$	1	$(4n+1)a<t<(4n+2)a$
		-1	$(4n+3)a<t<(4n+4)a$
			$n = 0,1,2,\cdots$
.41	$p^{-1}(e^{ap}-1)^{-2}$	n^2	$a<t<(n+1)a$
	$\cdot(e^{ap}+1)$		$n = 0,1,2,\cdots$

	$g(p) = \int_0^\infty f(t)e^{-pt}dt$	$f(t)$
5.42	$p^{-1}(e^{ap}-1)^{-3}$ $\cdot(e^{2ap}+4e^{ap}+1)$	n^3 $\quad na<t<(n+1)a$ $n = 0,1,2,\cdots$
5.43	$p^{-1}e^{-ap}(e^{ap}-1)^{-m}$	$\binom{n}{m}$ $\quad na<t<(n+1)a$ $n,m = 0,1,2,\cdots$
5.44	$p^{-1}(e^{bp}-a)^{-1}$	$(1-a)^{-1}(1-a^n)$ $\quad nb<t<(n+1)b$ $n = 0,1,2,\cdots$
5.45	$p^{-1}(e^{bp}-a)^{-1}$ $\cdot(e^{bp}-1)$	a^n $\quad nb<t<(n+1)b$ $n = 0,1,2,\cdots$
5.46	$p^{-1}(e^{bp}-a)^{-2}$ $\cdot(e^{bp}-1)$	na^{n-1} $\quad nb<t<(n+1)b$ $n = 0,1,2,\cdots$
5.47	$p^{-1}(e^{bp}-1)$ $\cdot[(e^{bp}-a)(e^{bp}-c)]^{-1}$	$\frac{a^n-c^n}{a-c}$ $\quad nb<t<(n+1)b$
5.48	$p^{-1}(e^{pb}-a)^{-3}$ $\cdot(e^{pb}-1)$	$\frac{1}{2}n(n-1)a^{n-2}$ $\quad nb<t<(n+1)b$ $n = 0,1,2,\cdots$

	$g(p) = \int_0^\infty f(t)e^{-pt}dt$	$f(t)$
5.49	$p^{-1}(e^{bp}-a)^{-3}$ $\cdot(e^{pb}-1)(e^{pb}+a)$	n^2a^{n-1} $\quad nb<t<(n+1)b$
5.50	$p^{-1}(e^p-1)(e^p-a-b)$ $\cdot[(e^p-a)(e^p-b)]^{-1}$	$-ab/(a-b)(a^{n-1}-b^{n-1})$ $n<t<n+1$ $n = 0,1,2,\cdots$
5.51	$p^{-1}[(e^p-a)(e^p-b)]^{-1}$ $\cdot(e^p-1)[(c-d)e^p-cb+ad]$	ca^n-db^n $n<t<(n+1)$ $n = 0,1,2,\cdots$
5.52	$p^{-1}e^{ap}(e^p-1)^{-n}$	$(a+1)_n$ $\quad n<t<n+1$ $n = 0,1,2,\cdots$
5.53	$p^{-1}(e^p-1)$ $\cdot(e^{2p}-2ae^p\cos b+a^2)^{-1}$	$\csc b.\ a^{n-1}\sin(bn)$ $n<t<(n+1)$ $n = 0,1,2,\cdots$
5.54	$p^{-1}(e^p-1)(e^p-a\cos b)$ $\cdot(e^{2p}-2ae^p\cos b+a^2)^{-1}$	$a^n\cos(bn)$ $\quad n<t<(n+1)$ $n = 0,1,2,\cdots$

	$g(p) = \int_0^\infty f(t)e^{-pt}dt$	$f(t)$
5.55	$p^{-2}(e^{ap}+1)^{-1}$	$\frac{1}{4}[1-(-1)^n](2t-a)+\frac{1}{2}(-1)^n an$ $na<t<(n+1)a, \quad n = 0,1,2,\cdots$
5.56	$p^{-2}(e^{ap}-1)^{-1}$	$nt-\frac{1}{2}an(n+1) \quad na<t<(n+1)a$ $n = 0,1,2,\cdots$
5.57	$p^{-2}(e^{bp}-a)^{-1}$	$(1-a)^{-1}\{(1-a^n)t-b(1-a)^{-1}[1-(n+1)a^n+na^{n+1}]\}$ $nb<t<(n+1)b, \quad n = 0,1,2,\cdots$
5.58	$p^{-2}(1-e^{-ap})^{-1}$ $\cdot(ap-1+e^{-ap})$	$a(n+1)-t \quad na<t<(n+1)a$ $n = 0,1,2,\cdots$
5.59	$p^{-2}(1-e^{ap})^{-1}$ $\cdot(ap+1-e^{ap})$	$t-na \quad na<t<(n+1)a$ $n = 0,1,2,\cdots$
5.60	$p^{-2}(1-e^{-2ap})^{-1}$ $\cdot[1-(1+ap)e^{-ap}]$	$t-2na \quad 2na<t<(2n+1)a$ $0 \quad (2n+1)a<t<(2n+2)a$ $n = 0,1,2,\cdots$
5.61	$p^{-2}(1+e^{-ap})^{-1}$ $\cdot(1-e^{-ap})$	$t-2na \quad 2na<t<(2n+1)a$ $2a(n+1)-t \quad (2n+1)a<t<(2n+2)a$ $n = 0,1,2,\cdots$

	$g(p) = \int_0^\infty f(t)e^{-pt}dt$	$f(t)$	
5.62	$p^{-2}(1-e^{-4ap})^{-1}$ $\cdot(1-e^{-ap})^{2}$	$t-2na$	$4na<t<(4n+1)a$
		$4n+2a-t$	$(4n+1)a<t<(4n+2)a$
		0	$(4n+2)a<t<(4n+4)a$
			$n = 0,1,2,\cdots$
5.63	$(p^2+a^2)^{-1}$ $[1+\exp(-2\pi mp/a)]^{-1}$	$a^{-1}\sin(at)$	$2n<at/(2\pi m)<2n+1$
	$m = 1,2,3,\cdots$	0	$(2n+1)<at/(2\pi m)<2n+2$
			$n = 0,1,2,\cdots$
5.64	$p(p^2+a^2)^{-1}$ $[1+\exp(-2\pi mp/a)]^{-1}$	$\cos(at)$	$2n<at/(2\pi m)<2n+1$
	$m = 1,2,3,\cdots$	0	$(2n+1)<at/(2\pi m)<2n+2$
			$n = 0,1,2,\cdots$
5.65	$p(p^2+a^2)^{-1}$ $(1+e^{-\pi p/a})$ $\cdot(1-e^{-\pi/a})^{-1}$	$\cos(at)$	$2n\pi<at<(2n+1)\pi$
		$-\cos(at)$	$(2n+1)\pi<at<(2n+2)\pi$
5.66	$e^{a/p}-1$	$(a/t)^{\frac{1}{2}} I_1[2(at)^{\frac{1}{2}}]$	
5.67	$1-e^{-a/p}$	$(a/t)^{\frac{1}{2}} J_1[2(at)^{\frac{1}{2}}]$	

	$g(p) = \int_0^\infty f(t)e^{-pt}dt$	$f(t)$
5.68	$p^{-1}(1-e^{-1/p})^{-1}$	$\sum_{n=0}^{\infty} J_0[2(nt)^{\frac{1}{2}}]$
5.69	$p^{-\frac{1}{2}}e^{a/p}$	$(\pi t)^{-\frac{1}{2}}\cosh[2(at)^{\frac{1}{2}}]$
5.70	$p^{-\frac{1}{2}}e^{-a/p}$	$(\pi t)^{-\frac{1}{2}}\cos[2(at)^{\frac{1}{2}}]$
5.71	$p^{-3/2}e^{a/p}$	$(\pi a)^{-\frac{1}{2}}\sinh[2(at)^{\frac{1}{2}}]$
5.72	$p^{-3/2}e^{-a/p}$	$(\pi a)^{-\frac{1}{2}}\sin[2(at)^{\frac{1}{2}}]$
5.73	$p^{-5/2}e^{a/p}$	$\pi^{-\frac{1}{2}}a^{-1}\{t^{\frac{1}{2}}\cosh[2(at)^{\frac{1}{2}}]-\frac{1}{2}a^{-\frac{1}{2}}\sinh[2(at)^{\frac{1}{2}}]\}$
5.74	$p^{-5/2}e^{-a/p}$	$\pi^{-\frac{1}{2}}a^{-1}\{\frac{1}{2}a^{-\frac{1}{2}}\sin[2(at)^{\frac{1}{2}}]-t^{\frac{1}{2}}\cos[2(at)^{\frac{1}{2}}]\}$
5.75	$p^{-\nu-1}e^{a/p}$ $\mathrm{Re}\ \nu > -1$	$(t/a)^{\frac{1}{2}\nu}I_\nu[2(at)^{\frac{1}{2}}]$
5.76	$p^{-\nu-1}e^{a/p}$ $\mathrm{Re}\ \nu > -1$	$(t/a)^{\frac{1}{2}\nu}J_\nu[2(at)^{\frac{1}{2}}]$
5.77	$p^{-1}e^{a/p}\log p$	$K_0(z)-\frac{1}{2}\log(t/a)I_0(z)$ $z = 2(at)^{\frac{1}{2}}$

	$g(p) = \int_0^\infty f(t)e^{-pt}dt$	$f(t)$
5.78	$p^{-1}e^{-a/p}\log p$	$-\frac{1}{2}J_0(z)\log(t/a)-\frac{1}{2}\pi Y_0(z)$ $z = 2(at)^{\frac{1}{2}}$
5.79	$p^{-\frac{1}{2}}e^{a/p}\log p$	$-\frac{1}{2}(\pi t)^{-\frac{1}{2}}[\log(t/a)\cosh z$ $+e^{z}Ei(-2z)+e^{-z}\overline{Ei}(2z)]$ $z = 2(at)^{\frac{1}{2}}$
5.80	$p^{-\frac{1}{2}}e^{-a/p}\log p$	$-(\pi t)^{-\frac{1}{2}}[\frac{1}{2}\log(t/a)\cos z$ $+\cos z\ Ci(2z)+\sin z\ Si(2z)]$ $z = 2(at)^{\frac{1}{2}}$
5.81	$p^{-3/2}e^{a/p}\log p$	$-\frac{1}{2}(\pi a)^{-\frac{1}{2}}[\log(t/a)\sinh z$ $+e^{z}\ Ei(-2z)-e^{-z}\ \overline{Ei}(2z)]$
5.82	$p^{-3/2}e^{-a/p}\log p$	$-(\pi a)^{-\frac{1}{2}}[\frac{1}{2}\log(t/a)\sin z$ $+\sin z\ Ci(2z)-\cos z\ Si(2z)]$ $z = 2(at)^{\frac{1}{2}}$
5.83	$p^{-2}e^{a/p}\log p$	$-(\frac{t}{a})^{\frac{1}{2}}[\frac{1}{2}\log(t/a)I_1(z)$ $+K_1(z)-z^{-1}I_0(z)]$ $z = 2(at)$

	$g(p) = \int_0^\infty f(t)e^{-pt}dt$	$f(t)$
5.84	$p^{-2}e^{-a/p}\log p$	$-\tfrac{1}{2}(t/a)^{\frac{1}{2}}[\log(t/a)J_1(z)$ $+\pi Y_1(z)+2z^{-1}J_0(z)]$ $z = 2(at)^{\frac{1}{2}}$
5.85	$e^{-ap^{\frac{1}{2}}}$	$\tfrac{1}{2}a\pi^{-\frac{1}{2}}t^{-3/2}\exp(-\tfrac{1}{4}a^2/t)$
5.86	$p^{\frac{1}{2}}e^{-ap^{\frac{1}{2}}}$	$\tfrac{1}{4}\pi^{-\frac{1}{2}}(a^2-2t)t^{-5/2}\exp(-\tfrac{1}{4}a^2/t)$
5.87	$p^{-\frac{1}{2}}e^{-ap^{\frac{1}{2}}}$	$(\pi t)^{-\frac{1}{2}}\exp(-\tfrac{1}{4}a^2/t)$
5.88	$pe^{-ap^{\frac{1}{2}}}$	$\tfrac{1}{4}a\pi^{-\frac{1}{2}}t^{-5/2}(\tfrac{1}{2}a^2/t-3)\exp(-\tfrac{1}{4}a^2/t)$
5.89	$p^{-1}e^{-ap^{\frac{1}{2}}}$	$\mathrm{Erfc}(\tfrac{1}{2}at^{-\frac{1}{2}})$
5.90	$p^{-1}(1-e^{ap^{\frac{1}{2}}})$	$\mathrm{Erf}(\tfrac{1}{2}at^{-\frac{1}{2}})$
5.91	$p^{3/2}e^{-ap^{\frac{1}{2}}}$	$\tfrac{1}{4}\pi^{-\frac{1}{2}}t^{-5/2}(3-\frac{3}{2}a^2/t + \tfrac{1}{4}a^4/t^2)$
5.92	$p^{-3/2}e^{-ap^{\frac{1}{2}}}$	$2(t/\pi)^{\frac{1}{2}}\exp(-\tfrac{1}{4}a^2/t)-a\ \mathrm{Erfc}(\tfrac{1}{2}at^{-\frac{1}{2}})$

	$g(p) = \int_0^\infty f(t)e^{-pt}dt$	$f(t)$
5.93	$p^{\frac{1}{2}n-\frac{1}{2}}e^{-(2ap)^{\frac{1}{2}}}$	$(2t)^{-\frac{1}{2}n}(\pi t)^{-\frac{1}{2}}e^{-\frac{1}{2}a/t}He_n[(t/a)^{-\frac{1}{2}}]$
5.94	$p^{\nu}e^{-ap^{\frac{1}{2}}}$	$2^{-\nu-\frac{1}{2}}\pi^{-\frac{1}{2}}t^{-\nu-1}\exp(-\frac{1}{8}a^2/t)D_{2\nu+1}[a(2t)^{-\frac{1}{2}}]$
5.95	$e^{-ap^{\frac{1}{2}}}(p^{\frac{1}{2}}+b)^{-1}$	$(\pi t)^{-\frac{1}{2}}\exp(-\frac{1}{4}a^2/t)-b\ \exp(ab+b^2t)$ $\cdot\mathrm{Erfc}(\frac{1}{2}at^{-\frac{1}{2}}+bt^{\frac{1}{2}})$
5.96	$p^{-\frac{1}{2}}e^{-ap^{\frac{1}{2}}}$ $\cdot(p^{\frac{1}{2}}+b)^{-1}$	$\exp(ab+b^2t)\mathrm{Erfc}(\frac{1}{2}at^{-\frac{1}{2}}+bt^{\frac{1}{2}})$
5.97	$p^{-3/2}(1+ap^{\frac{1}{2}})$ $\cdot e^{ap^{\frac{1}{2}}}$	$2(t/\pi)^{\frac{1}{2}}\exp(-\frac{1}{4}a^2/t)$
5.98	$p^{-1}(p^{\frac{1}{2}}+b)^{-1}$ $\cdot e^{-ap^{\frac{1}{2}}}$	$b^{-1}\mathrm{Erfc}(\frac{1}{2}at^{-\frac{1}{2}})-b^{-1}\exp(ab+b^2t)$ $\cdot\mathrm{Erfc}(\frac{1}{2}at^{-\frac{1}{2}}+bt^{\frac{1}{2}})$
5.99	$p^{\frac{1}{2}}(p^{\frac{1}{2}}+b)^{-1}$ $\cdot e^{-ap^{\frac{1}{2}}}$	$(\pi t)^{-\frac{1}{2}}(\frac{1}{2}a/t-b)\exp(-\frac{1}{4}a^2/t)$ $+b^2\exp(ab+b^2t)\mathrm{Erfc}(\frac{1}{2}at^{-\frac{1}{2}}+bt^{\frac{1}{2}})$

	$g(p) = \int_0^\infty f(t)e^{-pt}dt$	$f(t)$
5.100	$p(p^{\frac{1}{2}}+b)^{-1}$ $\cdot e^{-ap^{\frac{1}{2}}}$	$\pi^{-\frac{1}{2}}t^{-3/2}(b^2t-\frac{1}{2}-\frac{1}{2}ab+\frac{1}{4}a^2/t)\exp(-\frac{1}{4}a^2/t)$ $-b^3\exp(ab+b^2t)\mathrm{Erfc}(\frac{1}{2}at^{-\frac{1}{2}}+bt^{\frac{1}{2}})$
5.101	$p^{-3/2}(p^{\frac{1}{2}}+b)^{-1}$ $\cdot e^{-ap^{\frac{1}{2}}}$	$b^{-1}[2(t/\pi)^{\frac{1}{2}}\exp(-\frac{1}{4}a^2/t)-(a+b^{-1})\mathrm{Erfc}(\frac{1}{2}at^{-\frac{1}{2}})$ $+b^{-1}\exp(ab+b^2t)\mathrm{Erfc}(\frac{1}{2}at^{-\frac{1}{2}}+bt^{\frac{1}{2}})]$
5.102	$(p^{\frac{1}{2}}+b)^{-2}e^{-ap^{\frac{1}{2}}}$	$(2bt^2+ab+1)\exp(ab+b^2t)$ $\cdot\mathrm{Erfc}(\frac{1}{2}at^{-\frac{1}{2}}+bt^{\frac{1}{2}})-2b(t/\pi)^{\frac{1}{2}}\exp(-\frac{1}{4}a^2/t)$
5.103	$p^{-\frac{1}{2}}(p^{\frac{1}{2}}+b)^{-2}e^{-ap^{\frac{1}{2}}}$	$2(t/\pi)^{\frac{1}{2}}\exp(-\frac{1}{4}a^2/t)$ $-(2bt+a)\exp(ab+b^2t)\mathrm{Erfc}(\frac{1}{2}at^{-\frac{1}{2}}+bt^{\frac{1}{2}})$
5.104	$p^{-1}(p^{\frac{1}{2}}+b)^{-2}e^{-ap^{\frac{1}{2}}}$	$b^{-2}\mathrm{Erfc}(\frac{1}{2}at^{-\frac{1}{2}})-2b^{-1}(t/\pi)^{\frac{1}{2}}\exp(-\frac{1}{4}a^2/t)$ $+(2t-b^{-2}+a/b)\exp(ab+b^2t)$ $\cdot\mathrm{Erfc}(\frac{1}{2}at^{-\frac{1}{2}}+bt^{\frac{1}{2}})$
5.105	$p^{-\nu}\exp(-ap^{-\frac{1}{2}})$ Re $\nu > 0$	$\frac{1}{2}\pi^{-\frac{1}{2}}a^{\frac{1}{2}-\nu}t^{-3/2}$ $\cdot\int_0^\infty u^{\nu+\frac{1}{2}}\exp(-\frac{1}{4}u^2/t)J_{2\nu-1}[2(au)^{\frac{1}{2}}]du$

	$g(p) = \int_0^\infty f(t)e^{-pt}dt$	$f(t)$
5.106	$\exp(-3p^{1/3})$	$\pi^{-1}3^{1/2}t^{-3/2}K_{1/3}(2t^{-1/2})$
5.107	$p^{-1/3}\exp(-3p^{1/3})$	$3^{1/2}(\pi t)^{-1}K_{2/3}(2t^{-1/2})$
5.108	$p^{-2/3}\exp(-3p^{1/3})$	$\pi^{-1}(3/t)^{1/2}K_{1/3}(2t^{-1/2})$
5.109	$(p^2-a^2)^{-1/2}$ $\cdot\exp[-b(p^2+a^2)^{1/2}]$	$0 \quad t < b$ $J_0[a(t^2-b^2)^{1/2}] \quad t > b$
5.110	$(p^2-a^2)^{-1/2}$ $\cdot\exp[-b(p^2-a^2)^{1/2}]$	$0 \quad t < b$ $I_0[a(t^2-b^2)^{1/2}] \quad t > b$
5.111	$(p^2+a^2)^{-1/2}$ $\cdot\exp[bp-b(p^2 \; a^2)^{1/2}]$	$J_0[a(t^2+2bt)^{1/2}$
5.112	$(p^2-a^2)^{-1/2}$ $\cdot\exp[bp-b(p^2-a^2)^{1/2}]$	$I_0[a(t^2+2bt)^{1/2}]$
5.113	$z^{-1}(z-p)^{1/2}e^{-bz}$ $z = (p^2+a^2)^{1/2}$	$0 \quad t < b$ $(2/\pi)^{1/2}(t+b)^{-1/2}\sin[a(t^2-b^2)] \quad t > b$

	$g(p) = \int_0^\infty f(t)e^{-pt}dt$	$f(t)$
5.114	$z^{-1}(z+p)^{\frac{1}{2}}e^{-bz}$ $z = (p^2+a^2)^{\frac{1}{2}}$	$0 \qquad t < b$ $(2/\pi)^{\frac{1}{2}}(t+b)^{-\frac{1}{2}}\cos[a(t^2-b^2)^{\frac{1}{2}}] \quad t > b$
5.115	$s^{-1}(p-s)^{\frac{1}{2}}e^{-bs}$ $s = (p^2-a^2)^{\frac{1}{2}}$	$0 \qquad t < b$ $(2/\pi)^{\frac{1}{2}}(t+b)^{-\frac{1}{2}}\sinh[a(t^2-b^2) \quad t > b$
5.116	$s^{-1}(s+p)^{\frac{1}{2}}e^{-bs}$ $s = (p^2-a^2)^{-\frac{1}{2}}$	$0 \qquad t < b$ $(2/\pi)^{\frac{1}{2}}(t+b)^{-\frac{1}{2}}\cosh[a(t^2-b^2)^{\frac{1}{2}}] \quad t > b$
5.117	$z^{-1}(z-p)^{\frac{1}{2}}e^{b(p-z)}$ $z = (p^2+a^2)^{\frac{1}{2}}$	$(2/\pi)^{\frac{1}{2}}(t+2b)^{-\frac{1}{2}}\sin[a(t^2+2bt)^{\frac{1}{2}}]$
5.118	$z^{-1}(z+p)^{\frac{1}{2}}e^{b(p-z)}$ $z = (p^2+a^2)^{\frac{1}{2}}$	$(2/\pi)^{\frac{1}{2}}(t+2b)^{-\frac{1}{2}}\cos[a(t^2+2bt)^{\frac{1}{2}}]$
5.119	$s^{-1}(p+s)^{\frac{1}{2}}e^{b(p-s)}$ $s = (p^2-a^2)^{\frac{1}{2}}$	$(2/\pi)^{\frac{1}{2}}(t+2b)^{-\frac{1}{2}}\cosh[a(t^2+2bt)^{\frac{1}{2}}]$
5.120	$s^{-1}(p-s)^{\frac{1}{2}}e^{b(p-s)}$ $s = (p^2-a^2)^{\frac{1}{2}}$	$(2/\pi)^{\frac{1}{2}}(t+2b)^{-\frac{1}{2}}\sinh[a(t^2+2bt)^{\frac{1}{2}}]$

	$g(p) = \int_0^\infty f(t)e^{-pt}dt$	$f(t)$	
5.121	$1-\exp[-b(p^2+a^2)^{\frac{1}{2}}+bp]$	$ab(t^2+2bt)^{-\frac{1}{2}}J_1[a(t^2+2bt)^{\frac{1}{2}}]$	
5.122	$\exp[bp-b(p^2-a^2)^{\frac{1}{2}}]$ -1	$ab(t^2+2bt)^{-\frac{1}{2}}I_1[a(t^2+2bt)^{\frac{1}{2}}]$	
5.123	$e^{-bp}-p(p^2+a^2)^{-\frac{1}{2}}$ $\cdot\exp[-b(p^2+a^2)^{\frac{1}{2}}]$	0 $at(t^2-b^2)^{-\frac{1}{2}}J_1[a(t^2-b^2)^{\frac{1}{2}}]$	$t < b$ $t > b$
5.124	$-e^{-bp}+p(p^2-a^2)^{-\frac{1}{2}}$ $\cdot\exp[-b(p^2-a^2)^{\frac{1}{2}}]$	0 $at(t^2-b^2)^{\frac{1}{2}}I_1[a(t^2-b^2)^{\frac{1}{2}}]$	$t < b$ $t > b$
5.125	$[1-p(p^2+a^2)^{-\frac{1}{2}}]$ $\cdot\exp[-b(p^2+a^2)^{\frac{1}{2}}]$	0 $a(\frac{t-b}{t+b})^{\frac{1}{2}}J_1[a(t^2-b^2)^{\frac{1}{2}}]$	$t < b$ $t > b$
5.126	$[p(p^2-a^2)^{-\frac{1}{2}}-1]$ $\cdot\exp[-b(p^2-a^2)^{\frac{1}{2}}]$	0 $a(\frac{t-b}{t+b})^{\frac{1}{2}}I_1[a(t^2-b^2)^{\frac{1}{2}}]$	$t < b$ $t > b$
5.127	e^{-bp} $-e^{-b(p^2+a^2)^{\frac{1}{2}}}$	0 $ab(t^2-b^2)^{-\frac{1}{2}}J_1[a(t^2-b^2)^{\frac{1}{2}}]$	$t < b$ $t > b$

	$g(p) = \int_0^\infty f(t)e^{-pt}dt$	$f(t)$	
5.128	e^{-bp} $-e^{-b(p^2-a^2)^{\frac{1}{2}}}$	0 $-ab(t^2-b^2)^{-\frac{1}{2}}I_1[a(t^2-b^2)^{\frac{1}{2}}]$	$t < b$ $t > b$
5.129	$p^{-1}\exp[-a(b^2+p)^{\frac{1}{2}}]$	$\frac{1}{2}e^{-ab}\mathrm{Erfc}(\frac{1}{2}at^{-\frac{1}{2}}-bt^{\frac{1}{2}})$ $+\frac{1}{2}e^{ab}\mathrm{Erfc}(\frac{1}{2}at^{-\frac{1}{2}}+bt^{\frac{1}{2}})$	
5.130	$(p^2+a^2)^{-1}$ $\cdot\exp[-b(p^2+a^2)^{\frac{1}{2}}]$	0 $\int_b^t J_0[a(t-u)]J_0[a(u^2-b^2)^{\frac{1}{2}}]du$	$t < b$ $t > b$
5.131	$(p^2-a^2)^{-1}$ $\cdot\exp[-b(p^2-a^2)^{\frac{1}{2}}]$	0 $\int_b^t I_0[a(t-u)]I_0[a(u^2-b^2)^{\frac{1}{2}}]$	$t < b$ $t > b$
5.132	$\exp[-b(p^2+a^2)^{\frac{1}{2}}]$ $\cdot[b(p^2+a^2)^{-1}+(p^2+a^2)^{-3/2}]$	0 $a^{-1}(t^2-b^2)^{\frac{1}{2}}J_1[a(t^2-b^2)^{\frac{1}{2}}]$	$t < b$ $t > b$
5.133	$\exp[-b(p^2-a^2)^{\frac{1}{2}}]$ $\cdot[b(p^2-a^2)^{-1}+(p^2-a^2)^{-3/2}]$	0 $a^{-1}(t^2-b^2)^{\frac{1}{2}}I_1[a(t^2-b^2)^{\frac{1}{2}}]$	$t < b$ $t > b$

	$g(p) = \int_0^\infty f(t)e^{-pt}dt$	$f(t)$
5.134	$1-p(p^2+a^2)^{-\frac{1}{2}}$ $\cdot\exp\{-b[(p^2+a^2)^{\frac{1}{2}}-p]\}$	$a(t+b)(t^2+2bt)^{\frac{1}{2}}J_1[a(t^2+2bt)^{\frac{1}{2}}]$
5.135	$1-p(p^2-a^2)^{-\frac{1}{2}}$ $\cdot\exp\{b[p-(p^2-a^2)^{\frac{1}{2}}]\}$	$-a(t+b)(t^2+2bt)^{\frac{1}{2}}I_1[a(t^2+2bt)^{\frac{1}{2}}]$
5.136	$p\exp[-b(p^2+a^2)^{-\frac{1}{2}}]$ $\cdot[b(p^2+a^2)^{-1}+(p^2+a^2)^{-3/2}]$	0 $\quad t<b$ $t\,J_0[a(t^2-b^2)^{\frac{1}{2}}]$ $\quad t>b$
5.137	$p\exp[-b(p^2-a^2)^{\frac{1}{2}}]$ $\cdot[b(p^2-a^2)^{-1}+(p^2-a^2)^{-3/2}]$	0 $\quad t<b$ $t\,I_0[a(t^2-b^2)^{\frac{1}{2}}]$ $\quad t>b$
5.138	$[(p+a)(p+b)]^{-\frac{1}{2}}$ $\exp\{-c[(p+a)(p+b)]^{\frac{1}{2}}\}$	0 $\quad t<c$ $\exp[-\frac{1}{2}at-\frac{1}{2}bt)I_0[\frac{1}{2}(a-b)(t^2-c^2)^{\frac{1}{2}}]$ $\quad t>c$
5.139	$[(p+a)(p+b)]^{-\frac{1}{2}}$ $\cdot\{p+\frac{1}{2}(a+b)+[(p+a)(p+b)]^{\frac{1}{2}}\}^{-\nu}$ $\cdot\exp\{-b[(p+a)(p+b)]^{\frac{1}{2}}\}$ $\mathrm{Re}\,\nu>-1$	0 $\quad t<b$ $2^{\nu}(a-b)^{-\nu}\left(\frac{t-b}{t+b}\right)^{\frac{1}{2}\nu}$ $\cdot e^{-\frac{1}{2}at-\frac{1}{2}bt}I_\nu[\frac{1}{2}(a-b)(t^2-b^2)^{\frac{1}{2}}]$ $\quad t>b$

	$g(p) = \int_0^\infty f(t)e^{-pt}dt$	$f(t)$
5.140	$(p^2+a^2)^{-\frac{1}{2}} \exp[-b(p^2+a^2)^{\frac{1}{2}}]$ $\cdot[(p^2+a^2)^{\frac{1}{2}}-p]^{\nu}$	$0 \quad t < b$ $a^{\nu}\left(\frac{t-b}{t+b}\right)^{\frac{1}{2}\nu} J_{\nu}[a(t^2-b^2)^{\frac{1}{2}}] \quad t > b$
5.141	$(p^2-a^2)^{-\frac{1}{2}}\exp[-b(p^2-a^2)^{\frac{1}{2}}]$ $\cdot[p-(p^2-a^2)^{\frac{1}{2}}]^{\nu}$	$0 \quad t < b$ $a^{\nu}\left(\frac{t-b}{t+b}\right)^{\frac{1}{2}\nu} I_{\nu}[a(t^2-b^2)^{\frac{1}{2}}] \quad t > b$
5.142	$(p^2+a^2)^{-\frac{1}{2}}[(p^2+a^2)^{\frac{1}{2}}-p]^{\nu}$ $\cdot\exp\{-b[(p^2+a^2)^{\frac{1}{2}}-p]\}$	$a^{\nu}(1+2b/t)^{-\frac{1}{2}\nu}J_{\nu}[a(t^2+2bt)^{\frac{1}{2}}]$ $\mathrm{Re}\,\nu > -1$
5.143	$(p^2-a^2)^{-\frac{1}{2}}[p-(p^2-a^2)^{\frac{1}{2}}]^{\nu}$ $\cdot\exp\{b[p-(p^2-a^2)^{\frac{1}{2}}]\}$	$a^{\nu}(1+2b/t)^{-\frac{1}{2}\nu}I_{\nu}[a(t^2+2bt)^{\frac{1}{2}}]$ $\mathrm{Re}\,\nu > -1$
5.144	e^{-bp} $-\exp[-b(p^2-ia^2)^{\frac{1}{2}}]$	$0 \quad t < b$ $abe^{i3\pi/4}(t^2-b^2)^{-\frac{1}{2}}$ $\cdot\{\mathrm{ber}_1[a(t^2-b^2)^{\frac{1}{2}}]$ $+i\,\mathrm{bei}_1[a(t^2-b^2)^{\frac{1}{2}}]\} \quad t > b$
5.145	$(p^2-ia^2)^{-\frac{1}{2}}$ $\cdot\exp[-b(p^2-ia^2)^{\frac{1}{2}}]$	$0 \quad t < b$ $\mathrm{ber}[a(t^2-b^2)^{\frac{1}{2}}]+i\,\mathrm{bei}[a(t^2-b^2)^{\frac{1}{2}}]$ $t > b$

	$g(p) = \int_0^\infty f(t)e^{-pt}dt$	$f(t)$
5.146	$(p^2-ia^2)^{-\frac{1}{2}}[p-(p^2-ia^2)^{\frac{1}{2}}]^\nu$ $\cdot\exp[-b(p^2-ia^2)^{\frac{1}{2}}]$	$0 \quad t < b$ $(ia)^\nu e^{-i3\pi\nu/4}(\frac{t-b}{t+b})^{\frac{1}{2}\nu}$ $\cdot\{ber_\nu[a(t^2-b^2)^{\frac{1}{2}}]+i\ bei_\nu[a(t^2-b^2)^{\frac{1}{2}}]$ $\quad t > b$

2.6 Logarithmic Functions

	$g(p) = \int_0^\infty f(t)e^{-pt}dt$	$f(t)$
6.1	$p^{-1}\log p$	$-\gamma-\log t$
6.2	$p^{-n-1}\log p$ $n=1,2,3,\cdots$	$(n!)^{-1}t^n(1+\frac{1}{2}+\cdots+\frac{1}{n}-\gamma-\log t)$
6.3	$p^{-\frac{1}{2}}\log p$	$-(\pi t)^{-\frac{1}{2}}(\log t+\gamma+\log 4)$
6.4	$p^{-n-\frac{1}{2}}\log p$ $n=1,2,3,\cdots$	$\pi^{-\frac{1}{2}}2^{2n}(n!)[(2n)!]^{-1}t^{n-\frac{1}{2}}$ $\cdot[2(1+\frac{1}{3}+\cdots+\frac{1}{2n-1})-\gamma-\log 4-\log t]$
6.5	$p\log(1+a/p)-a$	$t^{-2}[(1+at)e^{-at}-1]$
6.6	$(p+\frac{1}{2}a)\log(1+a/p)-a$	$at^{-1}-\frac{1}{2}t^{-2}(2+at)(1-e^{-at})$
6.7	$p\log[(p+a)/(b+p)]$ $+ b-a$	$t^{-1}e^{-at}(a+t^{-1})-t^{-1}e^{-bt}(b+t^{-1})$
6.8	$p^{-\nu}\log p$ $\mathrm{Re}\,\nu > 0$	$[\Gamma(\nu)]^{-1}t^{\nu-1}[\psi(\nu)-\log t]$
6.9	$p^{-1}\log(p/a+1)$	$-\mathrm{Ei}(-at)$

	$g(p) = \int_0^\infty f(t)e^{-pt}dt$	$f(t)$
6.10	$p^{-1}\log(p/a-1)$	$-\overline{E}i(at)$
6.11	$p^{-1}\log(p/a+1)$	$-Ei(-at)$
6.12	$p^{-2}\log(p/a+1)$	$1-e^{-t}-t\ Ei(-t)$
6.13	$\log[(p+a)/(p-a)]$	$2t^{-1}\sin h(at)$
6.14	$\log[(p+b)/(p-a)]$	$t^{-1}(e^{-at}-e^{-bt})$
6.15	$p^{-1}\log[(p+a)/(p-a)]$	$2\ Shi(at)$
6.16	$\log(1-a/p)$	$t^{-1}(1-e^{-at})$
6.17	$(p^2+a^2)^{-1}\log p$	$a^{-1}\{\cos(at)Si(at)$ $+\sin(at)[\log a-Ci(at)]\}$
6.18	$p(p^2+a^2)^{-1}\log p$	$\cos(at)[\log a-Ci(at)]$ $-\sin(at)Si(at)]$
6.19	$(p+a)^{-1}\log(p+b)$ $b > a$	$e^{-at}\{\log(b-a)-Ei[-(b-a)t]\}$

	$g(p) = \int_0^\infty f(t)e^{-pt}dt$	$f(t)$
6.20	$(p^2+a^2)^{-1}\log(p/a)$	$a^{-1}[\cos(at)\mathrm{Si}(at)$ $-\sin(at)\mathrm{Ci}(at)]$
6.21	$p(p^2+a^2)^{-1}\log(p/a)$	$-\sin(at)\mathrm{Si}(at)$ $-\cos(at)\mathrm{Ci}(at)$
6.22	$p^{-1}(\log p)^2$	$(\gamma+\log t)^2-\pi^2/6$
6.23	$p^{-2}(\log p)^2$	$t[(1-\gamma-\log t)^2+1-\pi^2/6]$
6.24	$p^{-\nu}(\log p)^2$ $\quad \mathrm{Re}\,\nu > 0$	$[\Gamma(\nu)]^{-1}t^{\nu-1}[(\psi(\nu)-\log t)^2-\psi'(\nu)]$
6.25	$p^{-2}(\log p)^3$	$t[(1-\gamma-\log t)^3+\psi^{(n)}(1)-\frac{1}{2}\pi^2$ $-3(1-\pi^2/6)(\gamma+\log t)+5]$
6.26	$p^{-1}\log(p^2+a^2)$	$2\log a-2\,\mathrm{Ci}(at)$
6.27	$p^{-2}\log(p^2+a^2)$	$2t[\log a+(at)^{-1}\sin(at)$ $-\mathrm{Ci}(at)]$
6.28	$p^{-1}\log(p^2-a^2)$	$2\log a - 2\,\mathrm{Cih}(at)$

	$g(p) = \int_0^\infty f(t)e^{-pt}dt$	$f(t)$
6.29	$p^{-2}\log(p^2-a^2)$	$-2t\,\mathrm{Cih}(at)+2a^{-1}\sin h(at)$ $+\ 2t\log a$
6.30	$\log\left[\frac{p^2+ap+b}{p^2-ap+b}\right]$	$4t^{-1}\sin h(\frac{1}{2}at)\cos[t(b-a^2/4)^{\frac{1}{2}}]$
6.31	$p^{-1}e^{a/p}\log p$	$K_0[2(at)^{\frac{1}{2}}]-\frac{1}{2}\log(t/a)I_0[2(at)^{\frac{1}{2}}]$
6.32	$p^{-1}e^{-a/p}\log p$	$-\frac{1}{2}\pi Y_0[2(at)^{\frac{1}{2}}]-\frac{1}{2}\log(t/a)J_0[2(at)^{\frac{1}{2}}]$
6.33	$p^{-\frac{1}{2}}e^{a/p}\log p$	$-\frac{1}{2}(\pi t)^{-\frac{1}{2}}\{\log(t/a)\cosh[2(at)^{\frac{1}{2}}]$ $+\exp[2(at)^{\frac{1}{2}}]\mathrm{Ei}[-4(at)^{\frac{1}{2}}]$ $+\exp[-2(at)^{\frac{1}{2}}]\overline{\mathrm{Ei}}[4(at)^{\frac{1}{2}}]\}$
6.34	$p^{-\frac{1}{2}}e^{-a/p}\log p$	$-(\pi t)^{-\frac{1}{2}}\{\log(t/a)\cos[2(at)^{\frac{1}{2}}]$ $+\ \cos[2(at)^{\frac{1}{2}}]\mathrm{Ci}[4(at)^{\frac{1}{2}}]$ $+\ \sin[2(at)^{\frac{1}{2}}]\mathrm{Si}[4(at)^{\frac{1}{2}}]\}$
6.35	$p^{-3/2}e^{a/p}\log p$	$-\frac{1}{2}(\pi a)^{-\frac{1}{2}}\{\log(t/a)\sinh[2(at)^{\frac{1}{2}}]$ $+\ \exp[2(at)^{\frac{1}{2}}]\mathrm{Ei}[-4(at)^{\frac{1}{2}}]$ $-\ \exp[-2(at)^{\frac{1}{2}}]\overline{\mathrm{Ei}}[4(at)^{\frac{1}{2}}]\}$

	$g(p) = \int_0^\infty f(t)e^{-pt}dt$	$f(t)$
6.36	$p^{-3/2}\log p\ e^{-a/p}$	$-(\pi a)^{-1/2}\{\log(t/a)\sin[2(at)^{1/2}]$ $+ \sin[2(at)^{1/2}]\mathrm{Ci}[4(at)^{1/2}]$ $- \cos[2(at)^{1/2}]\mathrm{Si}[4(at)^{1/2}]\}$
6.37	$\log(1+a^2/p^2)$	$2t^{-1}[1-\cos(at)]$
6.38	$p\ \log(1+a^2/p^2)$	$2t^{-1}\{a\ \sin(at)-t^{-1}[1-\cos(at)]\}$
6.39	$\log(1-a^2/p^2)$	$2t^{-1}[1-\cosh(at)]$
6.40	$p\ \log(1-a^2/p^2)$	$2t^{-1}\{t^{-1}[\cosh(at)-1]-a\ \sin\ at\}$
6.41	$p^{-1}\log(1+a^2/p^2)$	$-2\ \mathrm{Ci}(at)$
6.42	$\log\ [\frac{(p+a)^2+c^2}{(p+b)^2+c^2}]$	$2t^{-1}\cos(ct)(e^{-bt} - e^{-at})$
6.43	$\log[(p^2+a^2)/(p^2+b^2)]$	$2t^{-1}[\cos(bt) - \cos(at)]$
6.44	$p\ \log[(p^2+a^2)/(p^2+b^2)]$	$2t^{-2}[\cos(at)+at\ \sin(at)$ $- \cos(bt)-bt\ \sin(bt)]$

	$g(p) = \int_0^\infty f(t)e^{-pt}dt$	$f(t)$
6.45	$\log[(p^2-a^2)/(p^2-b^2)]$	$2t^{-1}[\cosh(at) - \cosh(bt)]$
6.46	$(p^2+a^2)^{-1}\log(p^2+a^2)$	$-a^{-1}\sin(at)[\log(\frac{1}{2}t/a)+\gamma+Ci(2at)]$ $+ a^{-1} \cos(at)Si(2at)$
6.47	$p(p^2+a^2)^{-1}\log(p^2+a^2)$	$-\cos(at)[\log(\frac{1}{2}t/a)+\gamma+Ci(2at)]$ $- \sin(at)Si(2at)$
6.48	$\log\left(\frac{p^{\frac{1}{2}}+a}{p^{\frac{1}{2}}-a}\right)$	$t^{-1}e^{a^2t}\, Erf(at^{\frac{1}{2}})$
6.49	$(p^2-a^2)^{-1}\{a \log(p^2-a^2)$ $-p \log[(p+a)/(p-a)]\}$	$-2 \sinh(at)(\gamma+\log t)$
6.50	$(p^2-a^2)^{-1}\{p \log(p^2-a^2)$ $-a \log[(p+a)/(p-a)]\}$	$-2 \cosh(at)(\gamma+\log t)$
6.51	$p^{-\frac{1}{2}}\log[2p/a-1$ $+ (4p^2/a^2-4p/a)^{\frac{1}{2}}]$	$- (\pi t)^{-\frac{1}{2}}\overline{Ei}(at)$
6.52	$p^{-\frac{1}{2}}\log[(p/a)^{\frac{1}{2}}+(1+p/a)^{\frac{1}{2}}]$	$- \frac{1}{2}(\pi t)^{-\frac{1}{2}}Ei(-at)$

	$g(p) = \int_0^\infty f(t) e^{-pt} dt$	$f(t)$
6.53	$(p^2+a^2)^{-\frac{1}{2}} \log[a/p+(1+a^2/p^2)^{\frac{1}{2}}]$	$\frac{1}{2}\pi \mathbf{H}_0(at)$
6.54	$p^{-1}-pa^{-1}(p^2+a^2)^{-\frac{1}{2}}$ $\cdot \log[a/p+(1+a^2/p^2)^{\frac{1}{2}}]$	$\frac{1}{2}\pi \mathbf{H}_1(at)$
6.55	$(b+p)^{-\frac{1}{2}} \log[(p+a)^{\frac{1}{2}}+(p+b)^{\frac{1}{2}}]$	$\frac{1}{2}(\pi t)^{-\frac{1}{2}} e^{-bt}$ $\cdot \{\log(a-b)-Ei[-(a-b)t]\}$
6.56	$p^{-1} \log[p+(p^2+a^2)^{\frac{1}{2}}]$	$\log a + Ji_0(at)$
6.57	$(p^2+a^2)^{-\frac{1}{2}} \log[p/a+(1+p^2/a^2)^{\frac{1}{2}}]$	$-\frac{1}{2}\pi Y_0(at)$
6.58	$(p^2+a^2)^{-\frac{1}{2}} \log[p+(a^2+p^2)^{\frac{1}{2}}]$	$J_0(at) \log a - \frac{1}{2}\pi Y_0(at)$
6.59	$(p^2+a^2)^{-3/2} \log[p+(p^2+a^2)^{\frac{1}{2}}]$	$a^{-1} t[J_1(at) \log a - \frac{1}{2}\pi Y_1(at)]$ $-a^{-2} \cos(at)$
6.60	$(p^2+a^2)^{-\frac{1}{2}} \log(p^2+a^2)$	$-\frac{1}{2}\pi Y_0(at)$ $-J_0(at)[\gamma+\log(2t/a)]$

	$g(p) = \int_0^\infty f(t)e^{-pt}dt$	$f(t)$
6.61	$(p^2-a^2)^{-\frac{1}{2}}\log(p^2-a^2)$	$K_0(at)$ $-I_0(at)[\gamma+\log(2t/a)]$
6.62	$p(p^2+a^2)^{-3/2}\log[p+(p^2+a^2)^{\frac{1}{2}}]$	$t[J_0(at)\log a-\frac{1}{2}\pi Y_0(at)]$ $+ a^{-1}\sin(at)$
6.63	$(p^2-a^2)^{-\frac{1}{2}}\log[p+(p^2-a^2)^{\frac{1}{2}}]$	$I_0(at)\log a + K_0(at)$
6.64	$(p^2-a^2)^{-\frac{1}{2}}\log[p/a+(p^2/a^2-1)^{\frac{1}{2}}]$	$K_0(at)$
6.65	$(p^2-a^2)^{-3/2}\log[p+(p^2-a^2)^{\frac{1}{2}}]$	$a^{-1}t[I_1(at)\log a - K_1(at)]$ $+ a^{-2}\cosh(at)$
6.66	$p(p^2-a^2)^{-3/2}\log[p+(p^2-a^2)^{\frac{1}{2}}]$	$t[I_0(at)\log a + K_0(at)]$ $+ a^{-1}\sinh(at)$
6.67	$p(p^2+a^2)^{-\frac{1}{2}}\log[a/p+(1+a^2/p^2)^{\frac{1}{2}}]$	$a-\frac{1}{2}\pi a\mathbf{H}_1(at)$
6.68	$(p^2+1)^{-\frac{1}{2}}\log\left[\frac{p+(1+p^2)^{\frac{1}{2}}}{1+(1+p^2)^{\frac{1}{2}}}\right]$	$\frac{1}{2}\pi[\mathbf{H}_0(at) - Y_0(at)]$

	$g(p) = \int_0^\infty f(t)e^{-pt}dt$	$f(t)$
6.69	$p^{-1}\log^2[p/a+(1+p^2/a^2)^{\frac{1}{2}}]$	$Yi_0(at)$
6.70	$p^{-1}\log^2[p/a+(p^2/a^2-1)^{\frac{1}{2}}]$	$2Ki_0(at)-\pi^2/4$
6.71	$p^{-1}\log^2[p+(p^2+a^2)^{\frac{1}{2}}]$	$\log a[\log a+2Ji_0(at)]$ $- \pi Yi_0(at)$
6.72	$(p^2+a^2)^{-\frac{1}{2}}[p+(p^2+a^2)^{\frac{1}{2}}]^{-\frac{1}{2}}$ $\cdot\log[p/a+(1+p^2/a^2)^{\frac{1}{2}}]$	$-a^{-1}(\frac{1}{2}\pi t)^{-\frac{1}{2}}[\sin(at)\mathrm{Ci}(2at)$ $-\cos(at)\mathrm{Si}(2at)]$
6.73	$(p^2+a^2)^{-\frac{1}{2}}[p+(p^2+a^2)^{\frac{1}{2}}]^{\frac{1}{2}}$ $\cdot\log[p/a+(1+p^2/a^2)^{\frac{1}{2}}]$	$-(\frac{1}{2}\pi t)^{-\frac{1}{2}}[\cos(at)\mathrm{Ci}(2at)$ $+ \sin(at)\mathrm{Si}(2at)]$
6.74	$(\log p)^{-1}$	$\int_0^\infty [\Gamma(u)]^{-1}t^{u-1}\,du$
6.75	$[p\ \log(p/a)]^{-1}$	$\nu(at)$
6.76	$p^{-1}[\log(p/a)]^{-\alpha}$ $\mathrm{Re}\ \alpha > 0$	$[\Gamma(\alpha)]^{-1}\mu(at,\alpha-1)$

	$g(p) = \int_0^\infty f(t)e^{-pt}dt$	$f(t)$
.77	$p^{-\alpha-1}(\log p)^{-1}$ Re $\alpha \geqq 0$	$\nu(t,\alpha)$
.78	$[(p+a)(p+b)]^{-\frac{1}{2}}$ $\cdot\log[(p+a)^{\frac{1}{2}}-(p+b)^{\frac{1}{2}}]$	$\frac{1}{2}\exp[-\frac{1}{2}(a+b)t]$ $\cdot\{I_0[\frac{1}{2}(a-b)t]\log(a-b)$ $-K_0[\frac{1}{2}(a-b)t]\}$
79	$(p^2+a^2)^{-\frac{1}{2}}[(p^2+a^2)^{\frac{1}{2}}-p]^{\frac{1}{2}}$ $\cdot\log[(p^2+a^2)^{\frac{1}{2}}-p]$	$(\frac{1}{2}\pi t)^{-\frac{1}{2}}[\log a\ \sin(at)$ $+\sin(at)\text{Ci}(2at)$ $-\cos(at)\text{Si}(2at)]$
80	$(p^2+a^2)^{-\frac{1}{2}}[(p^2+a^2)^{\frac{1}{2}}+p]^{\frac{1}{2}}$ $\cdot\log[(p^2+a^2)^{\frac{1}{2}}-p]$	$(\frac{1}{2}\pi t)^{-\frac{1}{2}}[\log a\ \cos(at)$ $+\cos(at)\text{Ci}(2at)$ $+\sin(at)\text{Si}(2at)]$

(For other results see 5.77 - 5.84)

2.7 Trigonometric- and Inverse Functions

	$g(p) = \int_0^\infty f(t)e^{-pt}dt$	$f(t)$
7.1	$p^{-1}\sin(a/p)$	$\mathrm{bei}[2(at)^{\frac{1}{2}}]$
7.2	$p^{-1}\cos(ap)$	$\mathrm{ber}[2(at)^{\frac{1}{2}}]$
7.3	$p^{-\frac{1}{2}}\sin(a/p)$	$(\pi t)^{-\frac{1}{2}}\sinh[(2at)^{\frac{1}{2}}]\sin[2(at)^{\frac{1}{2}}]$
7.4	$p^{-\frac{1}{2}}\cos(a/p)$	$(\pi t)^{-\frac{1}{2}}\cosh[(2at)^{\frac{1}{2}}]\cos[(2at)^{\frac{1}{2}}]$
7.5	$p^{-3/2}\sin(a/p)$	$(\pi a)^{-\frac{1}{2}}\cosh[(2at)^{\frac{1}{2}}]\sin[(2at)^{\frac{1}{2}}]$
7.6	$p^{-3/2}\cos(a/p)$	$(\pi a)^{-\frac{1}{2}}\sinh[(2at)^{\frac{1}{2}}]\cos[(2at)^{\frac{1}{2}}]$
7.7	$p^{-\nu}\sin(a/p)$ $\mathrm{Re}\ \nu > -1$	$(t/a)^{\frac{1}{2}\nu-\frac{1}{2}}\{\sin(3\pi\nu/4+\pi/4)\mathrm{ber}_{\nu-1}[2(at)^{\frac{1}{2}}]$ $-\cos(3\pi\nu/4+\pi/4)\mathrm{bei}_{\nu-1}[2(at)^{\frac{1}{2}}]\}$
7.8	$p^{-\nu}\cos(a/p)$ $\mathrm{Re}\ \nu > 0$	$-(t/a)^{\frac{1}{2}\nu-\frac{1}{2}}\{\cos(3\pi\nu/4+\pi/4)\mathrm{ber}_{\nu-1}[2(at)^{\frac{1}{2}}$ $+\sin(3\pi\nu/4+\pi/4)\mathrm{bei}_{\nu-1}[2(at)^{\frac{1}{2}}]\}$
7.9	$p^{-\nu-1}\sin(\frac{1}{4}a^2/p+3\pi\nu/4)$ $\mathrm{Re}\ \nu > -1$	$(\frac{1}{2}a)^{-\nu}t^{\frac{1}{2}\nu}\mathrm{bei}_{\nu}(at^{\frac{1}{2}})$

	$g(p) = \int_0^\infty f(t)e^{-pt}dt$	$f(t)$
7.10	$p^{-\nu-1}\cos(\frac{1}{4}a^2/p+3\pi\nu/4)$ $\mathrm{Re}\,\nu > -1$	$(\frac{1}{2}a)^{-\nu}t^{\frac{1}{2}\nu}\mathrm{ber}_\nu(at^{\frac{1}{2}})$
7.11	$p^{-\frac{1}{2}}e^{-ap^{\frac{1}{2}}}\sin(bp^{\frac{1}{2}})$ $a \geqq b$	$(\pi t)^{-\frac{1}{2}}\exp[-\frac{1}{4}(a^2-b^2)/t]$ $\cdot\sin(\frac{1}{2}ab/t)$
7.12	$p^{-\frac{1}{2}}e^{-ap^{\frac{1}{2}}}\cos(bp^{\frac{1}{2}})$ $a \geqq b$	$(\pi t)^{-\frac{1}{2}}\exp[-\frac{1}{4}(a^2-b^2)/t]$ $\cdot\cos(\frac{1}{2}ab/t)$
7.13	$e^{-ap^{\frac{1}{2}}}\sin(bp^{\frac{1}{2}})$ $a \geqq b$	$\frac{1}{2}\pi^{-\frac{1}{2}}t^{-3/2}\exp[-\frac{1}{4}(a^2-b^2)/t]$ $\cdot[a\sin(\frac{1}{2}ab/t)-b\cos(\frac{1}{2}ab/t)]$
7.14	$e^{-ap^{\frac{1}{2}}}\cos(bp^{\frac{1}{2}})$ $a \geqq b$	$\frac{1}{2}\pi^{-\frac{1}{2}}t^{-3/2}\exp[-\frac{1}{4}(a^2-b^2)/t]$ $\cdot[b\sin(\frac{1}{2}ab/t)+a\cos(\frac{1}{2}ab/t)]$
7.15	$p^{-\nu}\cos(ap^{-\frac{1}{2}})$ $\mathrm{Re}\,\nu > 0$	$[\Gamma(\nu)]^{-1}t^{\nu-1}{}_0F_2(\ ;\nu,\frac{1}{2};-\frac{1}{4}a^2t)$
7.16	$p^{-\nu}\sin(ap^{-\frac{1}{2}})$ $\mathrm{Re}\,\nu > -\frac{1}{2}$	$a[\Gamma(\frac{1}{2}+\nu)]^{-1}t^{\nu-\frac{1}{2}}{}_0F_2(\ ;\nu+\frac{1}{2},3/2;-\frac{1}{4}a^2t)$

	$g(p) = \int_0^\infty f(t)e^{-pt}dt$	$f(t)$
7.17	$\int_0^{\frac{1}{2}\pi}(p^2+a^2\cos^2u)^{-\frac{1}{2}}\cos(2nu)\,du$	$\frac{1}{2}(-1)^n\pi J_n^2(\frac{1}{2}at)$
7.18	$\int_0^{\frac{1}{2}\pi}(p^2+a^2\cos^2u)^{-\mu-1}$ $\cdot\cos^{\mu+\frac{1}{2}}u\,\cos[(\mu-\frac{1}{2})u]\,du$ Re $\mu > -1$	$\pi[\Gamma(\mu+1)]^{-1}2^{-\mu-\frac{1}{2}}a^{-\mu-1}$ $\cdot t^{\mu}\sin(\frac{1}{2}at)J_\mu(\frac{1}{2}at)$
7.19	$\int_0^{\frac{1}{2}\pi}(p^2+a^2\cos^2u)^{-\mu}$ $\cdot\cos^{\mu-\frac{1}{2}}u\,\cos[(\pi+\frac{1}{2})u]\,du$ Re $\mu > 0$	$\pi[\Gamma(\mu)]^{-1}2^{\frac{1}{2}-\mu}a^{-\mu}$ $\cdot t^{\mu-1}\cos(\frac{1}{2}at)J_\mu(\frac{1}{2}at)$
7.20	$\int_0^{\frac{1}{2}\pi}(p^2+\cos^2u)^{-\mu-\frac{1}{2}}$ $\cdot\cos^{\mu}u\,\cos(\nu u)\,du$ Re $\mu > -\frac{1}{2}$	$\frac{1}{2}(2a)^{-\mu}\pi^{3/2}[\Gamma(\frac{1}{2}+\mu)]^{-1}$ $\cdot t^{\mu}J_{\frac{1}{2}\mu+\frac{1}{2}\nu}(\frac{1}{2}at)J_{\frac{1}{2}\mu-\frac{1}{2}\nu}(\frac{1}{2}at)$
7.21	$\int_0^{\frac{1}{2}\pi}(p^2+\sin^2u)^{-\frac{1}{2}}\sin u$ $\cdot[p+(p^2+\sin^2u)^{\frac{1}{2}}]^{-\nu}du$ Re $\nu > -1$	$(\frac{1}{2}\pi/t)^{\frac{1}{2}}\mathbf{H}_{\nu-\frac{1}{2}}(t)$

	$g(p) = \int_0^\infty f(t)e^{-pt}dt$	$f(t)$
7.22	$\int_0^{\frac{1}{2}\pi} (p^2-\sin^2 u)^{-\frac{1}{2}}\sin u$ $\cdot[p+(p^2-\sin^2 u)^{\frac{1}{2}}]^{-\nu}du$ Re $\nu > -1$	$(\frac{1}{2}\pi/t)^{\frac{1}{2}}\mathbf{L}_\nu(t)$
7.23	$\int_0^\pi [b^2+(p+ia\cos u)^2]^{-\mu}$ $\cdot\sin^{2\nu}u\,du$ Re $\mu > 0$	$2^{\mu+\nu-\frac{1}{2}}\Gamma(\frac{1}{2}+\nu)[\Gamma(2\mu)]^{-1}\pi^{\frac{1}{2}}a^{-\nu}b^{\frac{1}{2}-\mu}$ $\cdot t^{\mu-\nu-\frac{1}{2}}J_\nu(at)J_{\mu-\frac{1}{2}}(bt)$
7.24	$\csc(\pi\nu)[\int_0^\pi \exp(a\cos u)$ $\cdot\cos(pu)du-\pi I_p(a)]$	$\exp(-a\cosh t)$
7.25	$p^{-\nu-1}\exp[-\frac{1}{4}(a^2+b^2)/p]$ $\cdot\int_0^\pi \exp(\frac{1}{2}ab/p\cos u)$ $\cdot\sin^{2\nu}u\,du$	$2^{2\nu}\pi^{\frac{1}{2}}\Gamma(\frac{1}{2}+\nu)(ab)^{-\nu}$ $\cdot J_\nu(at^{\frac{1}{2}})J_\nu(bt^{\frac{1}{2}})$ Re $\nu > -\frac{1}{2}$
7.26	$\arctan(a/p)$	$t^{-1}\sin(at$
7.27	$p\,\text{arccot}\,p-1$	$t^{-2}(t\cos t-\sin t)$

	$g(p) = \int_0^\infty f(t)e^{-pt}dt$	$f(t)$
7.28	$p \arctan(a/p) - a$	$t^{-2}[at \cos(at) - \sin(at)]$
7.29	$p^{-1}\arctan(p/a)$	$-\mathrm{si}(at)$
7.30	$p^{-1}\mathrm{arccot}(p/a)$	$\mathrm{Si}(at)$
7.31	$(p^2+a^2)^{-1}\arctan(a/p)$	$\frac{1}{2}a^{-1}\cos(at)[\mathrm{Ci}(2at) - \gamma - \log(2at)]$ $+\frac{1}{2}a^{-1}\sin(at)\mathrm{Si}(2at)$
7.32	$p(p^2+a^2)^{-1}\arctan(a/p)$	$\frac{1}{2}\sin(at)[\gamma + \log(2at) - \mathrm{Ci}(2at)]$ $+\frac{1}{2}\cos(at)\mathrm{Si}(2at)$
7.33	$\log(p^2+a^2)\arctan(a/p)$	$-2t^{-1}\sin(at)(\gamma + \log t)$
7.34	$(p^2+1)^{-1}\{\arctan[(p-1)^{-1}]$ $+\frac{1}{2}p \log(2p^2-2p+2)\}$	$-\cos t \,\overline{\mathrm{Ei}}(t)$
7.35	$(p^2+1)^{-1}\{p \arctan[(p-1)^{-1}]$ $-\frac{1}{2}\log(p^2-2p+2)\}$	$\sin t \,\overline{\mathrm{Ei}}(t)$

	$g(p) = \int_0^\infty f(t)e^{-pt}dt$	$f(t)$
7.36	$(p^2+a^2)^{-1}\{a \arctan[a(p+1)^{-1}]$ $+\frac{1}{2}p \log[(p+1)^2+a^2]\}$	$-\cos(at)\mathrm{Ei}(-t)$
7.37	$(p^2+a^2)^{-1}\{p \arctan[a(p+1)^{-1}]$ $-\frac{1}{2}a \log[(p+1)^2+a^2]\}$	$\sin(at)\mathrm{Ei}(-t)$
7.38	$(p^2+a^2)^{-1}\{p \arctan(\frac{2bp}{b^2-a^2-p^2})$ $+\frac{1}{2}p \log[\frac{(b+a)^2+p^2}{(b-a)^2+p^2}]\}$	$-2b^{-1}\cos(at)\mathrm{si}(bt)$
7.39	$(p^2+a^2)^{-1}\{a \arctan(\frac{2ap}{p^2+b^2-a^2})$ $-\frac{1}{2}a \log[\frac{p^2+(b^2-a^2)^2+4a^2p^2}{b^2}]\}$	$2b^{-1}\mathrm{snn}(at)\mathrm{Ci}(bt)$
7.40	$(-p)^{-\frac{1}{2}}\arctan[\frac{p^2+2ap}{(p+a)^2}]$ $-2a < \mathrm{Re}\, p < 0$	$(\pi t)^{-\frac{1}{2}}\mathrm{Ei}(-2at)$
7.41	$\arctan[2ap/(p^2+b^2-a^2)]$	$2t^{-1}\sin(at)\cos(bt)$
7.42	$\arctan[(p^2-a^2)/2bp]$	$2t^{-1}\sin(bt)\cosh[t(a^2-b^2)^{\frac{1}{2}}]$

	$g(p) = \int_0^\infty f(t)e^{-pt}dt$	$f(t)$
7.43	$\arcsin[a(p^2+a^2)^{-\frac{1}{2}}]$	$t^{-1}\sin(at)$
7.44	$(p^2-a^2)^{-\frac{1}{2}}\arcsin(a/p)$	$\frac{1}{2}\pi\mathbf{L}_0(at)$
7.45	$p(p^2-a^2)^{-\frac{1}{2}}\arcsin(a/p)$	$\frac{1}{2}\pi a\mathbf{L}_1(at)+a$
7.46	$\arccos[p(p^2+a^2)^{-\frac{1}{2}}]$	$t^{-1}\sin(at)$
7.47	$(p^2-a^2)^{-\frac{1}{2}}\arccos(a/p)$	$K_0(at)$
7.48	$(p^2+a^2)^{-\frac{1}{2}}$ $\cdot\sin[b+\arctan(a/p)]$	$\sin(at+b)$
7.49	$(p^2+a^2)^{-\frac{1}{2}}$ $\cdot\cos[b+\arctan(a/p)]$	$\cos(at+b)$
7.50	$(p^2+a^2)^{-\frac{1}{2}\nu}$ $\cdot\sin[\nu \arctan(a/p)]$	$[\Gamma(\nu)]^{-1}t^{\nu-1}\sin(at)$ $\mathrm{Re}\ \nu > -1$

	$g(p) = \int_0^\infty f(t)e^{-pt}dt$	$f(t)$
7.51	$(p^2+a^2)^{-\frac{1}{2}\nu}$ $\cdot\cos[\nu \arctan(a/p)]$	$[\Gamma(\nu)]^{-1}t^{\nu-1}\cos(at)$ $\mathrm{Re}\ \nu > 0$
7.52	$p^{-\nu}\cos[2n \arcsin(p^{-\frac{1}{2}})]$ $n=0,1,2,\cdots,\ \mathrm{Re}\ \nu > 0$	$[\Gamma(\nu)]^{-1}t^{\nu-1}{}_2F_2(-n,n;\nu;\frac{1}{2};t)$
7.53	$p^{-\nu}\sin[(2n+1)\arcsin(p^{-\frac{1}{2}})]$ $n=0,1,2,\cdots,\ \mathrm{Re}\ \nu > 0$	$[\Gamma(\nu)]^{-1}(2n+1)t^{\nu-1}$ $\cdot{}_2F_2(-n,n+1;\nu;\frac{3}{2};t)$

2.8 Hyperbolic- and Inverse Functions

	$g(p) = \int_0^\infty f(t)e^{-pt}dt$	f(t)	
8.1	$p^{-1}\operatorname{sech}(ap)$	0	$0<t<a$
		2	$(4n-3)a<t<(4n-1)a$
		0	$(4n-1)a<t<(4n+1)a$
			$n = 1,2,3,\cdots$
8.2	$p^{-1}\operatorname{csch}(ap)$	0	$0<t<a$
		2n	$(2n-1)a<t<(2n+1)a$
			$n = 1,2,3,\cdots$
8.3	$p^{-1}[1-\operatorname{sech}(ap)]$	1	$0<t<a$
		-1	$(4n-3)a<t<(4n-1)a$
		1	$(4n-1)a<t<(4n+1)a$
			$n = 1,2,3,\cdots$
8.4	$p^{-1}\operatorname{sech}(2ap)\cosh(ap)$	1	$(4n-3)a<t<(4n-1)a$
		2	$(8n-5)a<t<(8n-3)a$
		0	otherwise
			$n = 1,2,3,\cdots$
8.5	$p^{-1}\operatorname{sech}(2ap)\sinh(ap)$	1	$(8n-7)a<t<(8n-5)a$
		-1	$(8n-3)a<t<(8n-1)a$
		0	otherwise
			$n = 1,2,3,\cdots$

	$g(p) = \int_0^\infty f(t)e^{-pt}dt$	$f(t)$
8.6	$p^{-1}\operatorname{sech}(ap)\tanh(ap)$	$0 \quad t < a$ $4n-2 \quad (4n-3)a<t<(4n-1)a$ $-4n \quad (4n-1)a<t<(4n+1)a$ $n = 1,2,3,\cdots$
8.7	$p^{-1}\tanh(ap)$	$(-1)^{n-1} \quad 2(n-1)a<t<2na$ $n = 1,2,3,\cdots$
8.8	$p^{-1}\coth(ap)$	$2n-1 \quad 2(n-1)a<t<2na$ $n = 1,2,3,\cdots$
8.9	$p^{-2}\operatorname{sech}(ap)$	$0 \quad t < a$ $t-(-1)^n(t-2na) \quad (2n-1)a<t<(2n+1)a$ $n = 1,2,3,\cdots$
8.10	$p^{-2}\operatorname{csch}(ap)$	$0 \quad t < a$ $2n(t-an) \quad (2n-1)a<t<(2n+1)a$ $n = 1,2,3,\cdots$
8.11	$(p^2+\frac{1}{4}\pi^2/a^2)^{-1}\operatorname{csch}(ap)$	$2a\pi^{-1}[\lvert\cos(\frac{1}{2}\pi t/a)\rvert-\cos(\frac{1}{2}\pi t/a)]$

	$g(p) = \int_0^\infty f(t)e^{-pt}dt$	$f(t)$
8.12	$p^{-2}\tanh(ap)$	$a-(-1)^n(2an+a-t)$ $\quad 2na<t<2(n+1)a$ $n = 0,1,2,\cdots$
8.13	$p^{-2}\coth(ap)$	$(2n+1)t-2an(n+1)$ $\quad 2na<t<2(n+1)a$ $n = 0,1,2,\cdots$
8.14	$(p^2+\frac{1}{4}\pi^2/a^2)^{-1}\coth(ap)$	$2a\pi^{-1}\lvert\sin(\frac{1}{2}\pi t/a)\rvert$
8.15	$p(p^2+\frac{1}{4}\pi^2/a^2)^{-1}\coth(ap)$	$\cos(\frac{1}{2}\pi t/a)$ $\quad 2an<t<(2n+1)2a$ $-\cos(\frac{1}{2}\pi t/a)$ $\quad (2n+1)2a<t<(2n+2)2a$ $n = 0,1,2,\cdots$
8.16	$\operatorname{csch}(ap)\sinh(bp)$ $b < a$	$\sum_{n=0}^{\infty} \delta[t-(a-b+2na)]$ $- \sum_{n=0}^{\infty} \delta[t-(a+b+2na)]$
8.17	$\operatorname{sech}(ap)\sinh(bp)$ $b < a$	$\sum_{n=0}^{\infty} (-1)^n\delta[t-(a-b+2na)]$ $- \sum_{n=0}^{\infty} (-1)^n\delta[t-(a+b+2na)]$

	$g(p) = \int_0^\infty f(t)e^{-pt}dt$	$f(t)$
8.18	$\mathrm{csch}(ap)\cosh(bp)$ $b < a$	$\sum_{n=0}^{\infty} \delta[t-(a-b+2na)]$ $+\sum_{n=0}^{\infty} \delta[t-(a+b+2na)]$
8.19	$\mathrm{sech}(ap)\cosh(bp)$ $b < a$	$\sum_{n=0}^{\infty} (-1)^n\delta[t-(a-b+2na)]$ $+\sum_{n=0}^{\infty} (-1)^n\delta[t-(a+b+2na)]$
8.20	$p^{-1}e^{bp}\mathrm{sech}(pa)$ $b < a$	$2\sum_{n=0}^{\infty} (-1)^n H[t-(a-b+2na)]$ $=1-4\pi^{-1}\sum_{n=0}^{\infty}(-1)^n(2n+1)^{-1}\cos[(n+\frac{1}{2})(t+b)\pi/a]$
8.21	$p^{-1}e^{-bp}\mathrm{sech}(pa)$ $b > -a$	$2\sum_{n=0}^{\infty} (-1)^n H[t-(a+b+2na)]$ $=1-4\pi^{-1}\sum_{n=0}^{\infty}(-1)^n(2n+1)^{-1}\cos[(n+\frac{1}{2})(t-b)\pi/a]$
8.22	$p^{-1}e^{bp}\mathrm{csch}(pa)$ $b < a$	$2\sum_{n=0}^{\infty} H[t-(a-b+2na)]$ $=(t+b)/a+2\pi^{-1}\sum_{n=1}^{\infty}(-1)^n n^{-1}\sin[n(t+b)\pi/a]$

	$g(p) = \int_0^\infty f(t)e^{-pt}dt$	$f(t)$
8.23	$p^{-1}e^{-bp}\operatorname{csch}(pa)$ $b > -a$	$2\sum_{n=0}^{\infty} H[t-(a+b+2na)]$ $=(t-b)/a+2\pi^{-1}\sum_{n=1}^{\infty}(-1)^n n^{-1}\sin[n(t-b)\pi/a]$
8.24	$e^{bp}\operatorname{csch}(pa)$ $b < a$	$2\sum_{n=0}^{\infty} \delta[t-(a-b+2na)]$
8.25	$e^{-bp}\operatorname{csch}(pa)$ $b > -a$	$2\sum_{n=0}^{\infty} \delta[t-(a+b+2na)]$
8.26	$e^{bp}\operatorname{sech}(pa)$ $b < a$	$2\sum_{n=0}^{\infty} (-1)^n\delta[t-(a-b+2na)]$
8.27	$e^{-bp}\operatorname{sech}(pa)$ $b > -a$	$2\sum_{n=0}^{\infty} (-1)^n\delta[t-(a+b+2na)]$

	$g(p) = \int_0^\infty f(t)e^{-pt}dt$	$f(t)$
8.28	$p^{-1}\operatorname{sech}(ap)\sinh(bp)$ $b < a$	$4\pi^{-1}\sum_{n=0}^{\infty}(-1)^n(2n+1)^{-1}$ $\cdot\sin[(n+\tfrac{1}{2})\pi b/a]\sin[(n+\tfrac{1}{2})\pi t/a]$ $=\sum_{n=0}^{\infty}(-1)^n H[t-(a-b+2na)]$ $-\sum_{n=0}^{\infty}(-1)^n H[t-(a+b+2na)]$
8.29	$p^{-1}\operatorname{csch}(ap)\sinh(bp)$ $b < a$	$b/a+2\pi^{-1}\sum_{n=1}^{\infty}(-1)^n n^{-1}\sin(n\pi b/a)$ $\cdot\cos(n\pi t/a)$ $=\sum_{n=0}^{\infty}H[t-(a-b+2na)]$ $-\sum_{n=0}^{\infty}H[t-(a+b+2na)]$
8.30	$p^{-1}\operatorname{sech}(ap)\cosh(bp)$ $b < a$	$1-4\pi^{-1}\sum_{n=0}^{\infty}(-1)^n(2n+1)^{-1}$ $\cdot\cos[(n+\tfrac{1}{2})\pi b/a]\cos[(n+\tfrac{1}{2})\pi t/a]$ $=\sum_{n=0}^{\infty}(-1)^n H[t-(a-b+2na)]$ $+\sum_{n=0}^{\infty}(-1)^n H[t-(a+b+2na)]$

	$g(p) = \int_0^\infty f(t)e^{-pt}dt$	$f(t)$
8.31	$p^{-1}\operatorname{csch}(ap)\cosh(bp)$ $b < a$	$2\pi^{-1}\sum_{n=1}^{\infty}(-1)^n n^{-1}\cos(n\pi b/a)$ $\cdot\sin(n\pi t/a) \quad + \quad t/a$ $= \sum_{n=0}^{\infty} H[t-(a-b+2na)]$ $+ \sum_{n=0}^{\infty} H[t-(a+b+2na)]$
8.32	$p^{-2}\operatorname{csch}(ap)\sinh(bp)$ $b \leq a$	$bt/a+2a\pi^{-2}\sum_{n=1}^{\infty}(-1)^n n^{-2}\sin(n\pi b/a)$ $\cdot\sin(n\pi t/a)$
8.33	$p^{-2}\operatorname{sech}(ap)\cosh(bp)$ $b \leq a$	$t-2a\pi^{-2}\sum_{n=0}^{\infty}(-1)^n(n+\frac{1}{2})^{-2}$ $\cdot\cos[(n+\frac{1}{2})\pi b/a]\sin[(n+\frac{1}{2})\pi t/a]$
8.34	$p^{-2}\operatorname{sech}(ap)\sinh(bp)$ $b \leq a$	$b-2a\pi^{-2}\sum_{n=0}^{\infty}(-1)^n(n+\frac{1}{2})^{-2}$ $\cdot\sin[(n+\frac{1}{2})\pi b/a]\cos[(n+\frac{1}{2})\pi t/a]$
8.35	$p^{-2}\operatorname{csch}(ap)\cosh(bp)$ $b \leq a$	$\frac{1}{2}b^2/a - \frac{1}{6}a - 2a\pi^{-2}$ $\cdot\sum_{n=1}^{\infty}(-1)^n n^{-2}\cos(n\pi b/a)\cos(n\pi t/a)$

	$g(p) = \int_0^\infty f(t)e^{-pt}dt$	$f(t)$
3.36	$p^{-3}\operatorname{csch}(ap)\cosh(bp)$	$(\frac{1}{2}b^2/a - \frac{1}{6}a)t - 2a^2\pi^{-3} \cdot \sum_{n=1}^{\infty}(-1)^n n^{-3}\cos(n\pi b/a)\sin(n\pi t/a)$
.37	$p^{-3}\operatorname{sech}(ap)\sinh(bp)$ $b \leqq a$	$bt - 2a^2\pi^{-3}\sum_{n=0}^{\infty}(-1)^n(n+\frac{1}{2})^{-3} \cdot \sin[(n+\frac{1}{2})\pi b/a]\sin[(n+\frac{1}{2})\pi t/a]$
.38	$p^{-\frac{1}{2}}\sinh(a/p)$	$\frac{1}{2}(\pi t)^{-\frac{1}{2}}\{\cosh[2(at)^{\frac{1}{2}}] - \cos[2(at)^{\frac{1}{2}}]\}$
.39	$p^{-\frac{1}{2}}\cosh(a/p)$	$\frac{1}{2}(\pi t)^{-\frac{1}{2}}\{\cosh[2(at)^{\frac{1}{2}}] + \cos[2(at)^{\frac{1}{2}}]\}$
.40	$p^{-3/2}\sinh(a/p)$	$\frac{1}{2}(\pi a)^{-\frac{1}{2}}\{\sinh[2(at)^{\frac{1}{2}}] - \sin[2(at)^{\frac{1}{2}}]\}$
.41	$p^{-3/2}\cosh(a/p)$	$\frac{1}{2}(\pi a)^{-\frac{1}{2}}\{\sinh[2(at)^{\frac{1}{2}}] + \sin[2(at)^{\frac{1}{2}}]\}$
.42	$p^{-5/2}\sinh(a/p)$	$\frac{1}{2}a^{-1}(t/\pi)^{\frac{1}{2}}\{\cosh[2(at)^{\frac{1}{2}}] + \cos[2(at)^{\frac{1}{2}}]\}$ $-\frac{1}{4}\pi^{-\frac{1}{2}}a^{-3/2}\{\sinh[2(at)^{\frac{1}{2}}] + \sin[2(at)^{\frac{1}{2}}]\}$
.43	$p^{-5/2}\cosh(a/p)$	$\frac{1}{2}a^{-1}(t/\pi)^{\frac{1}{2}}\{\cosh[2(at)^{\frac{1}{2}}] - \cos[2(at)^{\frac{1}{2}}]\}$ $-\frac{1}{4}\pi^{-\frac{1}{2}}a^{-3/2}\{\sinh[2(at)^{\frac{1}{2}}] - \sin[2(at)^{\frac{1}{2}}]\}$

	$g(p) = \int_0^\infty f(t)e^{-pt}dt$	$f(t)$
8.44	$p^{-\nu}\sinh(a/p)$ Re $\nu > -1$	$\frac{1}{2}(t/a)^{\frac{1}{2}\nu-\frac{1}{2}}\{I_{\nu-1}[2(at)^{\frac{1}{2}}]-J_{\nu-1}[2(at)^{\frac{1}{2}}]\}$
8.45	$p^{-\nu}\cosh(a/p)$ Re $\nu > 0$	$\frac{1}{2}(t/a)^{\frac{1}{2}\nu-\frac{1}{2}}\{I_{\nu-1}[2(at)^{\frac{1}{2}}]+J_{\nu-1}[2(at)^{\frac{1}{2}}]\}$
8.46	$\operatorname{sech}(ap^{\frac{1}{2}})$	$-a^{-2}[\frac{\partial}{\partial\nu}\theta_1(\frac{1}{2}\nu\|ta^{-2})]_{\nu=o}$
8.47	$\operatorname{csch}(ap^{\frac{1}{2}})$	$-a^{-2}[\frac{\partial}{\partial\nu}\theta_4(\frac{1}{2}\nu\|ta^{-2})]_{\nu=o}$
8.48	$p^{-\frac{1}{2}}\operatorname{sech}(ap^{\frac{1}{2}})$	$a^{-1}\hat{\theta}_2(\frac{1}{2}\|ta^{-2})$
8.49	$p^{-\frac{1}{2}}\operatorname{csch}(ap^{\frac{1}{2}})$	$a^{-1}\theta_4(0\|ta^{-2})$
8.50	$p^{-1}\operatorname{sech}(ap^{\frac{1}{2}})$	$1-a^{-1}\int_o^a \theta_1(\frac{1}{2}ua^{-1}\|ta^{-2})du$
8.51	$p^{-\frac{1}{2}}\tanh(ap^{\frac{1}{2}})$	$a^{-1}\theta_2(0\|ta^{-2})$
8.52	$p^{-\frac{1}{2}}\coth(ap^{\frac{1}{2}})$	$a^{-1}\theta_3(0\|ta^{-2})$

	$g(p) = \int_0^\infty f(t)e^{-pt}dt$	$f(t)$
8.53	$p^{-1}\tanh(ap^{\frac{1}{2}})$	$\int_0^1 \hat{\theta}_2(\frac{1}{2}u \mid ta^{-2})du$
8.54	$\sinh(\nu p^{\frac{1}{2}})\operatorname{csch}(ap^{\frac{1}{2}})$ $-a < \nu < a$	$a^{-1} \frac{\partial}{\partial\nu} \theta_4(\frac{1}{2}\nu a^{-1} \mid ta^{-2})$
8.55	$\cosh(\nu p^{\frac{1}{2}})\operatorname{sech}(ap^{\frac{1}{2}})$ $-a < \nu < a$	$a^{-1} \frac{\partial}{\partial\nu} \theta_1(\frac{1}{2}\nu a^{-1} \mid ta^{-2})$
8.56	$\sinh(\nu p^{\frac{1}{2}})\operatorname{sech}(ap^{\frac{1}{2}})$ $-a < \nu < a$	$a^{-1} \frac{\partial}{\partial\nu} \hat{\theta}_1(\frac{1}{2}\nu a^{-1} \mid ta^{-2})$
8.57	$\operatorname{csch}(ap^{\frac{1}{2}})\cosh(\nu p^{\frac{1}{2}})$	$-a^{-1} \frac{\partial}{\partial\nu} \hat{\theta}_4(\frac{1}{2}\nu a^{-1} \mid ta^{-2})$
8.58	$p^{-\frac{1}{2}}\cosh(\nu p^{\frac{1}{2}})\operatorname{sech}(ap^{\frac{1}{2}})$ $-a \leq \nu \leq a$	$a^{-1}\hat{\theta}_1(\frac{1}{2}\nu a^{-1} \mid ta^{-2})$
8.59	$p^{-\frac{1}{2}}\sinh(\nu p^{\frac{1}{2}})\operatorname{sech}(ap^{\frac{1}{2}})$ $-a \leq \nu \leq a$	$a^{-1}\theta_1(\frac{1}{2}\nu a^{-1} \mid ta^{-2})$
8.60	$p^{-\frac{1}{2}}\cosh(\nu p^{\frac{1}{2}})\operatorname{csch}(ap^{\frac{1}{2}})$ $-a \leq \nu \leq a$	$a^{-1}\theta_4(\frac{1}{2}\nu a^{-1} \mid ta^{-2})$

	$g(p) = \int_0^\infty f(t)e^{-pt}dt$	$f(t)$
8.61	$p^{-\frac{1}{2}}\sinh(\nu p^{\frac{1}{2}})\operatorname{csch}(ap^{\frac{1}{2}})$ $-a \leqq \nu \leqq a$	$-a^{-1}\hat{\theta}_4(\frac{1}{2}\nu a^{-1}\mid ta^{-2})$
8.62	$p^{-1}\sinh(\nu p^{\frac{1}{2}})\operatorname{csch}(ap^{\frac{1}{2}})$ $-a \leqq \nu \leqq a$	$a^{-1}\int_a^{\nu+a} \theta_3(\frac{1}{2}a^{-1}u\mid ta^{-2})du$ $= -a^{-1}\int_0^{\nu} \theta_4(\frac{1}{2}a^{-1}u\mid ta^{-2})du$
8.63	$p^{-1}\cosh(\nu p^{\frac{1}{2}})\operatorname{sech}(ap^{\frac{1}{2}})$ $-a \leqq \nu \leqq a$	$1+a^{-1}\int_a^{\nu} \theta_1(\frac{1}{2}a^{-1}u\mid ta^{-2})du$ $= 1-a^{-1}\int_0^{\nu+a} \theta_2(\frac{1}{2}a^{-1}u\mid ta^{-2})du$
8.64	$p^{-1}\sinh(\nu p^{\frac{1}{2}})\operatorname{sech}(ap^{\frac{1}{2}})$ $-a \leqq \nu \leqq a$	$-a^{-1}\int_0^{\nu} \hat{\theta}_1(\frac{1}{2}a^{-1}u\mid ta^{-2})du$ $= a^{-1}\int_a^{\nu+a} \hat{\theta}_2(\frac{1}{2}a^{-1}u\mid ta^{-2})du$
8.65	$p^{-1}\cosh(\nu p^{\frac{1}{2}})\operatorname{csch}(ap^{\frac{1}{2}})$ $-a \leqq \nu \leqq a$	$a^{-1}\int_0^{\nu} \hat{\theta}_4(\frac{1}{2}a^{-1}u\mid ta^{-2})du$ $+ a^{-1}\int_0^{t} [\frac{\partial}{\partial w}\hat{\theta}_4(\frac{1}{2}a^{-1}w\mid ua^{-2})du$

	$g(p) = \int_0^\infty f(t)e^{-pt}dt$	$f(t)$
.66	$p^{-\frac{1}{2}}[\tanh(p^{\frac{1}{2}}+a^{\frac{1}{2}})$ $+\tanh(p^{\frac{1}{2}}-a^{\frac{1}{2}})]$	$2e^{at}\theta_3(a^{\frac{1}{2}}t\|t)$
.67	$p^{-\frac{1}{2}}[\tanh(p^{\frac{1}{2}}+a^{\frac{1}{2}})$ $-\tanh(p^{\frac{1}{2}}-a^{\frac{1}{2}})]$	$2e^{at}\hat{\theta}_3(a^{\frac{1}{2}}t\|t)$
.68	$(p-ic)^{-1}\sinh(bp^{\frac{1}{2}})$ $\cdot\operatorname{csch}(ap^{\frac{1}{2}})$ $a > b > 0$	$e^{ict}\sinh[b(ic)^{\frac{1}{2}}]\operatorname{csch}[a(ic)^{\frac{1}{2}}]$ $+2\pi\sum_{n=1}^{\infty}(n^2\pi^2+ica^2)^{-1}n(-1)^n\sin(n\pi b/a)$ $\cdot\exp(-n^2\pi^2t/a$
.69	$(p-ic)^{-1}\cosh(bp^{\frac{1}{2}})$ $\operatorname{sech}(ap^{\frac{1}{2}})$ $a > b > 0$	$e^{ict}\cosh[b(ic)^{\frac{1}{2}}]\operatorname{sech}[a(ic)^{\frac{1}{2}}]$ $-2\pi\sum_{n=0}^{\infty}(-1)^n(n+\frac{1}{2})[(n+\frac{1}{2})^2\pi^2+ica^2]^{-1}$ $\cdot\cos[(n+\frac{1}{2})\pi b/a]\exp[-(n+\frac{1}{2})^2\pi^2t/a]$
.70	$p^{-\frac{1}{2}}\sinh(\frac{1}{2}ab/p)$ $\cdot\exp[-\frac{1}{4}(a^2+b^2)/p]$	$(\pi t)^{-\frac{1}{2}}\sin(at^{\frac{1}{2}})\sin(bt^{\frac{1}{2}})$
.71	$p^{-\frac{1}{2}}\cosh(\frac{1}{2}ab/p)$ $\cdot\exp[-\frac{1}{4}(a^2+b^2)/p]$	$(\pi t)^{-\frac{1}{2}}\cos(at^{\frac{1}{2}})\cos(bt^{\frac{1}{2}})$

	$g(p) = \int_0^\infty f(t)e^{-pt}dt$	$f(t)$
8.72	$(c^2+p^2)^{-\frac{1}{2}}\operatorname{sech}[a(c^2+p^2)^{\frac{1}{2}}]$ $\cdot\sinh[b(c^2+p^2)^{\frac{1}{2}}]$ $b \leqq a$	$2a^{-1}\sum_{n=0}^{\infty}(-1)^n\{c^2+[(n+\frac{1}{2})\pi/a]^2\}^{-\frac{1}{2}}$ $\cdot\sin[(n+\frac{1}{2})\pi b/a]$ $\cdot\sin\{t[c^2+(n+\frac{1}{2})^2\pi^2/a^2]^{\frac{1}{2}}\}$ $=\sum_{n=0}^{\infty}(-1)^n H[t-(a-b+2an)]$ $\cdot J_o\{c[t^2-(a-b+2an)^2]^{\frac{1}{2}}\}$ $-\sum_{n=0}^{\infty}(-1)^n H[t-(a+b+2an)]$ $\cdot J_o\{c[t^2-(a+b+2an)^2]^{\frac{1}{2}}\}$
8.73	$(c^2+p^2)^{-\frac{1}{2}}\operatorname{csch}[a(c^2+p^2)^{\frac{1}{2}}]$ $\cdot\sinh[b(c^2+p^2)^{\frac{1}{2}}]$ $b < a$	$\sum_{n=0}^{\infty} H(t-(a-b+2an)]$ $\cdot J_o\{c[t^2-(a-b+2an)^2]^{\frac{1}{2}}\}$ $-\sum_{n=0}^{\infty} H[t-(a+b+2an)^2]$ $\cdot J_o\{c[t^2-(a+b+2an)^2]^{\frac{1}{2}}\}$

	$g(p) = \int_0^\infty f(t)e^{-pt}dt$	$f(t)$
8.74	$(c^2+p^2)^{-\frac{1}{2}}\operatorname{sech}[a(c^2+p^2)^{\frac{1}{2}}]$ $\cdot\cosh[b(c^2+p^2)^{\frac{1}{2}}]$ $b < a$	$\sum_{n=0}^{\infty} (-1)^n H[t-(a-b+2an)]$ $\cdot J_o\{c[t^2-(a-b+2an)^2]^{\frac{1}{2}}\}$ $+ \sum_{n=0}^{\infty} (-1)^n H[t-(a+b+2an)]$ $\cdot J_o\{c[t^2-(a+b+2an)^2]^{\frac{1}{2}}\}$
8.75	$(c^2+p^2)^{-\frac{1}{2}}\operatorname{csch}[a(c^2+p^2)^{\frac{1}{2}}]$ $\cdot\cosh[b(c^2+p^2)^{\frac{1}{2}}]$ $b < a$	$a^{-1} \sum_{n=0}^{\infty} (-1)^n \varepsilon_n [c^2+(n\frac{\pi}{a})^2]^{-\frac{1}{2}}$ $\cdot\cos(n\pi b/a)\sin\{t[c^2+(n\pi/a)^2]^{\frac{1}{2}}\}$ $= \sum_{n=0}^{\infty} H[t-(a-b+2an)]$ $\cdot J_o\{c[t^2-(a\ b+2an)^2]^{\frac{1}{2}}\}$ $+ \sum_{n=0}^{\infty} H[t-(a+b+2an)]$ $\cdot J_o\{c[t^2-(a+b+2an)^2]^{\frac{1}{2}}\}$
8.76	$(c^2+p^2)^{-1}\operatorname{csch}[a(c^2+p^2)^{\frac{1}{2}}]$ $\cdot\sinh[b(c^2+p^2)^{\frac{1}{2}}]$ $b \leqq a$	$b(ac)^{-1}\sin(ct)+2\pi^{-1}$ $\cdot \sum_{n=1}^{\infty} (-1)^n n^{-1}\sin(n\pi b/a)$ $\cdot[c^2+(n\pi/a)^2]^{-\frac{1}{2}}\sin\{t[c^2+(n\pi/a)^2]^{\frac{1}{2}}\}$

	$g(p) = \int_0^\infty f(t)e^{-pt}dt$	$f(t)$
8.77	$(c^2+p^2)^{-1}\operatorname{sech}[a(p^2+c^2)^{\frac{1}{2}}]$ $\cdot\cosh[b(p^2+c^2)^{\frac{1}{2}}]$ $b \leqq a$	$c^{-1}\sin(ct)-4\pi^{-1}\sum_{n=0}^{\infty}(-1)^n(2n+1)^{-1}$ $\cdot\cos[(n+\frac{1}{2})\pi b/a][c^2+(n+\frac{1}{2})^2\pi^2/a^2]^{-\frac{1}{2}}$ $\cdot\sin\{t[c^2+(n+\frac{1}{2})^2\pi^2/a^2]^{\frac{1}{2}}\}$
8.78	$(c^2+p^2)^{-3/2}\operatorname{csch}[a(c^2+p^2)^{\frac{1}{2}}]$ $\cdot\cosh[b(p^2+c^2)^{\frac{1}{2}}]$ $b \leqq a$	$(\frac{1}{2}b^2/a-\frac{1}{6}a)c^{-1}\sin(ct)$ $-2a\pi^{-2}\sum_{n=1}^{\infty}(-1)^n n^{-2}\cos(n\pi b/a)$ $\cdot[c^2+(n\pi/a)^2]^{-\frac{1}{2}}\sin\{t[c^2+(n\pi/a)^2]^{\frac{1}{2}}\}$
8.79	$(c^2+p^2)^{-3/2}\operatorname{sech}[a(c^2+p^2)^{\frac{1}{2}}]$ $\cdot\sinh[b(p^2+c^2)^{\frac{1}{2}}]$ $b \leqq a$	$bc^{-1}\sin(ct)-2a\pi^{-2}\sum_{n=0}^{\infty}(-1)^n(n+\frac{1}{2})^{-2}$ $\cdot\sin[(n+\frac{1}{4})\pi b/a][c^2+(n+\frac{1}{2})^2\pi^2/a^2]^{-\frac{1}{2}}$ $\cdot\sin\{t[c^2+(n+\frac{1}{2})^2\pi^2/a^2]^{\frac{1}{2}}\}$
8.80	$\coth^{-1}(p/a)$	$t^{-1}\sin(at)$
8.81	$p^{-1}\sinh^{-1}(p/a)$	$-Ji_o(at)$
8.82	$p^{-1}[\sinh^{-1}(p/a)]^2$	$Yi_o(at)$

	$g(p) = \int_0^\infty f(t) e^{-pt} dt$	$f(t)$
8.83	$(p^2+a^2)^{-\frac{1}{2}} \sinh^{-1}(p/a)$	$-\frac{1}{2}\pi Y_o(at)$
8.84	$(p^2+a^2)^{-\frac{1}{2}} \sinh^{-1}(a/p)$	$\frac{1}{2}\pi \mathbf{H}_o(at)$
8.85	$(p^2-a^2)^{-\frac{1}{2}} \cosh^{-1}(p/a)$	$K_o(at)$
8.86	$(p^2-a^2)^{-1}[p \tanh^{-1}(a/p)$ $-\frac{1}{2}a \log(p^2-a^2)]$	$(\gamma+\log t)\sinh(at)$
8.87	$(p^2-a^2)^{-1}[\frac{1}{2}p \log(p^2-a^2)$ $-a \tanh^{-1}(a/p)]$	$(\gamma+\log t)\cosh(at)$
8.88	$(p^2-1)^{-\frac{1}{2}}$ $\cdot\sinh(\nu \cosh^{-1} p)$	$\pi^{-1}\sin(\pi\nu)K_\nu(t)$ $-1 < \mathrm{Re}\ \nu < 1$
8.89	$(p^2+a^2)^{-\frac{1}{2}}$ $\exp[-\nu \sinh^{-1}(p/a)]$	$J_\nu(at)$ $\mathrm{Re}\ \nu > -1$

2.9 Orthogonal Polynomials

	$g(p) = \int_0^\infty f(t)e^{-pt}dt$	$f(t)$
9.1	$p^{-n}T_n(1-1/p)$	$(-1)^n[(2_{n-1})!]^{-1}(2t)^{n-1}He_{2n}(t^{\frac{1}{2}})$
9.2	$p^{-n-1}U_n(1/p-1)$	$2^{n-1}[(2n-1)!]^{-1}t^{n-\frac{1}{2}}He_{2n-1}(t^{\frac{1}{2}})$
9.3	$p^{-\frac{1}{2}-\frac{1}{2}n}He_n[(2p)^{\frac{1}{2}}]$	$2^{\frac{1}{2}n-1}(\pi t)^{-\frac{1}{2}}[(1+it^{\frac{1}{2}})^n+(1-it^{\frac{1}{2}})^n]$
9.4	$p^{-n-\frac{1}{2}}He_{2n}[(2p)^{\frac{1}{2}}]$	$2^n(\pi t)^{-\frac{1}{2}}(1+t)^nT_n(\frac{1-t}{1+t})$
9.5	$p^{-n-2}He_{2n+1}[(2p)^{\frac{1}{2}}]$	$2^{n+3/2}(n+1)^{-1}(t/\pi)^{\frac{1}{2}}(t+1)^n$ $\cdot U_{n+1}(\frac{1-t}{1+t})$
9.6	$p^{-n-\frac{1}{2}}e^{-a/p}He_{2n}[(2a/p)^{\frac{1}{2}}]$	$\pi^{-\frac{1}{2}}(-2)^nt^{n-\frac{1}{2}}\cos[2(at)^{\frac{1}{2}}]$
9.7	$p^{-n-1}e^{-a/p}He_{2n+1}[(2a/p)^{\frac{1}{2}}]$	$(-1)^n2^{n+\frac{1}{2}}\pi^{-\frac{1}{2}}t^n\sin[2(at)^{\frac{1}{2}}]$
9.8	$p^{-\frac{1}{2}}(1-\frac{1}{2}b^{-2}/p)$ $\cdot He_n[a(b^2-\frac{1}{2}p^{-1})^{-\frac{1}{2}}]$	$(2\pi t)^{-\frac{1}{2}}[He_n(\frac{a+t^{\frac{1}{2}}}{b})$ $+He_n(\frac{a-t^{\frac{1}{2}}}{b})]$

	$g(p) = \int_0^\infty f(t)e^{-pt}dt$	$f(t)$
9.9	$p^{-1}P_n(1-a/p)$	${}_2F_2(-n,n+1;1,1;\frac{1}{2}at)$
9.10	$p^{-n-1}P_n(1-a/p)$	$(\frac{1}{2}a)^{n+1}t^n L_n(\frac{1}{2}at)/n!$
9.11	$p^{-\nu}P_n(1-a/p)$ $\operatorname{Re}\nu > 0$	$t^{n-1}\,{}_2F_2(-n,n+1;1;\nu;\frac{1}{2}at)$
9.12	$p^{-\frac{1}{2}}P_n(a/p)$	$(n!)^{-1}i^{-n}(\pi t)^{-\frac{1}{2}}He_n[(2at)^{\frac{1}{2}}]He_n[(-2at)^{\frac{1}{2}}]$
9.13	$p^{-\frac{1}{2}-\frac{1}{2}n}P_n[(2p)^{-\frac{1}{2}}]$	$(n!)^{-1}2^{\frac{1}{2}n}\pi^{-\frac{1}{2}}t^{\frac{1}{2}n-\frac{1}{2}}He_n(t^{\frac{1}{2}})$
9.14	$(p+b)^{-n-1}P_n(\frac{p+a}{p+b})$	$(n!)^{-1}t^n e^{-bt}L_n(\frac{1}{2}bt-\frac{1}{2}at)$
9.15	$(p+b)^{-\nu}P_n(\frac{p+a}{p+b})$	$[\Gamma(\nu)]^{-1}t^{\nu-1}e^{-bt}$ $\cdot\,{}_2F_2(-n,n+1;1;\nu;\frac{1}{2}bt-\frac{1}{2}at)$
9.16	$p^{-n-1}(p-a)^n$ $\cdot P_n[\frac{p^2-ap+b}{p(p-a)}]$	$L_n\{\frac{1}{2}[a-(a^2-2b)^{\frac{1}{2}}]\}L_n\{\frac{1}{2}[a+(a^2-2b)^{\frac{1}{2}}]\}$

	$g(p) = \int_0^\infty f(t)e^{-pt}dt$	$f(t)$
9.17	$p^{-n-1}(p-2a)^n$ $\cdot P_n[\frac{p^2+2(1-2a)(a^2-b^2)}{p(p-2a)}$	$L_n[(a-b)t]L_n[(a+b)t]$
9.18	$(p^2+a^2)^{-\frac{1}{2}n-\frac{1}{2}}$ $\cdot P_n[p(p^2+a^2)^{-\frac{1}{2}}]$	$(n!)^{-1}t^nJ_0(at)$
9.19	$(p^2-a^2)^{-\frac{1}{2}n-\frac{1}{2}}$ $\cdot P_n[p(p^2-a^2)^{-\frac{1}{2}}]$	$(n!)^{-1}t^nI_0(at)$
9.20	$(a+b+p)^{-\frac{1}{2}-\frac{1}{2}n}(a+b-p)^{\frac{1}{2}n}$ $\cdot P_n\{2(ab)^{\frac{1}{2}}[(a+b)^2-p^2]^{-\frac{1}{2}}\}$	$\pi^{-\frac{1}{2}}(n!)^{-1}t^{-\frac{1}{2}}e^{-at-bt}$ $\cdot He_n[2(at)^{\frac{1}{2}}]He_n[2(bt)^{\frac{1}{2}}]$
9.21	$p^{-n-2\nu}C_n^{\nu}(1-a/p)$	$\pi^{\frac{1}{2}}2^{1-n-4\nu}[\Gamma(\nu)\Gamma(\frac{1}{2}+\nu+n)]^{-1}a^{n+2\nu}$ $\cdot t^{n+2\nu-1}L_n^{\nu-\frac{1}{2}}(\frac{1}{2}at)$
9.22	$p^{-\mu}C_n^{\nu}(1-a/p)$ Re $\mu > 0$	$[nB(n,2\nu)\Gamma(\mu)]^{-1}t^{\mu-1}$ $\cdot {}_2F_2(-n,n+2\nu;\frac{1}{2}+\nu,\mu;\frac{1}{2}at)$

	$g(p) = \int_0^\infty f(t)e^{-pt}dt$	$f(t)$
9.23	$(p+b)^{-\mu}C_n^\nu(\frac{p+a}{p+b})$ $\mathrm{Re}\,\mu > 0,\ \mathrm{Re}\,\nu > 0$	$[nB(n,2\nu)\Gamma(\mu)]^{-1}t^{\mu-1}e^{-bt}$ $\cdot{}_2F_2(-n,n+2\nu;\tfrac{1}{2}+\nu,\mu;\tfrac{1}{2}bt-\tfrac{1}{2}at)$
9.24	$p^{-\nu-\frac{1}{2}n}C_n^\nu(p^{-\frac{1}{2}})$	$2^{\frac{1}{2}n}[n!\Gamma(\nu)]^{-1}t^{\nu+\frac{1}{2}n-1}He_n[(2t)^{\frac{1}{2}}]$
9.25	$p^{-n-1}L_n(p)$	$(n!)^{-1}(t+1)^nP_n(\frac{t-1}{t+1})$
9.26	$p^{-n-\nu}L_n^\alpha(p)$	$[\Gamma(n+\nu)]^{-1}(1+t)^nt^{\nu-1}$ $\cdot P_n^{\alpha,\nu-1}(\frac{t-1}{t+1})$
9.27	$p^{-\nu}L_n^\alpha(a/p)$ $\mathrm{Re}\,\nu > 0$	$(\alpha+1)_n[n!\Gamma(\nu)]^{-1}t^{\nu-1}$ $\cdot{}_1F_2(-n;\alpha+1,\nu;at)$
9.28	$p^{-n-\nu-1}e^{-a/p}L_n^\nu(a/p)$ $\mathrm{Re}(\nu+n) > -1$	$(n!)^{-1}a^{-\frac{1}{2}\nu}t^{\frac{1}{2}\nu+n}J_\nu[2(at)^{\frac{1}{2}}]$
9.29	$p^{-\nu-n-1}e^{-a/p}(p-1)^n$ $\cdot L_n^\nu[ap^{-1}(1-p)^{-1}]$	$(t/a)^{\frac{1}{2}\nu}L_n^\nu(t)J_\nu[2(at)^{\frac{1}{2}}]$ $\mathrm{Re}\,\nu > -1$

	$g(p) = \int_0^\infty f(t) e^{-pt} dt$	$f(t)$
9.30	$B(p, \frac{3}{2}+n) L_n^{p+\frac{1}{2}}(a)$	$(-2)^{-n}(2a)^{-\frac{1}{2}} He_{2n+1}[(2a^{\frac{1}{2}})(1-e^{-t})]^{\frac{1}{2}}$
9.31	$B(p+\frac{1}{2}, n+\frac{1}{2}) L_n^p(a)$	$(-2)^{-n}(e^t-1)^{-\frac{1}{2}} He_{2n}[(2a)^{\frac{1}{2}}(1-e^{-t})^{\frac{1}{2}}]$
9.32	$p^{-a-b-n-1} P_n^{a,b}(1-h/p)$	$[\Gamma(1+a+b+n)]^{-1}(\frac{1}{2}h)^{a+b+n} L_n^a(\frac{1}{2}ht)$
9.33	$O_n(p/a)$	$\frac{1}{2}a\{[at+(1+a^2t^2)^{\frac{1}{2}}]^n+[at-(1+a^2t^2)^{\frac{1}{2}}]^n\}$
9.34	$O_{2n}(p)$	$\frac{1}{2}\{[t+(1+t^2)^{\frac{1}{2}}]^{2n}+[t+(1+t^2)^{\frac{1}{2}}]^{-2n}\}$
9.35	$O_{2n+1}(p)$	$\frac{1}{2}\{[t+(1+t^2)^{\frac{1}{2}}]^{2n+1}-[t+(1+t^2)^{\frac{1}{2}}]^{-2n-1}\}$
9.36	$S_n(p/a)$	$(1+a^2t^2)^{-\frac{1}{2}}\{[at+(1+a^2t^2)^{\frac{1}{2}}]^n$ $-[at-(1+a^2t^2)^{\frac{1}{2}}]^n\}$

2.10 Gamma Function and Related Functions

	$g(p) = \int_0^\infty f(t)e^{-pt}dt$	$f(t)$	
10.1	$B(ap,\nu)$ $\mathrm{Re}\ \nu > 0$	$a^{-1}(1-e^{-t/a})^{\nu-1}$	
10.2	$B(\tfrac{1}{2}p/a-\nu,\ 2\nu+1)$ $\mathrm{Re}\ \nu > -\tfrac{1}{2}$	$2^{1+2\nu}a\ \sinh^{2\nu}(at)$	
10.3	$B(\nu,p-ia)+B(\nu,p+ia)$ $\mathrm{Re}\ \nu > 0$	$2(1-e^{-t})^{\nu-1}\cos(at)$	
10.4	$B(\nu,p-ia)-B(\nu,p+ia)$ $\mathrm{Re}\ \nu > -1$	$2i(1-e^{-t})^{\nu-1}\sin(at)$	
10.5	$b^{-p}B(p,a)$	0 $(1-be^{-t})^{a-1}$	$t < \log b$ $t > \log b$
10.6	$e^{-\pi p}[\Gamma(\nu+ip)\Gamma(\nu-ip)]^{-1}$ $\mathrm{Re}\ \nu > 3/2$	$\tfrac{1}{2}\pi^{-1}[\Gamma(2\nu-1)]^{-1}$ $\cdot(2\ \sin\tfrac{1}{2}t)^{2\nu-2}$ 0	 $t < 2\pi$ $t > 2\pi$
0.7	$e^{-ap}\Gamma(ap)$ $\cdot[\Gamma(1+ap+n)]^{-1}$	0 $(an!)^{-1}(1-be^{-t/a})^{n}$ $n = 0,1,2,\cdots$	$t < a \log b$ $t > a \log b$

	$(p) = \int_0^\infty f(t)e^{-pt}dt$	$f(t)$
10.8	$2^{-p}\Gamma(p)[\Gamma(\frac{1}{2}-\frac{1}{2}n+\frac{1}{2}p)$ $\cdot\Gamma(\frac{1}{2}+\frac{1}{2}n+\frac{1}{2}p)]^{-1}$	$\pi^{-1}(1-e^{-2t})^{-\frac{1}{2}}T_n(e^{-t})$ $n = 0,1,2,\cdots$
10.9	$2^p[\Gamma(1+p+\mu)]^{-1}$ $\cdot\Gamma(\frac{1}{2}+\frac{1}{2}p+\frac{1}{2}\nu)\Gamma(\frac{1}{2}p-\frac{1}{2}\nu)$	$2\pi^{\frac{1}{2}}(1-e^{-2t})^{\frac{1}{2}\mu}P_\nu^{-\mu}(e^t)$ $\mathrm{Re}\ \mu > -1$
10.10	$p^{-1}\Gamma(\frac{1}{2}+\frac{1}{2}ip/a)\Gamma(1-\frac{1}{2}ip/a)$ $\cdot[\Gamma(1+\nu+\frac{1}{2}ip/a)\Gamma(1+\nu-\frac{1}{2}ip/a)]^{-1}$	$[\Gamma(1+2\nu)]^{-1}2^{2\nu}\lvert\sin at\rvert^{2\nu}$ $\mathrm{Re}\ \nu > -\frac{1}{2}$
10.11	$\frac{\Gamma(p+a)\Gamma(p+b)}{\Gamma(p+c)\Gamma(p+d)}$ $\mathrm{Re}(c+d-a-b) > 0$	$[\Gamma(c+d-a-b)]^{-1}e^{-at}(1-e^{-t})^{c+d-a-b-1}$ $\cdot\,{}_2F_1(d-b,c-b;c+d-a-b;1-e^{-t})$
10.12	$\frac{\Gamma(p+a)}{\Gamma(p+b)}$ $\mathrm{Re}(b-a) > 0$	$[\Gamma(b-a)]^{-1}e^{-at}(1-e^{-t})^{b-a-1}$
10.13	$\log[\frac{\Gamma(p+a)\Gamma(p+b+c)}{\Gamma(p+a+c)\Gamma(p+b)}]$	$t^{-1}(1-e^{-t})^{-1}(e^{-at}-e^{-bt})(1-e^{-ct})$
10.14	$\log[\frac{\Gamma(ap+b)}{\Gamma(ap+c)}]$ $+(c-b)\psi(ap+d)$ $\mathrm{Re}\ a > 0$	$(1-e^{-t/a})^{-1}[t^{-1}(e^{-bt/a}-e^{-ct/a})$ $+a^{-1}(b-c)e^{-dt/a}]$

	$g(p) = \int_0^\infty f(t)e^{-pt}dt$	$f(t)$
10.15	$\log[(ap)^{\frac{1}{2}} \frac{\Gamma(ap)}{\Gamma(\frac{1}{2}+ap)}]$	$\frac{1}{2}t^{-1}\tanh(\frac{1}{4}t/a)$
0.16	$\Gamma(-\nu,ap)$ $\mathrm{Re}\ \nu > -1$	$0 \qquad t < a$ $a^{-\nu}[\Gamma(1+\nu)]^{-1}t^{-1}(t-a)^{\nu} \qquad t > a$
0.17	$e^{ap}\Gamma(-\nu,ap)$ $\mathrm{Re}\ \nu > -1$	$a^{-\nu}[\Gamma(1+\nu)]^{-1}(t+a)^{-1}t^{\nu}$
0.18	$\Gamma(-p,1)$	$\exp(-e^{t})$
0.19	$\gamma(p,1)$	$\exp(-e^{-t})$
0.20	$\Gamma(-p,a)-\Gamma(-p,b)$ $1 \leqq a < b$	$0 \qquad t < \log a$ $\exp(-e^{t}) \qquad \log a < t < \log b$ $0 \qquad t > \log b$
.21	$p^{-\nu}\Gamma(\nu,ap)$	$0 \qquad t < a$ $t^{\nu-1} \qquad t > a$
.22	$p^{-\nu}\gamma(\nu,ap)$	$t^{\nu-1} \qquad t < a$ $0 \qquad t > a$

	$g(p) = \int_0^\infty f(t)e^{-pt}dt$	$f(t)$
10.23	$p^{-\nu}e^{ap}\Gamma(\nu,ap)$	$(t+a)^{\nu-1}$
10.24	$p^{-\nu}e^{-ap}\gamma(\nu,-ap)$ Re $\nu > 0$	$(a-t)^{\nu-1}$ $\quad t < a$ 0 $\quad t > a$
10.25	$a^{p}\gamma(p,a)$	$\exp(-ae^{-t})$
10.26	$a^{-p}\Gamma(-p,a)$	$\exp(-ae^{t})$
10.27	$\gamma[\nu,\frac{1}{2}(p^2+a^2)^{\frac{1}{2}}-\frac{1}{2}p]$ Re $\nu > 0$	$(\frac{1}{2}a)^{\nu}t^{\frac{1}{2}\nu-1}(1+t)^{-\frac{1}{2}\nu}J_{\nu}[a(t^2+t)^{\frac{1}{2}}]$
10.28	$e^{-bp}\gamma[\nu,b(p^2+a^2)^{\frac{1}{2}}-bp]$ Re $\nu > 0$	0 $\quad t < b$ $(ab)^{\nu}(t-b)^{\frac{1}{2}\nu-1}(t+b)^{-\frac{1}{2}\nu}$ $\cdot J_{\nu}[a(t^2-b^2)^{\frac{1}{2}}]$ $\quad t > a$
10.29	$\gamma(\nu,a/p)$ Re $\nu > 0$	$a^{\frac{1}{2}\nu}t^{\frac{1}{2}\nu-1}J_{\nu}[2(at)^{\frac{1}{2}}]$

	$g(p) = \int_0^\infty f(t)e^{-pt}dt$	$f(t)$
10.30	$\gamma(\nu, ae^{i\pi}/p)$ $\operatorname{Re}\nu > 0$	$e^{i\pi\nu}a^{\frac{1}{2}\nu}t^{\frac{1}{2}\nu-1}I_\nu[2(at^{\frac{1}{2}})]$
10.31	$p^{\nu-1}\gamma(\nu, a/p)$ $\operatorname{Re}\nu > 0$	$a^{\frac{1}{2}\nu}\Gamma(\nu)t^{-\frac{1}{2}\nu}I_\nu[2(at)^{\frac{1}{2}}]$
10.32	$p^{\nu-1}\exp(-a/p)$ $\cdot\gamma(\nu, ae^{-i\pi}/p)$ $\operatorname{Re}\nu > 0$	$e^{-i\pi\nu}a^{\frac{1}{2}\nu}\Gamma(\nu)t^{-\frac{1}{2}\nu}J_\nu[2(at)^{\frac{1}{2}}]$
10.33	$p^{-\nu-1}e^{a/p}\Gamma(-\nu, a/p)$ $\operatorname{Re}\nu > -1$	$2[\Gamma(1+\nu)]^{-1}(t/a)^{\frac{1}{2}\nu}K_\nu[2(at)^{\frac{1}{2}}]$
10.34	$p^{\nu-3/2}e^{a/p}\Gamma(\nu, a/p)$ $\operatorname{Re}\nu < 3/2$	$\Gamma(\nu)(a/t)^{\frac{1}{2}\nu-\frac{1}{4}}\{[I_{\frac{1}{2}-\nu}[2(at)^{\frac{1}{2}}]$ $-\mathbf{L}_{\nu-\frac{1}{2}}[2(at)^{\frac{1}{2}}]\}$
10.35	$p^{\nu-1}e^{-a/p}\Gamma(\nu, e^{i\pi}a/p)$ $\operatorname{Re}\nu < 1$	$-\pi i[\Gamma(1-\nu)]^{-1}(a/t)^{\frac{1}{2}\nu}H_\nu^{(2)}[2(at)^{\frac{1}{2}}]$
10.36	$p^{\nu-1}e^{-a/p}\Gamma(\nu, e^{-i\pi}a/p)$ $\operatorname{Re}\nu < 1$	$\pi i[\Gamma(1-\nu)]^{-1}(a/t)^{\frac{1}{2}\nu}H_\nu^{(1)}[2(at)^{\frac{1}{2}}]$

	$g(p) = \int_0^\infty f(t)e^{-pt}dt$	$f(t)$
10.37	$p^{\nu-3/2}e^{a/p}\gamma(\nu,a/p)$	$\Gamma(\nu)(a/t)^{\frac{1}{2}\nu-\frac{1}{4}}\mathbf{L}_{\nu-\frac{1}{2}}[2(at)^{\frac{1}{2}}]$
10.38	$p^{\mu}e^{a/p}\gamma(\nu,a/p)$ $\mathrm{Re}(\nu,\mu) > 0$	$[\nu\Gamma(\nu-\mu)]^{-1}a^{\nu}t^{\nu-\mu-1}$ $\cdot\, {}_1F_2(1;\nu+1,\nu-\mu;at)$
10.39	$s^{-1}e^{bp}[(p-s)^{\nu}e^{-bs}\Gamma(-\nu,bp-bs)$ $-(p+s)^{\nu}e^{bs}\Gamma(-\nu,bp+bs)]$ $s = (p^2-a^2)^{\frac{1}{2}}$	$2a^{\nu}[\Gamma(1+\nu)]^{-1}t^{\frac{1}{2}\nu}$ $(t+2b)^{-\frac{1}{2}\nu}K_{\nu}[a(t^2+2bt)^{\frac{1}{2}}]$
10.40	$s^{-1}[(p-s)^{\nu}e^{-bs}\Gamma(-\nu,bp-bs)$ $-(p+s)^{\nu}e^{bs}\Gamma(-\nu,bp+bs)]$ $s = (p^2-a^2,^{\frac{1}{2}}$ $\mathrm{Re}\ \nu > -1$	$0 \quad t < b$ $2a^{\nu}[\Gamma(1+\nu)]^{-1}$ $\cdot\left(\frac{t-b}{t+b}\right)^{\frac{1}{2}\nu}K_{\nu}[a(t^2-b^2)^{\frac{1}{2}}]$ $t > b$
10.41	$B(p,\nu)[\psi(p+\nu)-\psi(p)]$ $\mathrm{Re}\ \nu > -1$	$t(1-e^{-t})^{\nu-1}$
10.42	$\psi(p+a)-\psi(p)$	$(1-e^{-t})^{-1}(1-e^{-at})$

	$g(p) = \int_0^\infty f(t)e^{-pt}dt$	$f(t)$
10.43	$\psi(\frac{1}{2}+\frac{1}{2}p)-\psi(\frac{1}{2}p)$	$2(1+e^{-t})^{-1}$
10.44	$\psi(b+p/a)$ $-\psi(c+p/a)$	$a(e^{-act}-e^{-abt})(1-e^{-at})^{-1}$
10.45	$p^{-1}[\psi(\frac{1}{2}+\frac{1}{2}p)$ $-\psi(\frac{1}{2}p)]$	$2\log(\frac{1}{2}+\frac{1}{2}e^{t})$
10.46	$\psi(a+b)+\psi(b+p)$ $-\psi(p)-\psi(p+a+b)$	$(1-e^{-t})^{-1}(1-e^{-at})(1-e^{-bt})$
10.47	$p^{-1}\psi(p/a)$	$-\gamma-\log(e^{at}-1)$
10.48	$\psi(ap+b)$ $-\psi(ap+c)$	$a^{-1}(1-e^{-t/a})^{-1}[\exp(-ct/a)-\exp(-bt/a)]$
10.49	$\psi(p-ia+1)$ $-\psi(p+ia+1)$	$-2i(e^{t}-1)\sin(at)$
10.50	$\psi(p-ia)-\psi(p+ia)$	$-2i(1-e^{-t})^{-1}\sin(at)$

	$g(p) = \int_0^\infty f(t)e^{-pt}dt$	$f(t)$
10.51	$\psi(\frac{1}{2}+\frac{1}{2}p) - \log(\frac{1}{2}p)$	$t^{-1}-\text{csch}\ t$
10.52	$\psi(\frac{3}{4}+\frac{1}{4}p/a) - \psi(\frac{1}{4}+\frac{1}{4}p/a)$	$2a\ \text{sech}(at)$
10.53	$\frac{1}{2}p[\psi(\frac{1}{2}+\frac{1}{4}p) - \psi(\frac{1}{4}p)]-1$	$\text{sech}^2 t$
10.54	$\psi(\frac{1}{2}p)-\log(\frac{1}{2}p) + p^{-1}$	$t^{-1}-\coth t$
10.55	$p^{-1}[\log(\frac{1}{2}p)-p^{-1} - \psi(\frac{1}{2}p)]$	$\log(\frac{\sinh t}{t})$
10.56	$(2p)^{-1}[\psi(\frac{1}{2}+\frac{1}{4}p) - \psi(\frac{1}{4}p)]-p^{-2}$	$\log(\cosh t)$
10.57	$p^{-1}\zeta(p)$	$n \quad \log n < t < \log(n+1)$ $n = 1,2,3,\cdots$

	$g(p) = \int_0^\infty f(t)e^{-pt}dt$	$f(t)$
10.58	$p^{-\nu}\zeta(p)$ $\mathrm{Re}\ \nu > 0$	$[\Gamma(\nu)]^{-1} \sum_{1\leq n\leq e^t} (t-\log n)^{\nu-1}$
10.59	$p^{-1}\zeta(p+a)$	$\sum_{1\leq n\leq e^t} n^{-a}$
10.60	$p^{-1} \frac{\zeta'(p)}{\zeta(p)}$	$-\psi(e^t)$
10.61	$2^{-p}\zeta(p-1)$	$\tanh[\frac{1}{2}\pi(e^{2t}-1)^{\frac{1}{2}}]$
10.62	$p^{-1}(1-2^{2-p})$ $\cdot\zeta(p-1)$	$n \quad \log(2n-1)<t<\log(2n+1)$
10.63	$\zeta(\nu,ap)$ $\mathrm{Re}\ \nu > 1$	$[\Gamma(\nu)]^{-1}a^{-\nu}t^{\nu-1}(1-e^{-t/a})^{-1}$
10.64	$\zeta(\nu,1+p/a)$ $\mathrm{Re}\ \nu > 1$	$a^{\nu}[\Gamma(\nu)]^{-1}t^{\nu-1}(e^{at}-1)^{-1}$

	$g(p) = \int_0^\infty f(t)e^{-pt}dt$	$f(t)$
10.65	$\sum_{k=0}^{\infty} z^k(k+1+p)^{-\nu}$ Re $\nu > 0$	$[\Gamma(\nu)]^{-1}t^{\nu-1}(e^t-z)^{-1}$
10.66	$\zeta(\nu,\frac{1}{2}+\frac{1}{2}p/a)$ Re $\nu > 1$	$\frac{1}{2}(2a)^{\nu}t^{\nu-1}\operatorname{csch}(at)$
10.67	$2^{-2p}[\zeta(\nu,\frac{1}{4}+\frac{1}{4}p/a)$ $-\zeta(\nu,\frac{3}{4}+\frac{1}{4}p/a)]$	$\frac{1}{2}[\Gamma(\nu)]^{-1}a^{\nu}t^{\nu-1}\operatorname{sech}(at)$ Re $\nu > 0$

2.11 Legendre Functions

	$g(p) = \int_0^\infty f(t)e^{-pt}dt$	$f(t)$
11.1	$p_\nu(p/a)$ $-1 < \mathrm{Re}\,\nu < 0$	$-(2a)^{\frac{1}{2}}\pi^{-3/2}\sin(\pi\nu)t^{-\frac{1}{2}}K_{\nu+\frac{1}{2}}(at)$
11.2	$q_\nu(p/a)$ $\mathrm{Re}\,\nu > -1$	$(\frac{1}{2}\pi a/t)^{\frac{1}{2}}I_{\nu+\frac{1}{2}}(at)$
11.3	$(p^2-a^2)^{-\frac{1}{2}\mu}p_\nu^\mu(p/a)$ $\mathrm{Re}(\mu+\nu)>-1,\ \mathrm{Re}(\mu-\nu)>0$	$(2a/\pi)^{\frac{1}{2}}[\Gamma(\mu-\nu)\Gamma(\mu+\nu+1)]^{-1}$ $\cdot t^{\mu-\frac{1}{2}}K_{\nu+\frac{1}{2}}(at)$
11.4	$(p^2-a^2)^{-\frac{1}{2}\mu}e^{-i\pi\mu}q_\nu^\mu(p/a)$ $\mathrm{Re}(\mu+\nu) > -1$	$(\frac{1}{2}\pi a)^{\frac{1}{2}}t^{\mu-\frac{1}{2}}I_{\nu+\frac{1}{2}}(at)$
11.5	$[(p+a)/(p-a)]^{-\frac{1}{2}\mu}$ $\cdot p_\nu^\mu(p/a)$ $-\frac{1}{2} < \mathrm{Re}\,\nu < \frac{1}{2}$	$-\pi^{-1}\sin(\pi\nu)t^{-1}W_{\mu,\frac{1}{2}+\nu}(2at)$
11.6	$[(p+a)/(p-a)]^{-\frac{1}{2}\mu}$ $\cdot e^{-i\pi\mu}q_\nu^\mu(p/a)$ $\mathrm{Re}\,\nu > -\frac{1}{2}$	$\frac{1}{2}[\Gamma(2+2\nu)]^{-1}\Gamma(1+\nu+\mu)$ $\cdot t^{-1}M_{\mu,\frac{1}{2}+\nu}(2at)$

	$g(p) = \int_0^\infty f(t) e^{-pt} dt$	$f(t)$
11.7	$q_\nu[(2ab)^{-1}(p^2+a^2+b^2)]$ $\mathrm{Re}\,\nu > -1$	$\pi (ab)^{\frac{1}{2}} J_{\nu+\frac{1}{2}}(at) J_{\nu+\frac{1}{2}}(bt)$
11.8	$p_\nu[(2ab)^{-1}(p^2+a^2+b^2)]$ $-1 < \mathrm{Re}\,\nu < 0$	$(ab)^{\frac{1}{2}} \sin(\pi\nu)$ $\cdot\ [J_{\nu+\frac{1}{2}}(bt) Y_{-\nu-\frac{1}{2}}(at)$ $+ J_{-\nu-\frac{1}{2}}(at)\ Y_{\nu+\frac{1}{2}}(bt)]$ $= (ab)^{\frac{1}{2}} \tan(\pi\nu)$ $\cdot\ [J_{\nu+\frac{1}{2}}(at) J_{\nu+\frac{1}{2}}(bt)$ $-J_{-\nu-\frac{1}{2}}(at) J_{-\nu-\frac{1}{2}}(bt)]$
11.9	$q_\nu[(2ab)^{-1}(p^2-a^2-b^2)]$ $\mathrm{Re}\,\nu > -\frac{1}{2}$	$\pi (ab)^{\frac{1}{2}} I_{\nu+\frac{1}{2}}(at) I_{\nu+\frac{1}{2}}(bt)$
11.10	$p_\nu[(2ab)^{-1}(p^2-a^2-b^2)]$ $-1 < \mathrm{Re}\,\nu < 0$	$(ab)^{\frac{1}{2}} \tan(\pi\nu)$ $[I_{\nu+\frac{1}{2}}(at) I_{\nu+\frac{1}{2}}(bt)$ $- I_{-\nu-\frac{1}{2}}(at) I_{-\nu-\frac{1}{2}}(bt)]$ $= -\pi^{-1} \sin(\pi\nu)(ab)^{\frac{1}{2}}$ $\cdot \{K_{\nu+\frac{1}{2}}(at)[I_{\nu+\frac{1}{2}}(bt)+I_{-\nu-\frac{1}{2}}(bt)]$ $+K_{\nu+\frac{1}{2}}(bt)[I_{\nu+\frac{1}{2}}(at)+I_{-\nu-\frac{1}{2}}(at)]\}$

	$g(p) = \int_0^\infty f(t)e^{-pt}dt$	$f(t)$
11.11	$q_n[(p/a)^{\frac{1}{2}}]$ $n = 1,2,3,\cdots$	$\frac{1}{2}a^{-\frac{1}{4}}\Gamma(1+\frac{1}{2}n)[\Gamma(\frac{3}{2}+n)]^{-1}$ $\cdot t^{-5/4}e^{\frac{1}{2}at}M_{-\frac{1}{4},\frac{1}{2}n+\frac{1}{4}}(at)$
11.12	$(2p+a)^{-\frac{1}{2}}(2p-a)^{\frac{1}{2}\mu}$ $\cdot p_\nu^\mu[(\frac{1}{2}+p/a)^{\frac{1}{2}}]$ $\mathrm{Re}(\nu+\mu)<1,\ \mathrm{Re}(\nu-\mu)>-1$	$a^{-\frac{1}{4}}2^{3/2\mu-\frac{1}{2}}[\Gamma(1+\frac{1}{2}\nu-\frac{1}{2}\mu)\Gamma(\frac{1}{2}-\frac{1}{2}\nu-\frac{1}{2}\mu)]^{-1}$ $\cdot t^{-\frac{1}{2}\mu-3/4}W_{\frac{1}{2}\mu+\frac{1}{4},\frac{1}{2}\nu+\frac{1}{4}}(at)$
11.13	$(2p+a)^{-\frac{1}{2}}(2p-a)^{-\frac{1}{2}\mu}$ $\cdot e^{-i\pi\mu}q_\nu^\mu[(\frac{1}{2}+p/a)^{\frac{1}{2}}]$ $\mathrm{Re}(\nu+\mu) > -1$	$2^{\frac{1}{2}\mu-3/2}a^{-\frac{1}{4}}\Gamma(\frac{1}{2}+\frac{1}{2}\nu+\frac{1}{2}\mu)[\Gamma(\frac{3}{2}+\nu)]^{-1}$ $\cdot t^{\frac{1}{2}\mu-3/4}M_{\frac{1}{4}-\frac{1}{2}\mu,\frac{1}{4}+\frac{1}{2}\nu}(at)$
11.14	$(2p-a)^{\frac{1}{2}\mu}p_\nu^\mu[(\frac{1}{2}+p/a)^{\frac{1}{2}}]$ $\mathrm{Re}(\nu+\mu)<0,\ \mathrm{Re}(\nu-\mu)>-1$	$2^{3/2-3/2\mu}a^{-\frac{1}{4}}$ $\cdot[\Gamma(\frac{1}{2}\mu-\frac{1}{2}\nu-\frac{1}{2})\Gamma(-\frac{1}{2}\nu-\frac{1}{2}\mu)]^{-1}$ $\cdot t^{-\frac{1}{2}\mu-5/4}W_{\frac{1}{2}\mu-\frac{1}{4},\frac{1}{2}\nu+\frac{1}{4}}(at)$
11.15	$(2p-a)^{-\frac{1}{2}\mu}e^{-i\pi\mu}$ $\cdot q_\nu^\mu[(\frac{1}{2}+p/a)^{\frac{1}{2}}]$ $\mathrm{Re}(\nu+\mu)>-1$	$2^{\frac{1}{2}\mu-1}a^{-\frac{1}{4}}\Gamma(1+\frac{1}{2}\nu+\frac{1}{2}\mu)[\Gamma\ \frac{3}{2}+\nu)]^{-1}$ $\cdot t^{\frac{1}{2}\mu-5/4}M_{-\frac{1}{4}-\frac{1}{2}\mu,\frac{1}{4}+\frac{1}{2}\nu}(at)$

	$g(p) = \int_0^\infty f(t)e^{-pt}dt$	$f(t)$
11.16	$(p-a)^{\frac{1}{2}\mu}(p+a)^{-\frac{1}{2}\nu-\frac{1}{2}}$ $\cdot P_\nu^\mu[(\frac{1}{2}+\frac{1}{2}p/a)^{-\frac{1}{2}}]$ $\mathrm{Re}(\nu-\mu) > -1$	$2^{\frac{1}{2}\nu-\frac{1}{2}\mu}\pi^{-\frac{1}{2}}[\Gamma(1+\nu-\mu)]^{-1}$ $\cdot t^{\frac{1}{2}(\nu-\mu-1)}D_{\nu+\mu}[2(at)^{\frac{1}{2}}]$
11.17	$(2p+a)^{-\frac{1}{2}}(2p-a)^{\frac{1}{2}\nu}$ $\cdot P_\nu^\mu\{[(\frac{2p+a}{2p-a})^{\frac{1}{2}}]\}$ $\mathrm{Re}(\nu+\mu) < 1$	$2^{\mu+\frac{1}{2}\nu-\frac{1}{2}}a^{-\frac{1}{2}}[\Gamma(\frac{1}{2}-\frac{1}{2}\mu-\frac{1}{2}\nu)\Gamma(1-\mu)]^{-1}$ $\cdot t^{-\frac{1}{2}\nu-1}M_{\frac{1}{2}+\frac{1}{2}\nu,-\frac{1}{2}\mu}(at)$
11.18	$(2p+a)^{-\frac{1}{2}}(2p-a)^{-\frac{1}{2}\nu-\frac{1}{2}}$ $\cdot e^{-i\pi\mu}Q_\nu^\mu\{[(\frac{2p+a}{2p-a})^{\frac{1}{2}}]\}$ $\mathrm{Re}(\pm\mu+\nu) > -2$	$2^{-2+\mu-\frac{1}{2}\nu}a^{-\frac{1}{2}}\Gamma(\frac{1}{2}+\frac{1}{2}\nu+\frac{1}{2}\mu)$ $\cdot[\Gamma(1-\frac{1}{2}\mu+\frac{1}{2}\nu)]^{-1}$ $\cdot t^{\frac{1}{2}\nu-\frac{1}{2}}W_{-\frac{1}{2}\nu,-\frac{1}{2}\nu}(at)$
11.19	$(2p-a)^{\frac{1}{2}\nu}P_\nu^\mu[(\frac{2p+a}{2p-a})^{\frac{1}{2}}]$ $\mathrm{Re}(\nu+\mu) < 0$	$2^{\frac{1}{2}\nu+\mu}a^{-\frac{1}{2}}[\Gamma(-\frac{1}{2}\mu-\frac{1}{2}\nu)\Gamma(1-\mu)]^{-1}$ $\cdot t^{-\frac{1}{2}\nu-3/2}M_{\frac{1}{2}\nu,-\frac{1}{2}\mu}(at)$
11.20	$(2p-a)^{-\frac{1}{2}-\frac{1}{2}\nu}e^{-i\pi\mu}$ $\cdot Q_\nu^\mu[(\frac{2p+a}{2p-a})^{\frac{1}{2}}]$ $\mathrm{Re}(\nu\pm\mu) > -1$	$2^{-3/2-\frac{1}{2}\nu+\mu}a^{-\frac{1}{2}}\Gamma(1+\frac{1}{2}\nu+\frac{1}{2}\mu)[\Gamma(\frac{1}{2}\mu-\frac{1}{2}-\frac{1}{2}\nu)]^{-1}$ $\cdot t^{\frac{1}{2}\nu-1}W_{-\frac{1}{2}-\frac{1}{2}\nu,\frac{1}{2}\mu}(at)$

	$g(p) = \int_0^\infty f(t)e^{-pt}dt$	$f(t)$
11.21	$p^{\nu}(a^2-p^2)^{\frac{1}{2}\nu}$ $\cdot p_{\nu}^{\nu}(a/p)$ $\mathrm{Re}\ \nu < 0$	$(\frac{1}{2}\pi)^{\frac{1}{2}}(a/t)^{\frac{1}{2}+\nu}[\Gamma(-2\nu)]^{-1}$ $\cdot[I_{-\nu-\frac{1}{2}}(at)-\mathbf{L}_{-\nu-\frac{1}{2}}(at)]$
11.22	$p^{\nu+1}(a^2-p^2)^{\frac{1}{2}\nu}p_{\nu}^{\nu}(a/p)$ $\mathrm{Re}\ \nu < -\frac{1}{2}$	$(\frac{1}{2}\pi)^{\frac{1}{2}}a[\Gamma(-2\nu)^{-1}a(a/t)^{\nu+\frac{1}{2}}$ $\cdot[I_{-\nu-3/2}(at)-\mathbf{L}_{-\nu-3/2}(at)]$
11.23	$p^{-\frac{1}{2}\nu-\frac{1}{2}}(p-a)^{\frac{1}{2}\mu}$ $\cdot p_{\nu}^{\mu}[(a/p)^{\frac{1}{2}}]$ $\mathrm{Re}\ \mu < 1,\ \mathrm{Re}(\nu-\mu) > -1$	$[\Gamma(\nu-\mu+1)]^{-1}(2t)^{\frac{1}{2}\nu-\frac{1}{2}\mu}(\pi t)^{-\frac{1}{2}}$ $\cdot e^{\frac{1}{2}at}D_{\mu+\nu}[(2at)^{\frac{1}{2}}]$
11.24	$(p^2+a^2)^{-\frac{1}{2}\nu-\frac{1}{2}}$ $\cdot P_{\nu}^{-\mu}[p(p^2+a^2)^{-\frac{1}{2}}]$ $\mathrm{Re}(\mu+\nu) > -1$	$[\Gamma(\nu+\mu+1)]^{-1}t^{\nu}J_{\mu}(at)$
11.25	$(p^2-a^2)^{-\frac{1}{2}\nu-\frac{1}{2}}$ $\cdot p_{\nu}^{-\mu}[p(p^2-a^2)^{-\frac{1}{2}}]$ $\mathrm{Re}(\nu+\mu) > -1$	$[\Gamma(\nu+\mu+1)]^{-1}t^{\nu}I_{\mu}(at)$

	$g(p) = \int_0^\infty f(t)e^{-pt}dt$	$f(t)$
11.26	$(p^2-a^2)^{-\frac{1}{2}\nu-\frac{1}{2}}e^{-i\pi\mu}$ $\cdot q_\nu^\mu[p(p^2-a^2)^{-\frac{1}{2}}]$ $\text{Re}(\nu\pm\mu) > -1$	$[\Gamma(1+\nu-\mu)]^{-1}t^\nu K_\mu(at)$
11.27	$(p+a)^{-\frac{1}{2}}(\frac{p-a}{p+a})^{\frac{1}{2}\nu}$ $\cdot p_\nu^\mu[p(p^2-a^2)^{-\frac{1}{2}}]$ $\text{Re}\ \mu < \frac{1}{2}$	$(\pi a)^{-\frac{1}{2}}[\Gamma(1-2\mu)]^{-1}$ $\cdot t^{-1}M_{\frac{1}{2}+\nu,-\mu}(at)$
11.28	$(p-a)^{-\frac{1}{2}}(\frac{p+a}{p-a})^{\frac{1}{2}\nu}$ $\cdot e^{-i\pi\mu}q_\nu^\mu[p(p^2-a^2)^{-\frac{1}{2}}]$ $-\frac{1}{2} < \text{Re}\ \mu < \frac{1}{2}$	$(\pi a)^{-\frac{1}{2}}\cos(\pi\mu)$ $\cdot t^{-1}W_{-\nu-\frac{1}{2},\mu}(at)$
11.29	$p^{-\frac{1}{2}}\{p_{-\frac{1}{4}}^{-\mu}[(1+a^2/p^2)^{\frac{1}{2}}]\}^2$ $\text{Re}\ \mu > -\frac{1}{4}$	$2^{2\mu}[\Gamma(\frac{1}{2}+2\mu)]^{-1}$ $\cdot t^{-\frac{1}{2}}J_\mu^2(\frac{1}{2}at)$
11.30	$p^{-\frac{1}{2}}p_{-\frac{1}{4}}^{\mu}[(1+a^2/p^2)^{\frac{1}{2}}]$ $\cdot p_{-\frac{1}{4}}^{-\mu}[(1+a^2/p^2)^{\frac{1}{2}}]$	$(\pi t)^{-\frac{1}{2}}J_\mu(\frac{1}{2}at)J_{-\mu}(\frac{1}{2}at)$

	$g(p) = \int_0^\infty f(t)e^{-pt}dt$	$f(t)$
11.31	$p^{-\frac{1}{2}}(p^2+a^2)^{-\frac{1}{2}}$ $\cdot P_{\frac{1}{4}}^{-\mu}[(1+a^2/p^2)^{\frac{1}{2}}]$ $\cdot P_{-\frac{1}{4}}^{-\mu}[(1+a^2/p^2)^{\frac{1}{2}}]$	$2^{\mu+3/2}[a\Gamma(\frac{3}{2}+2\mu)]^{-1}J_\mu^2(\frac{1}{2}at)$ Re $\mu > -\frac{3}{4}$
11.32	$p^{-\frac{1}{2}}P_{-\frac{1}{4}}^{\mu}[(1-a^2/p^2)^{\frac{1}{2}}]$ $\cdot P_{-\frac{1}{4}}^{-\mu}[(1-a^2/p^2)^{\frac{1}{2}}]$	$(\pi t)^{-\frac{1}{2}}I_\mu(\frac{1}{2}at)I_{-\mu}(\frac{1}{2}at)$
11.33	$p^{-\frac{1}{2}}\{P_{-\frac{1}{4}}^{-\mu}[(1-a^2/p^2)^{\frac{1}{2}}]\}^2$ Re $\mu > -\frac{1}{4}$	$2^{2\mu}[\Gamma(\frac{1}{2}+2\mu)]^{-1}t^{-\frac{1}{2}}I_\mu^2(\frac{1}{2}at)$
11.34	$p^{\frac{1}{2}}[P_\nu^{-\frac{1}{4}}(z)Q_\nu^{\frac{1}{4}}(z)$ $-P_\nu^{\frac{1}{4}}(z)Q_\nu^{-\frac{1}{4}}(z)]$ Re $\nu>-\frac{1}{4}$, $z = (1-p^2/a^2)^{\frac{1}{2}}$	$-a(\frac{1}{2}\pi)^{-\frac{1}{2}}t^{-\frac{1}{2}}I_{\nu+\frac{1}{2}}(\frac{1}{2}at)K_{\nu+\frac{1}{2}}(\frac{1}{2}at)$
11.35	$e^{-i\pi\mu}q_{p-\frac{1}{2}}^{\mu}(\cosh a)$ Re $\mu < \frac{1}{2}$	0 $t < a$ $(\frac{1}{2}\pi)^{\frac{1}{2}}[\Gamma(\frac{1}{2}-\mu)]^{-1}(\sinh a)^{\mu}$ $\cdot(\cosh t-\cosh a)^{-\mu-\frac{1}{2}}$ $t > a$

	$g(p) = \int_0^\infty f(t)e^{-pt}dt$	$f(t)$
11.36	$\Gamma(p-\nu)$ $\cdot P_\nu^{-p}(\coth a)$ $\mathrm{Re}\,\nu > -1$	$0 \qquad t < a$ $[\Gamma(1+\nu)]^{-1}(\sinh a)^{-\nu}$ $\cdot(\cosh t - \cosh a)^\nu \qquad t > a$
11.37	$2^p e^{i\pi p}\Gamma(p)$ $(z^2-1)^{\frac{1}{2}p} Q_{p-1}^{\nu-p}(z)$	$\frac{1}{2}\Gamma(\nu)e^{i\pi\nu}(z^2-1)^{\frac{1}{2}\nu}(1-e^{-t})^{-\frac{1}{2}}$ $\cdot\{[z+(1-e^{-t})^{\frac{1}{2}}]^{-\nu}+[z-(1-e^{-t})^{\frac{1}{2}}]^{-\nu}\}$

2.12 Bessel Functions

	$g(p) = \int_0^\infty f(t)e^{-pt}dt$	$f(t)$
12.1	$\sin(ap)J_0(ap$ $-\cos(ap)Y_0(ap)$	$\pi^{-1}(\frac{1}{2}t)^{-\frac{1}{2}}(t^2+4a^2)^{-\frac{1}{2}}$ $\cdot[t+(t^2+4a^2)^{\frac{1}{2}}]^{\frac{1}{2}}$
12.2	$\cos(ap)J_0(ap)$ $+\sin(ap)Y_0(ap)$	$\pi^{-1}(\frac{1}{2}t)^{-\frac{1}{2}}(t^2+4a^2)^{-\frac{1}{2}}$ $\cdot[(t^2+4a^2)^{\frac{1}{2}}-t]^{\frac{1}{2}}$
2.3	$\cos(ap)J_1(ap)$ $+\sin(ap)Y_1(ap)$	$-(\pi a)^{-1}(2t)^{-\frac{1}{2}}(t^2+4a^2)^{-\frac{1}{2}}$ $\cdot[(t^2+4a^2)^{\frac{1}{2}}-t]^{3/2}$
2.4	$\sin(ap)J_1(ap)$ $-\cos(ap)Y_1(ap)$	$(\pi a)^{-1}(2t)^{-\frac{1}{2}}(t^2+4a^2)^{-\frac{1}{2}}$ $\cdot[t+(t^2+4a^2)^{\frac{1}{2}}]^{3/2}$
2.5	$p^{-\nu}[\cos(ap+b)J_\nu(ap)$ $+\sin(ap+b)Y_\nu(ap)]$ $\mathrm{Re}\,\nu > -\frac{1}{2}$	$-2\pi^{-\frac{1}{2}}(2a)^{-\nu}[\Gamma(\frac{1}{2}+\nu)]^{-1}t^{\nu-\frac{1}{2}}$ $\cdot(t^2+4a^2)^{\frac{1}{2}\nu-\frac{1}{4}}$ $\cdot\sin[b+(\nu-\frac{1}{2})\,\mathrm{arccot}(\frac{1}{2}t/a)]$
2.6	$p^{-\nu}[\sin(ap+b)J_\nu(ap)$ $-\cos(ap+b)Y_\nu(ap)]$ $\mathrm{Re}\,\nu > -\frac{1}{2}$	$2\pi^{-\frac{1}{2}}(2a)^{-\nu}[\Gamma(\frac{1}{2}+\nu)]^{-1}t^{\nu-\frac{1}{2}}$ $\cdot(t^2+4a^2)^{\frac{1}{2}\nu-\frac{1}{4}}$ $\cdot\cos[b+(\nu-\frac{1}{2})\,\mathrm{arccot}(\frac{1}{2}t/a)]$

	$g(p) = \int_0^\infty f(t)e^{-pt}dt$	$f(t)$
12.7	$p^{\frac{1}{2}} J_{\nu+\frac{1}{4}}(\frac{1}{2}ap)J_{\nu-\frac{1}{4}}(\frac{1}{2}ap)$ $+Y_{\nu+\frac{1}{4}}(\frac{1}{2}ap)Y_{\nu-\frac{1}{4}}(\frac{1}{2}ap)]$	$2\pi^{-1}a^{-2\nu}(t^2+a^2)^{-\frac{1}{2}}(\frac{1}{2}\pi t)^{-\frac{1}{2}}$ $\cdot[t+(t^2+a^2)^{\frac{1}{2}}]^{2\nu}$
12.8	$p^{\frac{1}{2}}[J_{\nu+\frac{1}{4}}(\frac{1}{2}ap)Y_{\nu-\frac{1}{4}}(\frac{1}{2}ap)$ $-J_{\nu-\frac{1}{4}}(\frac{1}{2}ap)Y_{\nu+\frac{1}{4}}(\frac{1}{2}ap)]$	$2\pi^{-1}a^{2\nu}(\frac{1}{2}\pi t)^{-\frac{1}{2}}(t^2+a^2)^{-\frac{1}{2}}$ $\cdot[t+(t^2+a^2)^{\frac{1}{2}}]^{-2\nu}$
12.9	$p^{\frac{1}{2}}[J_{\frac{1}{4}+\nu}(\frac{1}{2}ap)J_{\frac{1}{4}-\nu}(\frac{1}{2}ap)$ $+Y_{\frac{1}{4}+\nu}(\frac{1}{2}ap)Y_{\frac{1}{4}-\nu}(\frac{1}{2}ap)]$	$2\pi^{-1}a^{-2\nu}(\frac{1}{2}\pi t)^{-\frac{1}{2}}(t^2+a^2)^{-\frac{1}{2}}$ $\cdot\{\sin[\pi(\nu+\frac{1}{4})][t+(t^2+a^2)^{\frac{1}{2}}]^{2\nu}$ $+\cos[\pi(\nu+\frac{1}{4})][(t^2+a^2)^{\frac{1}{2}}-t]^{2\nu}\}$
12.10	$p^{\frac{1}{2}}[J_{\frac{1}{4}+\nu}(\frac{1}{2}ap)Y_{\frac{1}{4}-\nu}(\frac{1}{2}ap)$ $-J_{\frac{1}{4}-\nu}(\frac{1}{2}ap)Y_{\frac{1}{4}+\nu}(\frac{1}{2}ap)]$	$2\pi^{-1}a^{-2\nu}(\frac{1}{2}\pi t)^{-\frac{1}{2}}(t^2+a^2)^{-\frac{1}{2}}$ $\cdot\{\sin[\pi(\nu+\frac{1}{4})][(t^2+a^2)^{\frac{1}{2}}-t]^{2\nu}$ $-\cos[\pi(\nu+\frac{1}{4})][(t^2+a^2)^{\frac{1}{2}}+t]^{2\nu}\}$
12.11	$J_0^2(ap)+Y_0^2(ap)$	$8\pi^{-2}(t^2+4a^2)^{-\frac{1}{2}}K[t(t^2+4a^2)^{-\frac{1}{2}}]$
12.12	$J_\nu^2(ap)+Y_\nu^2(ap)$	$2(\pi a)^{-1}P_{\nu-\frac{1}{2}}(1+\frac{1}{2}t^2/a^2)$

	$g(p) = \int_0^\infty f(t)e^{-pt}dt$	$f(t)$
12.13	$e^{-ap}J_0[a(b^2-p^2)]$	$\pi^{-1}(2at-t^2)^{-\frac{1}{2}}\cos[b(2at-t^2)^{\frac{1}{2}}]$ $t < 2a$ 0 $t > 2a$
12.14	$J_\nu^2(ap^{\frac{1}{2}})+Y_\nu^2(ap^{\frac{1}{2}})$	$2\pi^{-2}t^{-1}\exp(\frac{1}{2}a^2/t)K_\nu(\frac{1}{2}a^2/t)$
12.15	$p^{-1}e^{a/p}J_0(b/p)$	$I_0(At^{\frac{1}{2}})J_0(Bt^{\frac{1}{2}})$ $\begin{matrix}A\\B\end{matrix} = 2^{\frac{1}{2}}[(a^2+b^2)^{\frac{1}{2}}\pm a]^{\frac{1}{2}}$
12.16	$p^{-1}e^{a/p}Y_0(b/p)$	$I_0(At^{\frac{1}{2}})Y_0(Bt^{\frac{1}{2}})-2\pi^{-1}J_0(Bt^{\frac{1}{2}})K_0(At^{\frac{1}{2}})$ $\begin{matrix}A\\B\end{matrix} = 2^{\frac{1}{2}}[(a^2+b^2)^{\frac{1}{2}}\pm a]^{\frac{1}{2}}$
12.17	$p^{-1}e^{-a/p}J_0(b/p)$	$J_0(At^{\frac{1}{2}})I_0(Bt^{\frac{1}{2}})$ $\begin{matrix}A\\B\end{matrix} = 2^{\frac{1}{2}}[(a^2+b^2)^{\frac{1}{2}}\pm a]^{\frac{1}{2}}$
12.18	$p^{-1}e^{-a/p}Y_0(b/p)$	$I_0(Bt^{\frac{1}{2}})Y_0(At^{\frac{1}{2}})$ $-2\pi^{-1}J_0(At^{\frac{1}{2}})K_0(Bt^{\frac{1}{2}})$ $\begin{matrix}A\\B\end{matrix} = 2^{\frac{1}{2}}[(a^2+b^2)^{\frac{1}{2}}\pm a]^{\frac{1}{2}}$

	$g(p) = \int_0^\infty f(t)e^{-pt}dt$	$f(t)$
12.19	$p^{-1}e^{-a/p}J_\nu(b/p)$ $\operatorname{Re}\nu > -1$	$J_\nu(At^{\frac{1}{2}})I_\nu(Bt^{\frac{1}{2}})$ $\begin{matrix}A\\B\end{matrix} = 2^{\frac{1}{2}}[(a^2+b^2)^{\frac{1}{2}}\pm a]^{\frac{1}{2}}$
12.20	$p^{-1}e^{a/p}J_\nu(b/p)$ $\operatorname{Re}\nu > -1$	$I_\nu(At^{\frac{1}{2}})J_\nu(Bt^{\frac{1}{2}})$ $\begin{matrix}A\\B\end{matrix} = 2^{\frac{1}{2}}[(a^2+b^2)^{\frac{1}{2}}\pm a]^{\frac{1}{2}}$
12.21	$p^{-\frac{1}{2}}[\cos(a/p)J_\nu(a/p)$ $+\sin(a/p)Y_\nu(a/p)]$ $-\frac{1}{2} < \operatorname{Re}\nu < \frac{1}{2}$	$-4\pi^{-3/2}t^{-\frac{1}{2}}\cos(\pi\nu)$ $\cdot\{\sin(\frac{1}{2}\pi\nu)\operatorname{ker}_{2\nu}[2(2at)^{\frac{1}{2}}]$ $+\cos(\frac{1}{2}\pi\nu)\operatorname{kei}_{2\nu}[2(2at)^{\frac{1}{2}}]\}$
12.22	$p^{-\frac{1}{2}}[\sin(a/p)J_\nu(a/p)$ $-\cos(a/p)Y_\nu(a/p)]$ $-\frac{1}{2} < \operatorname{Re}\nu < \frac{1}{2}$	$4\pi^{-3/2}t^{-\frac{1}{2}}\cos(\pi\nu)$ $\cdot\{\cos(\frac{1}{2}\pi\nu)\operatorname{ker}_{2\nu}[2(2at)^{\frac{1}{2}}]$ $-\sin[\frac{1}{2}\pi\nu)\operatorname{kei}_{2\nu}[2(2at)^{\frac{1}{2}}]\}$
12.23	$p^{-1}e^{-a/p}Y_\nu(b/p)$ $\begin{matrix}A\\B\end{matrix} = 2^{\frac{1}{2}}[(a^2+b^2)^{\frac{1}{2}}\pm a]^{\frac{1}{2}}$ $-1 < \operatorname{Re}\nu < 1$	$Y_\nu(At^{\frac{1}{2}})I_{-\nu}(Bt^{\frac{1}{2}})$ $-2\pi^{-1}\cos(\pi\nu)J_\nu(At^{\frac{1}{2}})K_\nu(Bt^{\frac{1}{2}})$

	$g(p) = \int_0^\infty f(t)e^{-pt}dt$	$f(t)$
12.24	$p^{-1}e^{a/p}Y_\nu(b/p)$ $\genfrac{}{}{0pt}{}{A}{B} = 2^{\frac{1}{2}}[(a^2+b^2)^{\frac{1}{2}}\pm a]^{\frac{1}{2}}$ $-1 < \mathrm{Re}\,\nu < 1$	$Y_\nu(Bt^{\frac{1}{2}})I_{-\nu}(At^{\frac{1}{2}})$ $-2\pi^{-1}\cos(\pi\nu)J_\nu(Bt^{\frac{1}{2}})K_\nu(At^{\frac{1}{2}})$
12.25	$p^\mu J_\nu(a/p)$ $\mathrm{Re}(\nu-\mu) > 0$	$(\frac{1}{2}a)^\nu[\Gamma(\nu+1)\Gamma(\nu-\mu)]^{-1}t^{\nu-\mu-1}$ $\cdot {}_0F_3(\ ;1+\nu,\frac{1}{2}\nu-\frac{1}{2}\mu,\frac{1}{2}+\frac{1}{2}\nu-\frac{1}{2}\mu;-\frac{1}{16}a^2t^2)$
12.26	$p^{-\mu}J_\nu[(a/p)^{\frac{1}{2}}]$ $\mathrm{Re}(\nu+2\mu) > 0$	$(\frac{1}{4}a)^{\frac{1}{2}\nu}[\Gamma(\frac{1}{2}\nu+\mu)\Gamma(1+\nu)]^{-1}t^{\mu-1+\frac{1}{2}\nu}$ $\cdot {}_0F_2(\ ;\mu+\frac{1}{2}\nu,1+\nu;-\frac{1}{4}at)$
12.27	$\sin(ap^2)J_0(ap^2)$ $-\cos(ap^2)Y_0(ap^2)$	$(\pi a)^{-\frac{1}{2}}\{\cos[t^2/(16a)]J_0[t^2/(16a)]$ $+\sin[t^2/(16a)]Y_0[t^2/(16a)]\}$
12.28	$\cos(ap^2)J_0(ap^2)$ $+Y_0(ap^2)\sin(ap^2)$	$(\pi a)^{-\frac{1}{2}}\{\cos[t^2/(16a)]J_0[t^2/(16a)]$ $-\sin[t^2/(16a)]Y_0[t^2/(16a)]$
12.29	$(p^2+1)^{-\frac{1}{2}}\exp[-ap(1+p^2)^{-1}]$ $\cdot J_\nu[a(1+p^2)^{-1}]$	$J_\nu(t)J_{2\nu}[2(at)^{\frac{1}{2}}]$
12.30	$(p^2+1)^{-\frac{1}{2}}\exp[-ap(1+p^2)^{-1}]$ $\cdot Y_0[a(1+p^2)^{-1}]$	$2J_0(t)Y_0[2(at)^{\frac{1}{2}}]$ $+J_0[2(at)^{\frac{1}{2}}]Y_0(t)$

	$g(p) = \int_0^\infty f(t)e^{-pt}dt$	$f(t)$
12.31	$p^{-\nu}e^{-iap}H_\nu^{(1)}(ap)$ $\mathrm{Re}\ \nu > -\frac{1}{2}$	$-2i\pi^{-\frac{1}{2}}(2a)^{-\nu}[\Gamma(\frac{1}{2}+\nu)]^{-1}$ $\cdot(t^2-2iat)^{\nu-\frac{1}{2}}$
12.32	$p^{-\nu}e^{iap}H_\nu^{(2)}(ap)$ $\mathrm{Re}\ \nu > -\frac{1}{2}$	$2i\pi^{-\frac{1}{2}}(2a)^{-\nu}[\Gamma(\frac{1}{2}+\nu)]^{-1}$ $\cdot(t^2+2iat)^{\nu-\frac{1}{2}}$
12.33	$(p^2+a^2)^{-\frac{1}{2}\nu}e^{ir}$ $\cdot H_\nu^{(2)}[(p^2+a^2)^{\frac{1}{2}}]$ $\mathrm{Re}\ \nu > -\frac{1}{2}$	$i(2a/\pi)^{\frac{1}{2}}a^{-\nu}(t^2+2it)^{\frac{1}{2}\nu-\frac{1}{4}}$ $\cdot J_{\nu-\frac{1}{2}}[a(t^2+2it)^{\frac{1}{2}}]$
12.34	$p^{-\frac{1}{2}}H_\nu^{(1)}(ap)H_\nu^{(2)}(ap)$	$(2t/\pi)^{\frac{1}{2}}a^{-1}\ P_{\nu-\frac{1}{2}}^{\frac{1}{4}}[(1+t^2/4a^2)^{\frac{1}{2}}]$ $\cdot P_{\nu-\frac{1}{2}}^{-\frac{1}{4}}[(1+t^2/4a^2)^{\frac{1}{2}}]$
12.35	$p^{-\nu}H_\nu^{(1)}(ap^{\frac{1}{2}})H_\nu^{(2)}(ap^{\frac{1}{2}})$ $\mathrm{Re}\ \nu > -\frac{1}{2}$	$2[\Gamma(\frac{1}{2}+\nu)]^{-1}a^{-2\nu-1}t^{-\frac{1}{2}+\frac{3}{2}\nu}$ $\cdot e^{\frac{1}{2}a^2/t}W_{\frac{1}{2}\nu,\frac{1}{2}\nu}(a^2/t)$

	$g(p) = \int_0^\infty f(t)e^{-pt}dt$	$f(t)$
12.36	$p^{\frac{1}{2}}[e^{i\pi\nu}H^{(1)}_{\frac{1}{2}+\nu}(ap)H^{(2)}_{\frac{1}{2}-\nu}(ap)$ $+e^{-i\pi\nu}H^{(1)}_{\frac{1}{2}-\nu}(ap)H^{(2)}_{\frac{1}{2}+\nu}(ap)]$	$4\pi^{-3/2}(2a)^{-2\nu}t^{-\frac{1}{2}}(t^2+4a^2)^{-\frac{1}{2}}$ $\cdot\{[(4a^2+t^2)^{\frac{1}{2}}+t]^{2\nu}$ $+[(4a^2+t^2)^{\frac{1}{2}}-t]^{2\nu}\}$
12.37	$p^{-\frac{1}{2}}[e^{i\pi\nu}H^{(1)}_{\frac{1}{2}+\nu}(ap)H^{(2)}_{\frac{1}{2}-\nu}(ap)$ $-e^{-i\pi\nu}H^{(1)}_{\frac{1}{2}-\nu}(ap)H^{(2)}_{\frac{1}{2}+\nu}(ap)]$	$i4\pi^{-3/2}(2a)^{-2\nu}t^{-\frac{1}{2}}(t^2+4a^2)^{-\frac{1}{2}}$ $\cdot\{(4a^2+t^2)^{\frac{1}{2}}+t]^{2\nu}$ $-[(4a^2+t^2)^{\frac{1}{2}}-t]^{2\nu}\}$
12.38	$J_{\nu-p}(a)Y_{-\nu-p}(a)$ $-J_{-\nu-p}(a)Y_{\nu-p}(a)$ $-\frac{1}{2} < \mathrm{Re}\,\nu < \frac{1}{2}$	$2\pi^{-2}\sin(2\pi\nu)$ $\cdot K_{2\nu}[2a\sinh(\frac{1}{2}t)]$
12.39	$Y_p(a)\frac{\partial}{\partial p}J_p(a)-J_p(a)\frac{\partial}{\partial a}Y_p(a)$	$2\pi^{-1}K_0[2a\sinh(\frac{1}{2}t)]$
12.40	$\Gamma(p)(\frac{1}{2}a)^{-p}J_{\nu+p}(a)$	$(1-e^{-t})^{\frac{1}{2}\nu}J_\nu[a(1-e^{-t})^{\frac{1}{2}}]$
12.41	$\Gamma(\frac{1}{2}+p)(\frac{1}{2}a)^{-p}J_p(a)$	$\pi^{-\frac{1}{2}}(e^t-1)^{-\frac{1}{2}}\cos[a(1-e^{-t})^{\frac{1}{2}}]$

2.13 Modified Bessel Functions

	$g(p) = \int_0^\infty f(t)e^{-pt}dt$	$f(t)$
13.1	$e^{-ap}I_0(ap)$	$\pi^{-1}(2at-t^2)^{-\frac{1}{2}}$ $t < 2a$ 0 $t > 2a$
13.2	$e^{-ap}I_1(ap)$	$\pi^{-1}(1-t/a)(2at-t^2)^{-\frac{1}{2}}$ $t < 2a$ 0 $t > 2a$
13.3	$K_0(ap)$	0 $t < a$ $(t^2-a^2)^{-\frac{1}{2}}$ $t > a$
13.4	$e^{ap}K_0(ap)$	$(t^2+2at)^{-\frac{1}{2}}$
13.5	$p^{-1}K_0(ap)$	0 $t < a$ $\log[t/a+(t^2/a^2-1)^{\frac{1}{2}}]$ $t > a$
13.6	$p^{-1}e^{ap}K_0(ap)$	$2\log\{[2a)^{-\frac{1}{2}}[t^{\frac{1}{2}}+(t+2a)^{\frac{1}{2}}]\}$
13.7	$K_1(ap)$	0 $t < a$ $a^{-1}t(t^2-a^2)^{-\frac{1}{2}}$ $t > a$
13.8	$e^{ap}K_1(ap)$	$(1+t/a)(t^2+2at)^{-\frac{1}{2}}$

	$g(p) = \int_0^\infty f(t)e^{-pt}dt$	$f(t)$
13.9	$K_n(ap)$ $n = 0,1,2,\cdots$	$0 \qquad t < a$ $(t^2-a^2)^{-\frac{1}{2}}T_n(t/a) \qquad t > a$
13.10	$e^{ap}K_n(ap)$ $n = 0,1,2,\cdots$	$(t^2+2at)^{-\frac{1}{2}}T_n(1+t/a)$
13.11	$p^{-1}K_n(ap)$ $n = 1,2,3,\cdots$	$0 \qquad t < a$ $(na)^{-1}(t^2-a^2)^{\frac{1}{2}}U_n(t/a) \qquad t > a$
13.12	$p^{-1}e^{ap}K_n(ap)$ $n = 1,2,3,\cdots$	$(na)^{-1}(t^2+2at)^{\frac{1}{2}}U_n(1+t/a)$
13.13	$p^{-\frac{1}{2}}e^{-p}I_{n+\frac{1}{2}}(p)$ $n = 0,1,2,\cdots$	$(2\pi)^{-\frac{1}{2}}P_n(1-t) \qquad t < 2$ $0 \qquad t > 2$
13.14	$p^{-\frac{1}{2}}e^{ap}K_{n+\frac{1}{2}}(p)$ $n = 0,1,2,\cdots$	$(\frac{1}{2}\pi/a)^{\frac{1}{2}}P_n(1+t/a)$

	$g(p) = \int_0^\infty f(t)e^{-pt}dt$	$f(t)$
13.15	$p^{-\nu}e^{-ap}I_{\nu+n}(ap)$	$(-1)^n\pi^{-1}(2/a)^\nu n!\Gamma(\nu)[\Gamma(2\nu+n)]^{-1}$ $\cdot[t(2a-t)]^{\nu-\frac{1}{2}}C_n^\nu(t/a-1)$ $t < 2a$ 0 $t > 2a$
13.16	$\log(bp)K_0(ap)$	0 $t < a$ $-(t^2-a^2)^{-\frac{1}{2}}[\gamma-\log(\frac{1}{2}ab)+\log(t^2-a^2)]$ $t > a$
13.17	$e^{ap}\log(bp)K_0(ap)$	$-(t^2+2at)^{-\frac{1}{2}}[\gamma-\log(\frac{1}{2}ab)$ $+ \log(t^2+2at)]$
13.18	$p^{-\nu}e^{-ap}I_\nu(ap)$ $\mathrm{Re}\ \nu > -\frac{1}{2}$	$\pi^{-\frac{1}{2}}(2a)^{-\nu}[\Gamma(\frac{1}{2}+\nu)]^{-1}(2at-t^2)^{\nu-\frac{1}{2}}$, $t < 2a$ 0 $t > 2a$
13.19	$p^\nu e^{-ap}I_\nu(ap)$	$\pi^{-3/2}(2a)^\nu\Gamma(\frac{1}{2}+\nu)\cos(2\pi\nu)(2at-t^2)^{-\nu-\frac{1}{2}}$ $t < 2a$ $-\pi^{-3/2}(2a)^\nu\Gamma(\frac{1}{2}+\nu)\sin(2\pi\nu)(t^2-2at)^{-\nu-}$ $t > 2a$

	$g(p) = \int_0^\infty f(t)e^{-pt}dt$	$f(t)$
13.20	$p^{-\nu}\operatorname{csch}(ap)I_\nu(ap)$ $\operatorname{Re}\nu>-\frac{1}{2}$, $n=0,1,\cdots$	$2\pi^{-\frac{1}{2}}(2a)^{-\nu}[\Gamma(\frac{1}{2}+\nu)]^{-1}(t-2na)^{\nu-\frac{1}{2}}$ $\cdot[2a(n+1)-t]^{\nu-\frac{1}{2}}, 2na<t<2a(n+1)$
13.21	$p^{-\nu}K_\nu(ap)$	$0 \quad t<a$ $\pi^{\frac{1}{2}}(2a)^{-\nu}[\Gamma(\frac{1}{2}+\nu)]^{-1}(t^2-a^2)^{\nu-\frac{1}{2}}$, $t>a$
13.22	$e^{ap}p^{-\nu}K_\nu(ap)$ $\operatorname{Re}\nu>-\frac{1}{2}$	$\pi^{\frac{1}{2}}(2a)^{-\nu}[\Gamma(\frac{1}{2}+\nu)]^{-1}(t^2+2at)^{\nu-\frac{1}{2}}$
13.23	$p^{-\mu}K_\nu(ap)$	$0 \quad t<a$ $(\frac{1}{2}\pi/a)^{\frac{1}{2}}(t^2-a^2)^{\frac{1}{2}\mu-\frac{1}{4}}P_{\nu-\frac{1}{2}}^{\frac{1}{2}-\mu}(t/a), \quad t>a$
13.24	$p^{-\mu}e^{ap}K_\nu(ap)$	$(\frac{1}{2}\pi/a)^{\frac{1}{2}}(t^2+2at)^{\frac{1}{2}\mu-\frac{1}{4}}P_{\nu-\frac{1}{2}}^{\frac{1}{2}-\mu}(1+t/a)$
13.25	$K_\nu(ap)$	$0 \quad t<a$ $\frac{1}{2}(t^2-a^2)^{-\frac{1}{2}}\{[t/a+(t^2/a^2-1)^{\frac{1}{2}}]^\nu$ $+[t/a-(t^2/a^2-1)^{\frac{1}{2}}]^\nu\} \quad t>a$
13.26	$e^{ap}K_\nu(ap)$	$\frac{1}{2}(t^2+2at)^{-\frac{1}{2}}\{[1+t/a+(t^2/a^2+2t/a)^{\frac{1}{2}}]^\nu$ $+[1+t/a-(t^2/a^2+2t/a)^{\frac{1}{2}}]^\nu\}$

	$g(p) = \int_0^\infty f(t)e^{-pt}dt$	$f(t)$
13.27	$p^{-1}K_\nu(ap)$	$0 \quad t < a$ $\frac{1}{2}\nu^{-1}\{[t/a+(t^2/a^2-1)^{\frac{1}{2}}]^\nu$ $-[t/a-(t^2/a^2-1)^{\frac{1}{2}}]^\nu\} \quad t > a$
13.28	$e^{ap}p^{-1}K_\nu(ap)$	$2^{-\nu-1}a^{-\nu}\nu^{-1}\{[(t+2a)^{\frac{1}{2}}+t^{\frac{1}{2}}]^{2\nu}$ $-[(t+2a)^{-\frac{1}{2}}-t^{\frac{1}{2}}]^{2\nu}\}$
13.29	$p^{\frac{1}{2}}K_{\frac{1}{4}}(ap)$ $\cdot[I_{\frac{1}{4}}(ap)+I_{-\frac{1}{4}}(ap)]$	$(\pi t)^{-\frac{1}{2}}(4a^2-t^2)^{-\frac{1}{2}} \quad t < 2a$ $0 \quad t > 2a$
13.30	$p^{\frac{1}{2}}K_{\frac{1}{4}}(ap)K_{3/4}(ap)$	$0 \quad t < 2a$ $a^{-1}(\frac{1}{2}\pi t)^{\frac{1}{2}}(t^2-4a^2) \quad t > 2a$
13.31	$p^{\frac{1}{2}}e^{2ap}K_{\frac{1}{4}}(ap)K_{3/4}(ap)$	$(\frac{1}{2}\pi)^{\frac{1}{2}}a^{-1}(t+2a)(t^2+4at)$
13.32	$p^{3/2}[K^2_{3/4}(ap)-K^2_{\frac{1}{4}}(ap)]$	$0 \quad t < 2a$ $(2\pi)^{\frac{1}{2}}t^{-3/2}(t^2-4a^2)^{-3/2} \quad t > 2a$
13.33	$e^{2ap}p^{3/2}$ $[K^2_{3/4}(ap)-K^2_{\frac{1}{4}}(ap)]$	$(2\pi)^{\frac{1}{2}}(t+2a)^{-3/2}(t^2+4at)$

	$g(p) = \int_0^\infty f(t)e^{-pt}dt$	$f(t)$
13.34	$p^{\frac{1}{2}}[I_{\frac{1}{4}}(\frac{1}{2}ap)I_{-\frac{3}{4}}(ap)$ $-I_{-\frac{1}{4}}(ap)I_{\frac{3}{4}}(ap)]$	$2^{\frac{1}{2}}\pi^{-\frac{3}{2}}a^{-1}t^{\frac{1}{2}}(4a^2-t^2)^{-\frac{1}{2}}$ $t < 2a$ 0 $t > 2a$
13.35	$p^{\frac{1}{2}}K_{\nu+\frac{1}{4}}(ap)K_{\nu-\frac{1}{4}}(ap)$	0 $t < 2a$ $(2a)^{-2\nu}(\frac{1}{2}\pi/t)^{\frac{1}{2}}(t^2-4a^2)^{-\frac{1}{2}}$ $\cdot\{[t+(t^2-a^2)^{\frac{1}{2}}]^{2\nu}+[t-(t^2-a^2)^{\frac{1}{2}}]^{2\nu}\}$ $t > 2a$
13.36	$p^{\frac{1}{2}}e^{2ap}K_{\nu+\frac{1}{4}}(ap)$ $\cdot K_{\nu-\frac{1}{4}}(ap)$	$(2a)^{-2\nu}(\frac{1}{2}\pi)^{\frac{1}{2}}(t+2a)^{-\frac{1}{2}}(t^2+4at)^{-\frac{1}{2}}$ $\cdot\{[t+2a+(t^2+4at)^{\frac{1}{2}}]^{2\nu}$ $+[t+2a-(t^2+4at)^{\frac{1}{2}}]^{2\nu}\}$
13.37	$p^{\frac{1}{2}}[I_{\nu-\frac{1}{4}}(ap)I_{-\nu-\frac{1}{4}}(ap)$ $-I_{\nu+\frac{1}{4}}(ap)I_{-\nu+\frac{1}{4}}(ap)]$	$(\frac{1}{2}\pi)^{-\frac{3}{2}}t^{-\frac{1}{2}}(4a^2-t^2)^{-\frac{1}{2}}$ $\cdot\cos[2\nu\arccos(\frac{1}{2}t/a)]$ $t < 2a$ 0 $t > 2a$
13.38	$K_\nu(ap)K_\nu(bp)$	0 $t < a+b$ $\frac{1}{2}\pi(ab)^{-\frac{1}{2}}P_{\nu-\frac{1}{2}}[(2ab)^{-1}(t^2-a^2-b^2)]$ $t > a+b$

	$g(p) = \int_0^\infty f(t)e^{-pt}dt$	$f(t)$
13.39	$I_\nu(bp)K_\nu(ap)$ $a > b$	$0 \qquad t < a-b$ $\frac{1}{2}(ab)^{-\frac{1}{2}}P_{\nu-\frac{1}{2}}[(2ab)^{-1}(a^2+b^2-t^2)]$ $a-b < t < a+b$
13.40		$\pi^{-1}(ab)^{-\frac{1}{2}}\cos(\pi\nu)$ $\cdot q_{\nu-\frac{1}{2}}[(2ab)^{-1}(t^2-a^2-b^2)]$ $t > a+b$
13.41	$p^{-1}K_0(ap^{\frac{1}{2}})$	$-\frac{1}{2}Ei(-\frac{1}{4}a^2/t)$
13.42	$p^{\frac{1}{2}\nu}K_\nu(ap^{\frac{1}{2}})$	$a^\nu(2t)^{-\nu-1}\exp(-\frac{1}{4}a^2/t)$
13.43	$p^{\frac{1}{2}\nu-1}K_\nu(ap^{\frac{1}{2}})$	$\frac{1}{2}(\frac{1}{2}a)^{-\nu}\Gamma(\nu,\frac{1}{4}a^2/t)$
13.44	$p^{-\frac{1}{2}}K_\nu(ap^{\frac{1}{2}})$	$\frac{1}{2}(\pi t)^{-\frac{1}{2}}\exp(-\frac{1}{8}a^2/t)K_{\frac{1}{2}\nu}(\frac{1}{8}a^2/t)$
13.45	$p^\mu K_\nu(ap^{\frac{1}{2}})$	$a^{-1}t^{-\mu-\frac{1}{2}}\exp(-\frac{1}{8}a^2/t)W_{\frac{1}{2}+\mu,\frac{1}{2}\nu}(\frac{1}{4}a^2/t)$
13.46	$p^{n+\frac{1}{2}\nu}K_\nu(ap^{\frac{1}{2}})$ $n = 0,1,2,\cdots$	$\frac{1}{2}(-1)^n n!(\frac{1}{2}a)^\nu t^{-n}\exp(-\frac{1}{4}a^2/t)L_n^\nu(\frac{1}{4}a^2/t)$

	$g(p) = \int_0^\infty f(t)e^{-pt}dt$	$f(t)$
13.47	$p^{\mu}K_{\nu}(ap^{\frac{1}{2}})$	$a^{-1}t^{-\mu-\frac{1}{2}}\exp(-\frac{1}{8}a^2/t)W_{\frac{1}{2}+\mu,\frac{1}{2}\nu}(\frac{1}{4}a^2/t)$
13.48	$e^{-ap}I_0[a(b^2+p^2)^{\frac{1}{2}}]$	$\pi^{-1}(2at-t^2)^{-\frac{1}{2}}\cosh[b(2at-t^2)^{\frac{1}{2}}]$ $t < 2a$ 0 $t > 2a$
13.49	$a^{-ap}I_0[a(p^2-b^2)^{\frac{1}{2}}]$	$\pi^{-1}(2at-t^2)^{-\frac{1}{2}}\cos[b(2at-t^2)^{\frac{1}{2}}]$ $t < 2a$ 0 $t > 2a$
13.50	$K_0[a(p^2+b^2)^{\frac{1}{2}}]$	0 $t < a$ $(t^2-a^2)^{-\frac{1}{2}}\cos[b(t^2-a^2)^{\frac{1}{2}}]$ $t > a$
13.51	$K_0[a(p^2-b^2)^{\frac{1}{2}}]$	0 $t < a$ $(t^2-a^2)^{-\frac{1}{2}}\cosh[b(t^2-a^2)^{\frac{1}{2}}]$ $t > a$
13.52	$e^{ap}K_0[a(p^2+b^2)^{\frac{1}{2}}]$	$(t^2+2at)^{-\frac{1}{2}}\cos[b(t^2+2at)^{\frac{1}{2}}]$
13.53	$e^{ap}K_0[a(p^2-b^2)^{\frac{1}{2}}]$	$(t^2+2at)^{-\frac{1}{2}}\cosh[b(t^2+2at)^{\frac{1}{2}}]$
13.54	$\arctan(b/p)$ $\cdot K_0[a(p^2+b^2)^{\frac{1}{2}}]$	0 $t < a$ $(t^2-a^2)^{-\frac{1}{2}}\sin[b(t^2-a^2)^{\frac{1}{2}}]$ $\cdot\log[t/a+(t^2/a^2-1)^{\frac{1}{2}}]$ $t > a$

	$g(p) = \int_0^\infty f(t)e^{-pt}dt$	$f(t)$
13.55	$e^{ap}\arctan(b/p)$ $\cdot K_0[a(p^2+b^2)^{\frac{1}{2}}]$	$(t^2+2at)^{-\frac{1}{2}}\sin[b(t^2+2at)^{\frac{1}{2}}]$ $\cdot\log[1+t/a+(t^2/a^2+2t/a)^{\frac{1}{2}}]$
13.56	$\log[(p+b)/(p-b)]$ $\cdot K_0[a(p^2-b^2)^{\frac{1}{2}}]$	$0 \quad t < a$ $2(t^2-a^2)^{-\frac{1}{2}}\sinh[b(t^2-a^2)^{\frac{1}{2}}]$ $\cdot\log[t/a+(t^2/a^2-1)^{\frac{1}{2}}] \quad t > a$
13.57	$e^{ap}\log[(p+b)/(p-b)]$ $\cdot K_0[a(p^2-b^2)^{\frac{1}{2}}]$	$2(t^2+2at)^{-\frac{1}{2}}\sinh[b(t^2+2at)^{\frac{1}{2}}]$ $\cdot\log[1+t/a+(t^2/a^2+2t/a)^{\frac{1}{2}}]$
13.58	$(p^2+b^2)^{-\frac{1}{2}}K_1[a(p^2+b^2)^{\frac{1}{2}}]$	$0 \quad t < a$ $(ab)^{-1}\sin[b(t^2-a^2)^{\frac{1}{2}}] \quad t > a$
13.59	$(p^2-b^2)^{-\frac{1}{2}}K_1[a(p^2-b^2)^{\frac{1}{2}}]$	$0 \quad t < a$ $(ab)^{-1}\sinh[b(t^2-a^2)^{\frac{1}{2}}] \quad t > a$
13.60	$e^{ap}(p^2+b^2)^{-\frac{1}{2}}K_1[a(p^2+b^2)^{\frac{1}{2}}]$	$(ab)^{-1}\sin[b(t^2+2at)^{\frac{1}{2}}]$
13.61	$e^{ap}(p^2-b^2)^{-\frac{1}{2}}K_1[a(p^2-b^2)^{\frac{1}{2}}]$	$(ab)^{-1}\sinh[b(t^2+2at)^{\frac{1}{2}}]$

	$g(p) = \int_0^\infty f(t)e^{-pt}dt$	$f(t)$
13.62	$(p^2+a^2)^{-\frac{1}{2}\nu}$ $\cdot K_\nu[b(p^2+a^2)^{\frac{1}{2}}]$ Re $\nu > -\frac{1}{2}$	$0 \qquad t < b$ $(\frac{1}{2}\pi a)^{\frac{1}{2}}(ab)^{-\nu}(t^2-b^2)^{\frac{1}{2}\nu-\frac{1}{4}}$ $\cdot J_{\nu-\frac{1}{2}}[a(t^2-b^2)^{\frac{1}{2}}] \qquad t > b$
13.63	$(p^2-a^2)^{-\frac{1}{2}\nu}$ $\cdot K_\nu[b(p^2-a^2)^{\frac{1}{2}}]$ Re $\nu > -\frac{1}{2}$	$0 \qquad t < b$ $(\frac{1}{2}\pi a)^{\frac{1}{2}}(ab)^{-\nu}(t^2-b^2)^{\frac{1}{2}\nu-\frac{1}{4}}$ $\cdot I_{\nu-\frac{1}{2}}[a(t^2-b^2)^{\frac{1}{2}}] \qquad t > b$
13.64	$e^{bp}(p^2+a^2)^{-\frac{1}{2}\nu}$ $\cdot K_\nu[b(p^2+a^2)^{\frac{1}{2}}]$ Re $\nu > -\frac{1}{2}$	$(\frac{1}{2}\pi a)^{\frac{1}{2}}(ab)^{-\nu}(t^2+2bt)^{\frac{1}{2}\nu-\frac{1}{4}}$ $\cdot J_{\nu-\frac{1}{2}}[a(t^2+2bt)^{\frac{1}{2}}]$
13.65	$e^{bp}(p^2-a^2)^{-\frac{1}{2}\nu}$ $\cdot K_\nu[b(p^2-a^2)^{\frac{1}{2}}]$ Re $\nu > -\frac{1}{2}$	$(\frac{1}{2}\pi a)^{\frac{1}{2}}(ab)^{-\nu}(t^2+2bt)^{\frac{1}{2}\nu-\frac{1}{4}}$ $\cdot I_{\nu-\frac{1}{2}}[a(t^2+2bt)^{\frac{1}{2}}]$
13.66	$(\frac{p+b}{p-b})^{\frac{1}{2}\nu}$ $\cdot K_\nu[a(p^2-b^2)^{\frac{1}{2}}]$	$\frac{1}{2}a^{-\nu}(t^2-a^2)^{-\frac{1}{2}}$ $\cdot\{[t+(t^2-a^2)^{\frac{1}{2}}]^\nu \exp[b(t^2-a^2)^{\frac{1}{2}}]$ $+[t-(t^2-a^2)^{\frac{1}{2}}]^\nu \exp[-b(t^2-a^2)^{\frac{1}{2}}]\} \qquad t > a$ $0 \qquad t < a$

	$g(p) = \int_0^\infty f(t)e^{-pt}dt$	$f(t)$
13.67	$e^{ap}\left(\frac{p+b}{p-b}\right)^{\frac{1}{2}\nu}$ $\cdot K_\nu[a(p^2-b^2)^{\frac{1}{2}}]$	$\frac{1}{2}a^{-\nu}(t^2+2at)^{-\frac{1}{2}}$ $\cdot\{[t+a+(t^2+2at)^{\frac{1}{2}}]^\nu \exp[b(t^2+2at)^{\frac{1}{2}}]$ $+[t+a-(t^2+2at)^{\frac{1}{2}}]^\nu \exp[-b(t^2+2at)^{\frac{1}{2}}]$
13.68	$e^{ap}\sin[\nu\ \arctan(b/p)]$ $\cdot K_\nu[a(p^2+b^2)^{\frac{1}{2}}]$	$\frac{1}{2}a^{-\nu}(t^2+2at)^{-\frac{1}{2}}\sin[b(t^2+2at)^{\frac{1}{2}}]$ $\cdot\{[a+t+(t^2+2at)^{\frac{1}{2}}]^\nu$ $-[a+t-(t^2+2at)^{\frac{1}{2}}]^\nu\}$
13.69	$e^{ap}\cos[\nu\ \arctan(b/p)$ $\cdot K_\nu[a(p^2+b^2)^{\frac{1}{2}}]$	$\frac{1}{2}a^{-\nu}(t^2+2at)^{-\frac{1}{2}}\cos[b(t^2+2at)^{\frac{1}{2}}]$ $\cdot\{[a+t+(t^2+2at)^{\frac{1}{2}}]^\nu$ $+[a+t-(t^2+2at)^{\frac{1}{2}}]^\nu\}$
13.70	$\sin[\nu\ \arctan(b/p)]$ $\cdot K_\nu[a(p^2+b^2)^{\frac{1}{2}}]$	$0 \quad t < a$ $\frac{1}{2}a^{-\nu}(t^2-a^2)^{-\frac{1}{2}}\sin[b(t^2-a^2)^{\frac{1}{2}}]$ $\cdot\{[t+(t^2-a^2)^{\frac{1}{2}}]^\nu-[t-(t^2-a^2)^{\frac{1}{2}}]^\nu\}$ $t > a$
13.71	$\cos[\nu\ \arctan(b/p)]$ $\cdot K_\nu[a(p^2+b^2)^{\frac{1}{2}}]$	$0 \quad t < a$ $\frac{1}{2}a^{-\nu}(t^2-a^2)^{-\frac{1}{2}}\cos[b(t^2-a^2)^{\frac{1}{2}}]$ $\{[t+(t^2-a^2)^{\frac{1}{2}}]+[t-(t^2-a^2)^{\frac{1}{2}}]^\nu\}$ $t>a$

	$g(p) = \int_0^\infty f(t)e^{-pt}dt$	$f(t)$
13.72	$p^{-\frac{1}{2}}[J_0(ap^{\frac{1}{2}})K_1(ap^{\frac{1}{2}}) + J_1(ap^{\frac{1}{2}})K_0(ap^{\frac{1}{2}})]$	$a^{-1}J_0(\frac{1}{2}a^2/t)$
13.73	$p^{-\frac{1}{2}}[Y_0(ap^{\frac{1}{2}})K_1(ap^{\frac{1}{2}}) + Y_1(ap^{\frac{1}{2}})K_0(ap^{\frac{1}{2}})]$	$a^{-1}Y_0(\frac{1}{2}a^2/t)$
13.74	$p^{-\frac{1}{2}}[aJ_0(bp^{\frac{1}{2}})K_1(ap^{\frac{1}{2}}) + bJ_1(bp^{\frac{1}{2}})K_0(ap^{\frac{1}{2}})]$	$\exp[-\frac{1}{4}(a^2-b^2)/t]J_0(\frac{1}{2}ab/t)$ $a > b$
13.75	$p^{-\frac{1}{2}}[aY_0(bp^{\frac{1}{2}})K_1(ap^{\frac{1}{2}}) + bY_1(bp^{\frac{1}{2}})K_0(ap^{\frac{1}{2}})]$	$\exp[-\frac{1}{4}(a^2-b^2)/t]Y_0(\frac{1}{2}ab/t)$ $a > b$
13.76	$p^{-\frac{1}{2}}[aI_0(bp^{\frac{1}{2}})K_1(ap^{\frac{1}{2}}) - bI_1(bp^{\frac{1}{2}})K_0(ap^{\frac{1}{2}})]$	$\exp[-\frac{1}{4}(a^2+b^2)/t]I_0(\frac{1}{2}ab/t)$
13.77	$p^{-\frac{1}{2}}[aK_0(bp^{\frac{1}{2}})K_1(ap^{\frac{1}{2}}) + bK_1(bp^{\frac{1}{2}})K_0(ap^{\frac{1}{2}})]$	$\exp[-\frac{1}{4}(a^2+b^2)/t]K_0(\frac{1}{2}ab/t)$

	$g(p) = \int_0^\infty f(t)e^{-pt}dt$	$f(t)$
13.78	$e^{ap}K_{\frac{1}{4}}(z_1)I_{\frac{1}{4}}(z_2)$ $z_{\frac{1}{2}} = \frac{1}{2}a[(p^2+b^2)^{\frac{1}{2}}\pm p]$	$(\frac{1}{2}\pi b)^{-\frac{1}{2}}(t^2+2at)^{-\frac{3}{4}}\sin[b(t^2+2at)^{\frac{1}{2}}]$
13.79	$e^{ap}K_{\frac{1}{4}}(z_1)I_{-\frac{1}{4}}(z_2)$ $z_{\frac{1}{2}} = \frac{1}{2}a[(p^2+b^2)^{\frac{1}{2}}\pm p]$	$(\frac{1}{2}\pi b)^{-\frac{1}{2}}(t^2+2at)^{-\frac{3}{4}}\cos[b(t^2+2at)^{\frac{1}{2}}]$
13.80	$K_{\frac{1}{4}}(z_1)I_{\frac{1}{4}}(z_2)$ $z_{\frac{1}{2}} = \frac{1}{2}a[(p^2+b^2)^{\frac{1}{2}}\pm p]$	$0 \qquad t < a$ $(\frac{1}{2}\pi b)^{-\frac{1}{2}}(t^2-a^2)^{-\frac{3}{4}}\sin[b(t^2-a^2)^{\frac{1}{2}}]$ $t > a$
13.81	$K_{\frac{1}{4}}(z_1)I_{-\frac{1}{4}}(z_2)$ $z_{\frac{1}{2}} = \frac{1}{2}a[(p^2+b^2)^{\frac{1}{2}}\pm p]$	$0 \qquad t < a$ $(\frac{1}{2}\pi b)^{-\frac{1}{2}}(t^2-a^2)^{-\frac{3}{4}}\cos[b(t^2-a^2)^{\frac{1}{2}}]$ $t > a$
13.82	$p^{-\frac{1}{2}}e^{ap}K_{\frac{1}{4}}(z_1)K_{\frac{1}{4}}(z_2)$ $z_{\frac{1}{2}} = \frac{1}{2}a[(b^2+p^2)^{\frac{1}{2}}\pm p]$	$(2\pi)^{\frac{1}{2}}(t+a)^{-\frac{1}{2}}(t^2+2at)^{-\frac{1}{2}}$ $\cdot\cos[b(t^2+2at)^{\frac{1}{2}}]$
13.83	$p^{-\frac{1}{2}}K_{\frac{1}{4}}(z_1)K_{\frac{1}{4}}(z_2)$ $z_{\frac{1}{2}} = \frac{1}{2}a[(b^2+p^2)^{\frac{1}{2}}\pm p]$	$0 \qquad t < a$ $(2\pi)^{-\frac{1}{2}}(t^2-a^2)^{-\frac{1}{2}}t^{-\frac{1}{2}}$ $\cdot\cos[b(t^2-a^2)^{\frac{1}{2}}] \qquad t > a$

	$g(p) = \int_0^\infty f(t)e^{-pt}dt$	$f(t)$
13.84	$e^{ap}K_{\frac{1}{4}}(z_1)I_{\frac{1}{4}}(z_2)$ $z_{\frac{1}{2}} = \frac{1}{2}a[p\pm(p^2-b^2)^{\frac{1}{2}}]$	$(\frac{1}{2}\pi b)^{-\frac{1}{2}}(t^2+2at)^{-\frac{3}{4}}\sinh[b(t^2+2at)^{\frac{1}{2}}]$
13.85	$e^{ap}K_{\frac{1}{4}}(z_1)I_{-\frac{1}{4}}(z_2)$ $z_{\frac{1}{2}} = \frac{1}{2}a[p\pm(p^2-b^2)^{\frac{1}{2}}]$	$(\frac{1}{2}\pi b)^{-\frac{1}{2}}(t^2+2at)^{-\frac{3}{4}}\cosh[b(t^2+2at)^{\frac{1}{2}}]$
13.86	$K_{\frac{1}{4}}(z_1)I_{\frac{1}{4}}(z_2)$ $z_{\frac{1}{2}} = \frac{1}{2}a[p\pm(p^2-b^2)^{\frac{1}{2}}]$	$0 \qquad t < a$ $(\frac{1}{2}\pi b)^{-\frac{1}{2}}(t^2-a^2)^{-\frac{3}{4}}\sinh[b(t^2-a^2)^{\frac{1}{2}}]$ $t > a$
13.87	$K_{\frac{1}{4}}(z_1)I_{-\frac{1}{4}}(z_2)$ $z_{\frac{1}{2}} = \frac{1}{2}a[p\pm(p^2-b^2)^{\frac{1}{2}}]$	$(\frac{1}{2}\pi b)^{-\frac{1}{2}}(t^2-a^2)^{-\frac{3}{4}}\cosh[b(t^2-a^2)^{\frac{1}{2}}]$ $t > a$ $0 \qquad t < a$
13.88	$e^{ap}K_{\frac{1}{4}}(z_1)K_{\frac{1}{4}}(z_2)$ $z_{\frac{1}{2}} = \frac{1}{2}a[p\pm(p^2-b^2)^{\frac{1}{2}}]$	$(\pi/b)^{\frac{1}{2}}(t^2+2at)^{-\frac{3}{2}}$ $\cdot\exp[-b(t^2+2at)^{\frac{1}{2}}]$
13.89	$K_{\frac{1}{4}}(z_1)K_{\frac{1}{4}}(z_2)$ $z_{\frac{1}{2}} = \frac{1}{2}a[p\pm(p^2-b^2)]$	$0 \qquad t < a$ $(\pi/b)^{\frac{1}{2}}(t^2-a^2)^{-\frac{3}{4}}\exp[-b(t^2-a^2)^{\frac{1}{2}}]$ $t > a$

	$g(p) = \int_0^\infty f(t)e^{-pt}dt$	$f(t)$
13.90	$p^{\frac{1}{2}}K_{\frac{1}{4}}(z_1)K_{\frac{1}{4}}(z_2)$ $z_{\frac{1}{2}} = \frac{1}{2}a[b\pm(b^2-p^2)^{\frac{1}{2}}]$	$(t/\pi)^{-\frac{1}{2}}(a^2+t^2)^{-\frac{1}{2}}$ $\cdot\exp[-b(a^2+t^2)^{\frac{1}{2}}]$
13.91	$e^{ap}K_0(z_1)K_0(z_2)$ $z_{\frac{1}{2}} = \frac{1}{2}a[(p^2+b^2)^{\frac{1}{2}}\pm p]$	$-\pi(t^2+2at)^{-\frac{1}{2}}Y_0[b(t^2+2at)^{\frac{1}{2}}]$
13.92	$K_0(z_1)K_0(z_2)$ $z_{\frac{1}{2}} = \frac{1}{2}a[(p^2+b^2)^{\frac{1}{2}}\pm p]$	$0 \quad t < a$ $(t^2-a^2)^{-\frac{1}{2}}Y_0[b(t^2-a^2)^{\frac{1}{2}}] \quad t > a$
13.93	$p^{-\nu}[K_\nu(ap^{\frac{1}{2}})]^2$	$\frac{1}{2}\pi^{\frac{1}{2}}a^{\nu-1}t^{-\frac{1}{2}-\frac{3}{2}\nu}$ $\cdot\exp(-\frac{1}{2}a^2/t)W_{\frac{1}{2}\nu,\frac{1}{2}\nu}(a^2/t)$
13.94	$J_\nu(bp^{\frac{1}{2}})K_\nu(ap^{\frac{1}{2}})$ $a \geqq b$	$\frac{1}{2}t^{-1}\exp[-\frac{1}{4}(a^2-b^2)/t]J_\nu(\frac{1}{2}ab/t)$
13.95	$Y_\nu(bp^{\frac{1}{2}})K_\nu(ap^{\frac{1}{2}})$ $a \geqq b$	$\frac{1}{2}t^{-1}\exp[-\frac{1}{4}(a^2-b^2)/t]Y_\nu(\frac{1}{2}ab/t)$
13.96	$I_\nu(bp^{\frac{1}{2}})K_\nu(ap^{\frac{1}{2}})$	$\frac{1}{2}t^{-1}\exp[-\frac{1}{4}(a^2+b^2)/t]I_\nu(\frac{1}{2}ab/t)$

	$g(p) = \int_0^\infty f(t)e^{-pt}dt$	$f(t)$
13.97	$K_\nu(bp^{\frac{1}{2}})K_\nu(ap^{\frac{1}{2}})$	$\frac{1}{2}t^{-1}\exp[-\frac{1}{4}(a^2+b^2)/t]K_\nu(\frac{1}{2}ab/t)$
13.98	$p^{\frac{1}{2}}K_{\nu-\frac{1}{2}}(ap)^{\frac{1}{2}}K_{\nu+\frac{1}{2}}(ap^{\frac{1}{2}})$	$\frac{1}{2}(\frac{1}{2}\pi)^{\frac{1}{2}}(at)^{-1}e^{-\frac{1}{2}a^2t}$ $\cdot W_{\frac{1}{2},\nu}(a^2/t)$
13.99	$K_\nu[p^{\frac{1}{2}}+(p-1)^{\frac{1}{2}}]$ $\cdot K_\nu[p^{\frac{1}{2}}-(p-1)^{\frac{1}{2}}]$	$\frac{1}{2}t^{-1}\exp(\frac{1}{2}t-1/t)K_\nu(\frac{1}{2}t)$
13.100	$e^{2ap}I_\nu(z_1)K_\nu(z_2)$ $z_{\frac{1}{2}} = a[(p^2+b^2)^{\frac{1}{2}}\mp p]$	$(t^2+4at)^{-\frac{1}{2}}J_{2\nu}[b(t^2+4at)^{\frac{1}{2}}]$
13.101	$I_\nu(z_1)K_\nu(z_2)$ $z_{\frac{1}{2}} = a[(p^2+b^2)^{\frac{1}{2}}\mp p]$	$0 \qquad t < 2a$ $(t^2-4a^2)^{-\frac{1}{2}}J_{2\nu}[b(t^2-4a^2)^{\frac{1}{2}}]$ $t > 2a$
13.102	$e^{2ap}K_\nu(z_1)K_\nu(z_2)$ $z_{\frac{1}{2}} = a[p\mp(p^2-b^2)^{\frac{1}{2}}]$ $-\frac{1}{2} < \mathrm{Re}\,\nu < \frac{1}{2}$	$2\cos(\pi\nu)$ $\cdot(t^2+4at)^{-\frac{1}{2}}K_{2\nu}[b(t^2+4at)^{\frac{1}{2}}]$

	$g(p) = \int_0^\infty f(t)e^{-pt}dt$	$f(t)$
13.103	$K_\nu(z_1)K_\nu(z_2)$ $z_{\frac{1}{2}} = a[p\mp(p^2-a^2)^{\frac{1}{2}}]$ $-\frac{1}{2} < \mathrm{Re}\,\nu < \frac{1}{2}$	$0 \qquad t < 2a$ $2\cos(\pi\nu)(t^2-4a^2)^{-\frac{1}{2}}K_{2\nu}[b(t^2-4a^2)^{\frac{1}{2}}]$ $t > 2a$
13.104	$I_\nu(z_1)K_\nu(z_2)$ $z_{\frac{1}{2}} = a[p\mp(p^2-a^2)^{\frac{1}{2}}]$ $\mathrm{Re}\,\nu > -\frac{1}{2}$	$0 \qquad t < 2a$ $(t^2-4a^2)^{-\frac{1}{2}}I_{2\nu}[b(t^2-4a^2)^{\frac{1}{2}}]$ $t > 2a$
13.105	$e^{2ap}I_\nu(z_1)K_\nu(z_2)$ $z_{\frac{1}{2}} = a[p\mp(p^2-b^2)^{\frac{1}{2}}]$ $\mathrm{Re}\,\nu > -\frac{1}{2}$	$(t^2+4at)^{-\frac{1}{2}}I_{2\nu}[b(t^2+4at)^{\frac{1}{2}}]$
13.106	$p^{-\frac{1}{2}}e^{a/p}I_{\frac{1}{4}}(a/p)$	$\pi^{-1}(2a)^{-\frac{1}{4}}t^{-3/4}\sinh[(8at)^{\frac{1}{2}}]$
13.107	$p^{-\frac{1}{2}}e^{-a/p}I_{\frac{1}{4}}(a/p)$	$\pi^{-1}(2a)^{-\frac{1}{4}}t^{-3/4}\sin[(8at)^{\frac{1}{2}}]$
13.108	$p^{-\frac{1}{2}}e^{a/p}I_{-\frac{1}{4}}(a/p)$	$\pi^{-1}(2a)^{-\frac{1}{4}}t^{-3/4}\cosh[(8at)^{\frac{1}{2}}]$
13.109	$p^{-\frac{1}{2}}e^{-a/p}I_{-\frac{1}{4}}(a/p)$	$\pi^{-1}(2a)^{-\frac{1}{4}}t^{-3/4}\cos[(8at)^{\frac{1}{2}}]$

	$g(p) = \int_0^\infty f(t)e^{-pt}dt$	$f(t)$
13.110	$p^{-\frac{1}{2}}e^{a/p}I_{3/4}(a/p)$	$-\pi^{-1}(2at)^{-3/4}\{\frac{1}{2}t^{-\frac{1}{2}}\sinh[(8at)^{\frac{1}{2}}] - (2a)^{\frac{1}{2}}\cosh[(8at)^{\frac{1}{2}}]\}$
13.111	$p^{-\frac{1}{2}}e^{-a/p}I_{3/4}(a/p)$	$\pi^{-1}(2at)^{-3/4}\{\frac{1}{2}t^{-\frac{1}{2}}\sin[(8at)^{\frac{1}{2}}] - (2a)^{\frac{1}{2}}\cos[(8at)^{\frac{1}{2}}]\}$
13.112	$p^{-\frac{1}{2}}e^{a/p}I_{-3/4}(a/p)$	$\pi^{-1}(2at)^{-3/4}\{(2a)^{\frac{1}{2}}\sinh[(8at)^{\frac{1}{2}}] - \frac{1}{2}\cosh[(8at)^{\frac{1}{2}}]\}$
13.113	$p^{-\frac{1}{2}}e^{-a/p}I_{-3/4}(a/p)$	$-\pi^{-1}(2at)^{-3/4}\{(2a)^{\frac{1}{2}}\sin[(8at)^{\frac{1}{2}}] - \frac{1}{2}\cos[(8at)^{\frac{1}{2}}]\}$
13.114	$p^{-3/2}e^{-a/p}I_{-\frac{1}{4}}(a/p)$	$(\frac{1}{2}\pi a)^{-\frac{1}{2}}C[8at)^{\frac{1}{2}}]$
13.115	$p^{-3/2}e^{-a/p}I_{\frac{1}{4}}(a/p)$	$(\frac{1}{2}\pi a)^{-\frac{1}{2}}S[(8at)^{\frac{1}{2}}]$
13.116	$p^{-\frac{1}{2}}e^{a/p}K_{\frac{1}{4}}(a/p)$	$a^{-\frac{1}{2}}(2t)^{-3/4}\exp[-(8at)^{\frac{1}{2}}]$
13.117	$p^{-3/2}e^{-a/p}K_{\frac{1}{4}}(a/p)$	$(\pi/a)^{\frac{1}{2}}\{C[(8at)^{\frac{1}{2}}]-S[(8at)^{\frac{1}{2}}]\}$

	$g(p) = \int_0^\infty f(t)e^{-pt}dt$	$f(t)$
13.118	$p^{-3/2}e^{a/p}[K_1(a/p) - K_0(a/p)]$	$(\pi a/8)^{1/2}K_1[(8at)^{1/2}]$
13.119	$p^{-1}e^{-a/p}K_0(b/p)$ $\begin{matrix}A\\B\end{matrix} = (a+b)^{1/2}\pm(a-b)^{1/2}$, $a \geqq b$	$-\frac{1}{2}\pi[J_0(At^{1/2})Y_0(Bt^{1/2}) + J_0(Bt^{1/2})Y_0(At^{1/2})]$
13.120	$p^{-1}e^{a/p}K_0(b/p)$ $\begin{matrix}A\\B\end{matrix} = (a+b)^{1/2}\pm(a-b)^{1/2}$, $a \geqq b$	$[I_0(At^{1/2})K_0(Bt^{1/2}) + I_0(Bt^{1/2})K_0(At^{1/2})]$
13.121	$I_\nu(a/p)$ $\mathrm{Re}\ \nu > 0$	$(\frac{1}{2}t/a)^{-1/2}\{\mathrm{ber}_\nu[(2at)^{1/2}]\mathrm{bei}'_\nu[(2at)^{1/2}] + \mathrm{bei}_\nu[(2at)^{1/2}]\mathrm{ber}'_\nu[(2at)^{1/2}]\}$
13.122	$pI_\nu(a/p)$ $\mathrm{Re}\ \nu > 1$	$\frac{1}{4}a^2\{[\mathrm{ber}'_\nu(2at)^{1/2}]^2 + [\mathrm{bei}'_\nu(2at)^{1/2}]^2\}$
13.123	$p^{-1}I_\nu(a/p)$ $\mathrm{Re}\ \nu > -1$	$\mathrm{ber}^2_\nu[(2at)^{1/2}] + \mathrm{bei}^2_\nu[(2at)^{1/2}]$ $= -J_\nu[(-2iat)^{1/2}]J_\nu[(2iat)^{1/2}]$

	$g(p) = \int_0^\infty f(t)e^{-pt}dt$	$f(t)$
13.124	$p^{-2}I_\nu(a/p)$	$(2t/a)^{\frac{1}{2}}\{\mathrm{ber}_\nu[(2at)^{\frac{1}{2}}]\mathrm{bei}'_\nu[(2at)^{\frac{1}{2}}]$ $-\mathrm{bei}_\nu[(2at)^{\frac{1}{2}}]\mathrm{ber}'_\nu[(2at)^{\frac{1}{2}}]\}$
13.125	$p^n I_\nu(a/p)$ $n=0,1,2,\cdots;\ \mathrm{Re}\ \nu>n$	$\frac{d^{n+1}}{dt^{n+1}}\{\mathrm{ber}^2_\nu[(2at)^{\frac{1}{2}}]+\mathrm{bei}^2_\nu[(2at)^{\frac{1}{2}}]\}$
13.126	$p^{-\mu}I_\nu(a/p)$ $\mathrm{Re}(\mu+\nu)>0$	$(\frac{1}{2}a)^\nu[\Gamma(1+\nu)\Gamma(\mu+\nu)]^{-1}t^{\mu+\nu-1}$ $\cdot {}_0F_3(\ ;\nu+1,\frac{1}{2}\mu+\frac{1}{2}\nu,\frac{1}{2}+\frac{1}{2}\mu+\frac{1}{2}\nu;a^2t^2/16)$
13.127	$p^{-\frac{1}{2}}e^{a/p}K_\nu(a/p)$ $-\frac{1}{2}<\mathrm{Re}\ \nu<\frac{1}{2}$	$2(\pi t)^{-\frac{1}{2}}\cos(\pi\nu)K_{2\nu}[(8at)^{\frac{1}{2}}]$
13.128	$p^{-\frac{1}{2}}e^{-a/p}K_\nu(a/p)$ $-\frac{1}{2}<\mathrm{Re}\ \nu<\frac{1}{2}$	$-(\pi/t)^{\frac{1}{2}}\{\sin(\pi\nu)J_{2\nu}[(8at)^{\frac{1}{2}}]$ $+\cos(\pi\nu)Y_{2\nu}[(8at)^{\frac{1}{2}}]\}$
13.129	$p^{-\frac{1}{2}}e^{-a/p}[\tan(\pi\nu)I_\nu(a/p)$ $+\pi^{-1}\sec(\pi\nu)K_\nu(a/p)]$	$-(\pi t)^{-\frac{1}{2}}Y_{2\nu}[(8at)^{\frac{1}{2}}]$ $-\frac{1}{2}<\mathrm{Re}\ \nu<\frac{1}{2}$

	$g(p) = \int_0^\infty f(t)e^{-pt}dt$	$f(t)$
13.130	$p^{-1}e^{a/p}K_\nu(a/p)$ $-1 < \text{Re}\,\nu < 1$	$2\pi^{-1}\sin(\pi\nu)K_\nu^2[(2at)^{\frac{1}{2}}]$ $+2K_\nu[(2at)^{\frac{1}{2}}]I_\nu[(2at)^{\frac{1}{2}}]$
13.131	$p^{-1}e^{-a/p}K_\nu(a/p)$ $-1 < \text{Re}\,\nu < 1$	$-\frac{1}{2}\pi\{\sin(\pi\nu)J_\nu^2[(2at)^{\frac{1}{2}}]$ $-\sin(\pi\nu)Y_\nu^2[(2at)^{\frac{1}{2}}]$ $+2\cos(\pi\nu)J_\nu[(2at)^{\frac{1}{2}}]Y_\nu[(2at)^{\frac{1}{2}}]\}$
13.132	$p^{-1}e^{a/p}I_\nu(a/p)$ $\text{Re}\,\nu > -1$	$I_\nu^2[(2at)^{\frac{1}{2}}]$
13.133	$p^{-1}e^{-a/p}I_\nu(a/p)$	$J_\nu^2[(2at)^{\frac{1}{2}}]$
13.134	$p^{-\frac{1}{2}}e^{a/p}I_\nu(a/p)$ $\text{Re}\,\nu > -\frac{1}{2}$	$(\pi t)^{-\frac{1}{2}}I_{2\nu}[(8at)^{\frac{1}{2}}]$
13.135	$p^{-\frac{1}{2}}e^{-a/p}I_\nu(a/p)$ $\text{Re}\,\nu > -\frac{1}{2}$	$(\pi t)^{-\frac{1}{2}}J_{2\nu}[(8at)^{\frac{1}{2}}]$

	$g(p) = \int_0^\infty f(t)e^{-pt}dt$	$f(t)$
13.136	$p^{-\frac{1}{2}}e^{-a/p}[\tan(\pi\nu)I_\nu(a/p)$ $+\pi^{-1}\sec(\pi\nu)K_\nu(a/p)]$ $-\frac{1}{2} < \mathrm{Re}\ \nu < \frac{1}{2}$	$-(\pi t)^{-\frac{1}{2}}Y_{2\nu}[(8at)^{\frac{1}{2}}]$
13.137	$p^{-1}e^{-a/p}I_\nu(b/p)$ $\mathrm{Re}\ \nu > -1;\ {A \atop B} = (a+b)^{\frac{1}{2}} \pm (a-b)^{\frac{1}{2}}$ $a > b$	$J_\nu(At^{\frac{1}{2}})J_\nu(Bt^{\frac{1}{2}})$
13.138	$p^{-1}e^{a/p}I_\nu(b/p)$ $\mathrm{Re}\ \nu > -1;\ {A \atop B} = (a+b)^{\frac{1}{2}} \pm (a-b)^{\frac{1}{2}}$ $a > b$	$I_\nu(At^{\frac{1}{2}})I_\nu(Bt^{\frac{1}{2}})$
13.139	$p^{-1}e^{-a/p}K_\nu(b/p)$ ${A \atop B} = (a+b)^{\frac{1}{2}} \pm (a-b)^{\frac{1}{2}}$ $-1 < \mathrm{Re}\ \nu < 1$	$-\frac{1}{2}\pi\{\sin(\pi\nu)[J_\nu(At^{\frac{1}{2}})J_\nu(Bt^{\frac{1}{2}})$ $-Y_\nu(At^{\frac{1}{2}})Y_\nu(Bt^{\frac{1}{2}})]$ $+\cos(\pi\nu)[J_\nu(At^{\frac{1}{2}})Y_\nu(Bt^{\frac{1}{2}})$ $+J_\nu(Bt^{\frac{1}{2}})Y_\nu(At^{\frac{1}{2}})]\}$

	$g(p) = \int_0^\infty f(t)e^{-pt}dt$	$f(t)$
13.140	$p^{-1}e^{a/p}K_\nu(b/p)$ $\begin{matrix}A\\B\end{matrix} = (a+b)^{\frac{1}{2}} \pm (a-b)^{\frac{1}{2}}$ $-1 < \mathrm{Re}\,\nu < 1$	$2\pi^{-1}\sin(\pi\nu)K_\nu(At^{\frac{1}{2}})K_\nu(Bt^{\frac{1}{2}})$ $+K_\nu(At^{\frac{1}{2}})I_\nu(Bt^{\frac{1}{2}})+I_\nu(At^{\frac{1}{2}})K_\nu(Bt^{\frac{1}{2}})$
13.141	$p^{-\frac{1}{2}}e^{-a/p}[e^{i\pi\nu}I_\nu(a/p)$ $+i\pi^{-1}K_\nu(a/p]$	$\cos(\pi\nu)(\pi t)^{\frac{1}{2}}H_{2\nu}^{(2)}[(8at)^{\frac{1}{2}}]$
13.142	$p^{-\frac{1}{2}}e^{-a/p}[e^{-i\pi\nu}I_\nu(a/p)$ $-i\ \pi^{-1}K_\nu(a/p)]$	$\cos(\pi\nu)(\pi t)^{-\frac{1}{2}}H_{2\nu}^{(1)}[(8at)^{\frac{1}{2}}]$
13.143	$p^{-2}\exp(-1/s^2)I_\nu(1/s^2)$	$2^{\frac{1}{2}-\nu}[\Gamma(1+2\nu)\Gamma(1+\nu)]^{-1}t^{2\nu}$ $\cdot {}_0F_2(\ ;1+\nu,1+2\nu;-\frac{1}{2}t^2)$
13.144	$p^{-3/2}e^{a/p}[I_{\nu-\frac{1}{2}}(a/p)$ $-I_{\nu+\frac{1}{2}}(a/p)]$	$(\frac{1}{2}a\pi)^{-\frac{1}{2}}I_{2\nu}[(8at)^{\frac{1}{2}}]$
13.145	$e^{ap^2}K_0(ap^2)$	$(\frac{1}{2}\pi/a)^{\frac{1}{2}}\exp(-\frac{1}{16}t^2/a)I_0(\frac{1}{16}t^2/a)$

	$g(p) = \int_0^\infty f(t)e^{-pt}dt$	$f(t)$
13.146	$p^{\frac{1}{2}}e^{ap^2}K_{\frac{1}{4}}(ap^2)$	$(2at)^{-\frac{1}{2}}\exp(-\frac{1}{8}t^2/a)$
13.147	$p^{-\frac{1}{2}}e^{ap^2}K_{\frac{1}{4}}(ap^2)$	$(8a)^{-\frac{1}{4}}\gamma(\frac{1}{4},\frac{1}{8}t^2/a)$
13.148	$p^{-2\nu}e^{ap^2}K_{\nu}(ap^2)$ $\mathrm{Re}\ \nu > -\frac{1}{2}$	$2\pi^{\frac{1}{2}}[\Gamma(1+2\nu)]^{-1}(8a)^{\frac{1}{2}\nu}t^{\nu-1}$ $\cdot\exp(-\frac{1}{16}t^2/a)M_{-\frac{3}{2}\nu,\frac{1}{2}\nu}(\frac{1}{8}t^2/a)$
13.149	$(\frac{1}{2}a)^{-p}[\Gamma(\frac{1}{2}+p)]^{-1}I_p(a)$	$\pi^{-\frac{1}{2}}(e^t-1)^{-\frac{1}{2}}\cosh[a(1-e^{-t})^{\frac{1}{2}}]$
13.150	$(\frac{1}{2}a)^{p}[\Gamma(\frac{1}{2}+p)]^{-1}K_p(a)$	$\frac{1}{2}\pi^{-\frac{1}{2}}(1-e^{-t})^{-\frac{1}{2}}\cos[a(e^t-1)^{\frac{1}{2}}]$
13.151	$(\frac{1}{2}a)^{p}[\Gamma(1+p)]^{-1}K_{\nu-p}(a)$ $\mathrm{Re}\ \nu > -1$	$\frac{1}{2}(e^t-1)^{\frac{1}{2}\nu}J_{\nu}[a(e^t-1)^{\frac{1}{2}}]$
13.152	$J_p(a)K_p(a)$	$\frac{1}{2}J_0[a(2\sinh t)^{\frac{1}{2}}]$
13.153	$I_{\nu+p}(a)K_{\nu-p}(a)$ $\mathrm{Re}\ \nu > -\frac{1}{2}$	$\frac{1}{2}J_{2\nu}[2a\sinh(\frac{1}{2}t)]$

2.14 Functions Related to Bessel Functions and Kelvin Functions

	$g(p) = \int_0^\infty f(t)e^{-pt}dt$	$f(t)$
14.1	$p^{-1}[\mathbf{H}_0(ap)-Y_0(ap)]$	$2\pi^{-1}\log[t/a+(1+t^2/a^2)^{\frac{1}{2}}]$
14.2	$\mathbf{H}_1(ap)-Y_1(ap)-2\pi^{-1}$	$2(\pi a)^{-1}t(t^2+a^2)^{-\frac{1}{2}}$
14.3	$p^{-2}[\mathbf{H}_0(ap)-Y_0(ap)]$	$2\pi^{-1}\{a-(t^2+a^2)^{\frac{1}{2}}$ $+t/a\log[t/a+(1+t^2/a^2)^{\frac{1}{2}}]\}$
14.4	$p^{-1}[I_0(ap)-\mathbf{L}_0(ap)]$	$2\pi^{-1}\arcsin(t/a)$ $t<a$; 1 $t>a$
14.5	$p^{-\nu}[\mathbf{H}_\nu(ap)-Y_\nu(ap)]$	$2\pi^{-\frac{1}{2}}(2a)^{-\nu}[\Gamma(\frac{1}{2}+\nu)]^{-1}(t^2+a^2)^{\nu-\frac{1}{2}}$
14.6	$p^{-\nu}[I_\nu(ap)-\mathbf{L}_\nu(ap)]$ Re $\nu>-\frac{1}{2}$	$2\pi^{-\frac{1}{2}}(2a)^{-\nu}[\Gamma(\frac{1}{2}+\nu)]^{-1}(a^2-t^2)^{\nu-\frac{1}{2}}$ $t<a$; 0 $t>a$
14.7	$\mathbf{J}_\nu(ap)-J_\nu(ap)$	$\pi^{-1}\sin(\pi\nu)a^\nu(t^2+a^2)^{-\frac{1}{2}}$ $\cdot[(t^2+a^2)^{\frac{1}{2}}+t]^{-\nu}$

	$g(p) = \int_0^\infty f(t)e^{-pt}dt$	$f(t)$
14.8	$p^{-1}[J_\nu(ap)-\mathbf{J}_\nu(ap)]$	$(\pi\nu)^{-1}a^\nu\sin(\pi\nu)$ $\cdot\{[(t^2+a^2)^{\frac{1}{2}}+t]^{-\nu}-a^{-\nu}\}$
14.9	$\mathbf{E}_\nu(ap)+Y_\nu(ap)$	$-\pi^{-1}a^{-\nu}(t^2+a^2)^{-\frac{1}{2}}\{[(a^2+t^2)^{\frac{1}{2}}+t]^\nu$ $+\cos(\pi\nu)[(a^2+t^2)^{\frac{1}{2}}-t]^\nu\}$
14.10	$p^{-\frac{1}{2}}[\mathbf{H}_0(ap^{\frac{1}{2}})-Y_0(ap^{\frac{1}{2}})]$	$\pi^{-3/2}t^{-\frac{1}{2}}\exp(\frac{1}{8}a^2/t)K_0(\frac{1}{8}a^2/t)$
14.11	$p^{-\frac{1}{2}}[I_0(ap^{\frac{1}{2}})-\mathbf{L}_0(ap^{\frac{1}{2}})]$	$(\pi t)^{-\frac{1}{2}}\exp(-\frac{1}{8}a^2/t)I_0(\frac{1}{8}a^2/t)$
14.12	$Y_1(ap^{\frac{1}{2}})-\mathbf{H}_{-1}(ap^{\frac{1}{2}})$	$-\frac{1}{4}a(\pi t)^{-3/2}\exp(\frac{1}{8}a^2/t)$ $\cdot[K_1(\frac{1}{8}a^2/t)-K_0(\frac{1}{8}a^2/t)]$
14.13	$p^{\frac{1}{2}\nu}[\mathbf{H}_\nu(ap^{\frac{1}{2}})-Y_\nu(ap^{\frac{1}{2}})]$ $\mathrm{Re}\,\nu<\frac{1}{2}$	$\pi^{-1}\cos(\pi\nu)(\frac{1}{2}a)^\nu t^{-\nu-1}$ $\cdot\exp(\frac{1}{4}a^2/t)\mathrm{Erfc}(\frac{1}{2}at^{-\frac{1}{2}})$
14.14	$p^{\frac{1}{2}\nu}[\mathbf{L}_\nu(ap^{\frac{1}{2}})-I_\nu(ap^{\frac{1}{2}})]$ $\mathrm{Re}\,\nu<\frac{1}{2}$	$i\pi^{-1}\cos(\pi\nu)(\frac{1}{2}a)^\nu t^{-\nu-1}$ $\exp(\frac{1}{4}a^2/t)\mathrm{Erf}(\frac{1}{2}iat^{-\frac{1}{2}})$

	$g(p) = \int_0^\infty f(t) e^{-pt} dt$	$f(t)$
14.15	$p^{-\frac{1}{2}\nu-\frac{1}{2}}[\mathbf{H}_\nu(ap^{\frac{1}{2}}) - Y_\nu(ap^{\frac{1}{2}})]$ $-2 < \mathrm{Re}\ \nu < \frac{1}{2}$	$2a^{-1}\pi^{-\frac{1}{2}}[\Gamma(\frac{1}{2}+\nu)]^{-1} t^{-\frac{1}{2}\nu}$ $\cdot \exp(\frac{1}{8}a^2/t) W_{\frac{1}{2}\nu,\frac{1}{2}\nu}(\frac{1}{4}a^2/t)$
14.16	$p^{-1}\mathbf{H}_0(a/p)$	$I_0[(2at)^{\frac{1}{2}}] Y_0[(2at)^{\frac{1}{2}}]$ $+2\pi^{-1} J_0[(2at)^{\frac{1}{2}}] K_0[(2at)^{\frac{1}{2}}]$
14.17	$p^{-\mu}\mathbf{H}_\nu(a/p)$ $\mathrm{Re}(\mu+\nu) > -1$	$\pi^{-\frac{1}{2}} a (\frac{1}{2}a)^\nu [\Gamma(\frac{3}{2}+\nu)\Gamma(\nu+\mu+1)]^{-1}$ $\cdot t^{\mu+\nu} {}_1F_4(1; \frac{3}{2}, \frac{3}{2}+\nu, \frac{1}{2}\mu+\frac{1}{2}\nu+\frac{1}{2}, 1+\frac{1}{2}\mu+\frac{1}{2}\nu;$ $-\frac{1}{16}a^2t^2)$
14.18	$p^{-\mu}\mathbf{L}_\nu(a/p)$ $\mathrm{Re}(\mu+\nu) > -1$	$\pi^{-\frac{1}{2}} a (\frac{1}{2}a)^\nu [\Gamma(\frac{3}{2}+\nu)\Gamma(\nu+\mu+1)]^{-1} t^{\mu+\nu}$ $\cdot {}_1F_4(1; \frac{3}{2}, \frac{3}{2}+\nu, \frac{1}{2}\mu+\frac{1}{2}\nu+\frac{1}{2}, 1+\frac{1}{2}\mu+\frac{1}{2}\nu;$ $\frac{1}{16}a^2t^2)$
14.19	$p^{\frac{1}{2}}[\mathbf{H}_{\frac{1}{4}}(p^2/a) - Y_{\frac{1}{4}}(p^2/a)]$	$a(\pi/t)^{-\frac{1}{2}} J_{-\frac{1}{4}}(at^2/4)$
14.20	$p^{\frac{1}{2}}[\mathbf{H}_{-\frac{1}{4}}(p^2/a) - Y_{-\frac{1}{4}}(p^2/a)]$	$a(\pi/t)^{-\frac{1}{2}} J_{\frac{1}{4}}(at^2/4)$

	$g(p) = \int_0^\infty f(t)e^{-pt}dt$	$f(t)$
14.21	$p^{3/2}[\mathbf{H}_{-1/4}(p^2/a) - Y_{-1/4}(p^2/a)]$	$\frac{1}{2}a^2\pi^{-1/2}t^{3/2}J_{-3/4}(at^2/4)$
14.22	$p^{3/2}[\mathbf{H}_{-3/4}(p^2/a) - Y_{-3/4}(p^2/a)]$	$-\frac{1}{2}a^2\pi^{-1/2}t^{3/2}J_{-1/4}(at^2/4)$
14.23	$\csc(\pi p)[\mathbf{J}_p(a) - J_p(a)]$	$\pi^{-1}\exp(-a\sinh t)$
14.24	$\Gamma(\frac{1}{2}+p)(\frac{1}{2}a)^{-p}\mathbf{H}_p(a)$	$\pi^{-1/2}(e^t-1)^{-1/2}\sin[a(1-e^{-t})^{1/2}]$
14.25	$\Gamma(\frac{1}{2}+p)(\frac{1}{2}a)^{-p}\mathbf{L}_p(a)$	$\pi^{-1/2}(e^t-1)^{-1/2}\sinh[a(1-e^{-t})^{1/2}]$
14.26	$\Gamma(\frac{1}{2}-p)(\frac{1}{2}a)^p$ $\cdot[I_p(a) - \mathbf{L}_{-p}(a)]$	$\pi^{-1/2}(1-e^{-t})^{-1/2}\sin[a(e^t-1)^{1/2}]$
14.27	$V_\nu(2p,0)$ $\mathrm{Re}\,\nu > 0$	$[\pi(1+t^2)]^{-1}t^{\nu-1}\sin(\pi\nu)$
14.28	$a^{-p}\Gamma(p)U_p(2a,0)$	$\cos[a(1-e^{-t})]$
14.29	$a^{-p}\Gamma(p)U_{p+1}(2a,0)$	$\sin[a(1-e^{-t})]$

	$g(p) = \int_0^\infty f(t)e^{-pt}dt$	$f(t)$
14.30	$p^{-1}S_{o,\nu}(p/a)$	$\frac{1}{2}\nu^{-1}\{[at+(1+a^2t^2)^{\frac{1}{2}}]^{\nu}-[(1+a^2t^2)^{\frac{1}{2}}-at]^{\nu}\}$
14.31	$p^{-1}S_{1,\nu}(p/a)$	$\frac{1}{2}\{[at+(1+a^2t^2)]^{\nu}+[(1+a^2t^2)^{\frac{1}{2}}-at]^{\nu}\}$
14.32	$S_{o,\nu}(p/a)$	$\frac{1}{2}a(1+a^2t^2)^{-\frac{1}{2}}\{[at+(1+a^2t^2)^{\frac{1}{2}}]^{\nu}$ $+[(1+a^2t^2)^{\frac{1}{2}}-at]^{\nu}\}$
14.33	$S_{-1,\nu}(p/a)$	$\frac{1}{2}a\nu^{-1}(1+a^2t^2)^{-\frac{1}{2}}\{[(1+a^2t^2)^{\frac{1}{2}}+at]^{\nu}$ $-[(1+a^2t^2)^{\frac{1}{2}}-at]^{\nu}\}$
14.34	$p^{-2}S_{2,o}(p/a)$	$(a^2+t^2)^{\frac{1}{2}}-t\ \log[t/a+(1+t^2/a^2)^{\frac{1}{2}}]$
14.35	$p^{-2}S_{2,\nu}(p)$	$1+(\nu-1/\nu)\int_0^t \sinh(\nu\ \sinh^{-1}u)du$ $=1+\frac{1}{2}(\nu-1/\nu)\int_0^t \{[(1+u^2)^{\frac{1}{2}}+u]^{\nu}$ $-[(1+u^2)^{\frac{1}{2}}-u]^{\nu}\}du$
14.36	$p^{-\frac{1}{2}\nu}S_{\mu,\nu}(ap^{\frac{1}{2}})$ $\mathrm{Re}(\mu-\nu) < 1$	$2^{\mu+\nu-1}a^{-\nu}[\Gamma(\frac{1}{2}+\frac{1}{2}\nu-\frac{1}{2}\mu)]^{-1}t^{\nu-1}$ $\cdot\exp(\frac{1}{4}a^2/t)\Gamma(\frac{1}{2}+\frac{1}{2}\nu+\frac{1}{2}\mu,\frac{1}{4}a^2/t)$

	$g(p) = \int_0^\infty f(t)e^{-pt}dt$	$f(t)$
14.37	$p^{\frac{1}{2}}S_{\nu,\frac{1}{2}}(ap)$ $\mathrm{Re}\,\nu < \frac{1}{2}$	$[\Gamma(\frac{1}{2}-\nu)]^{-1}a^{\nu+1}t^{-\frac{1}{2}-\nu}(t^2+a^2)^{-1}$
14.38	$p^{-\nu}S_{\nu,\mu}(p)$ $\mu+\nu\neq 0,-1,-2,\cdots$	$\pi^{\frac{1}{2}}2^{-\mu}\Gamma(\frac{1}{2}+\frac{1}{2}\nu-\frac{1}{2}\mu)[\Gamma(\frac{1}{2}\nu+\frac{1}{2}\mu)]^{-1}$ $\cdot(1+t^2)^{\frac{1}{2}\nu-\frac{1}{2}}\{2\pi^{-1}\cos[\frac{\pi}{2}(\nu+\mu)]$ $\cdot Q^{\mu}_{\nu-1}[t(1+t^2)^{-\frac{1}{2}}]+\sin[\frac{\pi}{2}(\nu+\mu)]$ $\cdot P^{\mu}_{\nu+1}[t(1+t^2)^{-\frac{1}{2}}]\}$
14.39	$p^{-\frac{1}{2}\nu-\frac{1}{2}}S_{\nu,\frac{1}{2}}[a(2p)^{\frac{1}{2}}]$ $\mathrm{Re}\,\nu > -\frac{1}{2}$	$2^{\frac{1}{2}\nu}a^{-\frac{1}{4}}t^{\frac{1}{2}\nu-\frac{1}{4}}$ $\cdot\exp(\frac{1}{4}a^2/t)D_{\nu}(a/t)$
14.40	$p^{-\mu-\frac{1}{2}}S_{2\mu,2\nu}[2(ap)^{\frac{1}{2}}]$ $\mathrm{Re}(\mu\pm\nu) > -\frac{1}{2}$	$2^{2\mu-1}a^{-\frac{1}{2}}t^{\mu}$ $\cdot\exp(\frac{1}{2}a^2/t)W_{\mu,\nu}(a/t)$
14.41	$p^{-\frac{1}{2}}S_{o,\nu}(ap^{\frac{1}{2}})$ $-\frac{1}{2} < \mathrm{Re}\,\nu < \frac{1}{2}$	$\frac{1}{2}\pi^{-\frac{1}{2}}t^{-\frac{1}{2}}\exp(\frac{1}{8}\,a^2/t)$ $\cdot K_{\frac{1}{2}\nu}(\frac{1}{8}\,a^2/t)$

	$g(p) = \int_0^\infty f(t)e^{-pt}dt$	$f(t)$
14.42	$p^{\frac{1}{2}}S_{0,\frac{1}{3}}(ap^{3/2})$	$a^{-1}\exp(-\frac{4}{27}t^3/a)$
14.43	$p^{-\frac{1}{2}}[e^{i\pi/4}S_{0,\frac{1}{3}}(ae^{i3\pi/4}p^{-\frac{1}{2}})$ $+e^{-i\pi/4}S_{0,\frac{1}{3}}(ae^{-i3\pi/4}p^{-\frac{1}{2}})]$	$(\frac{1}{2}a)^{-1/3}t^{-2/3}\cos[3(\frac{1}{2}a)^{2/3}t^{1/3}]$
14.44	$ip^{-\frac{1}{2}}[e^{i\pi/4}S_{0,\frac{1}{3}}(ae^{i3\pi/4}p^{-\frac{1}{2}})$ $-e^{-i\pi/4}S_{0,\frac{1}{3}}(ae^{-i3\pi/4}p^{-\frac{1}{2}})]$	$(\frac{1}{2}a)^{-1/3}t^{-2/3}\sin[3(\frac{1}{2}a)^{3/2}t^{1/3}]$
14.45	$p^{-\frac{1}{2}}\mathrm{ker}_\nu(ap^{\frac{1}{2}})$	$-\frac{1}{4}(t/\pi)^{-\frac{1}{2}}$ $\cdot[\cos(\frac{1}{4}\pi\nu+\frac{1}{8}a^2/t)Y_{\frac{1}{2}\nu}(\frac{1}{8}a^2/t)$ $+\sin(\frac{1}{4}\pi\nu+\frac{1}{8}a^2/t)J_{\frac{1}{2}\nu}(\frac{1}{8}a^2/t)]$
14.46	$p^{-\frac{1}{2}}\mathrm{kei}_\nu(ap^{\frac{1}{2}})$	$\frac{1}{4}(t/\pi)^{-\frac{1}{2}}$ $\cdot[\sin(\frac{1}{4}\pi\nu+\frac{1}{8}a^2/t)Y_{\frac{1}{2}\nu}(\frac{1}{8}a^2/t)$ $-\cos(\frac{1}{4}\pi\nu+\frac{1}{8}a^2/t)J_{\frac{1}{2}\nu}(\frac{1}{8}a^2/t)]$
14.47	$p^{-\frac{1}{2}\nu}\mathrm{ker}_\nu(ap^{\frac{1}{2}})$ Re $\nu > -1$	$\frac{1}{2}(\frac{1}{2}a)^\nu t^{\nu-1}\cos(\frac{1}{2}\pi\nu+\frac{1}{4}a^2/t)$

	$g(p) = \int_0^\infty f(t)e^{-pt}dt$	$f(t)$
14.48	$p^{-\frac{1}{2}\nu}\mathrm{kei}_\nu(ap^{\frac{1}{2}})$ $\mathrm{Re}\ \nu > -1$	$-\frac{1}{2}(\frac{1}{2}a)^\nu t^{\nu-1}\sin(\frac{1}{2}\pi\nu+\frac{1}{4}a^2/t)$
14.49	$p^{-1}\mathrm{ker}(ap^{\frac{1}{2}})$	$-\frac{1}{2}\mathrm{Ci}(\frac{1}{4}a^2/t)$
14.50	$p^{-1}\mathrm{kei}(ap^{\frac{1}{2}})$	$\frac{1}{2}\mathrm{si}(\frac{1}{4}a^2/t)$
14.51	$\mathrm{ker}_\nu(ap^{\frac{1}{2}})\mathrm{kei}_\nu(ap^{\frac{1}{2}})$	$-\frac{1}{8}\pi t^{-1}$ $\cdot[\cos(\frac{1}{2}\pi\nu+\frac{1}{2}a^2/t)J_\nu(\frac{1}{2}a^2/t)$ $-\sin(\frac{1}{2}\pi\nu+\frac{1}{2}a^2/t)Y_\nu(\frac{1}{2}a^2/t)$
14.52	$\mathrm{kei}_\nu^2(ap^{\frac{1}{2}})-\mathrm{ker}_\nu^2(ap^{\frac{1}{2}})$	$\frac{1}{4}\pi t^{-1}$ $\cdot[\sin(\frac{1}{2}\pi\nu+\frac{1}{2}a^2/t)J_\nu(\frac{1}{2}a^2/t)$ $+\cos(\frac{1}{2}\pi\nu+\frac{1}{2}a^2/t)Y_\nu(\frac{1}{2}a^2/t)]$

2.15 Special Cases of Whittaker Functions

	$g(p) = \int_0^\infty f(t)e^{-pt}dt$	$f(t)$	
15.1	$\mathrm{Ei}(-ap)$	0 $-t^{-1}$	$t < a$ $t > a$
15.2	$\mathrm{Ei}[-b(p-a)]$	0 $-t^{-1}e^{at}$	$t < b$ $t > b$
15.3	$\mathrm{Ei}[-b(p+a)]$	0 $-t^{-1}e^{-at}$	$t < b$ $t > b$
15.4	$p^{-1}\,\mathrm{Ei}(-ap)$	0 $\log(a/t)$	$t < a$ $t > a$
15.5	$e^{ap}\mathrm{Ei}(-ap)$	$-(t+a)^{-1}$	
15.6	$p^{-1}e^{ap}\mathrm{Ei}(-ap)$	$-\log(1+t/a)$	
15.7	$e^{-ap}\overline{\mathrm{Ei}}(ap)$	$-(t-a)^{-1}$ Cauchy principal value	
15.8	$e^{ap}[\mathrm{Ei}(-ap-cp)$ $-\mathrm{Ei}(-ap-bp)]$	0 $(t+a)^{-1}$ 0	$t < b$ $b < t < c$ $t > c$

	$g(p) = \int_0^\infty f(t)e^{-pt}dt$	$f(t)$
15.9	$e^{-ap}Ei(-ap-bp)$	$0 \quad t < b$ $(t+a)^{-1} \quad t > b$
15.10	$e^{-ap}Ei(ap-bp)$ $-e^{ap}Ei(-ap-bp)$	$0 \quad a < t < b$ $-2a(t^2-a^2)^{-1} \quad t > b$
15.11	$e^{-ap}\,\overline{Ei}(ap)$ $-e^{ap}Ei(-ap)$	$-2a(t^2-a^2)^{-1}$ Cauchy principal value
15.12	$e^{ap}[Ei(-ap)]^2$	$0 \quad t < a$ $2(t+a)^{-1}\log(t/a) \quad t > a$
15.13	$e^{(a+b)p}$ $\cdot Ei(-ap)Ei(-bp)$	$(t+a+b)^{-1}\log[(t+a)(t+b)(ab)^{-1}]$
15.14	$\overline{Ei}(p/a)Ei(-p/a)$	$t^{-1}\log\|1-a^2t^2\|$
15.15	$Ei(-ap)Ei(-bp)$	$0 \quad t < a+b$ $t^{-1}\log[(ab)^{-1}(t-a)(t-b)]$ $t > a+b$

	$g(p) = \int_0^\infty f(t)e^{-pt}dt$	$f(t)$
15.16	$p^{-1}e^{a/p}Ei(-a/p)$	$-2K_o[2(at)^{\frac{1}{2}}]$
15.17	$p^{-1}e^{-a/p}\overline{Ei}(a/p)$	$\pi Y_o[2(at)^{\frac{1}{2}}]$
15.18	$\exp(ap^2)Ei(-ap^2)$	$i(\pi a)^{-\frac{1}{2}}e^{-\frac{1}{4}t^2/a}Erf(\frac{1}{2}ia^{\frac{1}{2}}t)$
15.19	$p^{-\frac{1}{2}}Ei(-a/p)$	$2(\pi t)^{-\frac{1}{2}}Ci[2(at)^{\frac{1}{2}}]$
15.20	$p^{-\frac{1}{2}}\overline{Ei}(a/p)$	$(\pi t)^{-\frac{1}{2}}\{Ei[-2(at)^{\frac{1}{2}}]+\overline{Ei}(at)^{\frac{1}{2}}]\}$
15.21	$p^{-1}Ei(-a/p)$	$2Ji_o[2(at)^{\frac{1}{2}}]$
15.22	$p^{-\nu}Ei(-a/p)$ $Re\ \nu > 0$	$2t^{\nu-1}\int_\infty^{(at)^{\frac{1}{2}}} u^{-\nu} J_{\nu-1}(2u)du$
15.23	$p^{-\nu}e^{a/p}Ei(-a/p)$ $Re\ \nu > 0$	$t^{\nu-1}\int_\infty^{at} u^{-\frac{1}{2}-\frac{1}{2}\nu}J_{\nu-1}[2(u-at)^{\frac{1}{2}}]du$
15.24	$p^{-\frac{1}{2}}e^{a/p}Ei(-a/p)$	$(\pi t)^{-\frac{1}{2}}\{\exp[2(at)^{\frac{1}{2}}]Ei[-2(at)^{\frac{1}{2}}]$ $+\exp[-2(at)^{\frac{1}{2}}]\overline{Ei}[2(at)^{\frac{1}{2}}]\}$

	$g(p) = \int_0^\infty f(t)e^{-pt}dt$	$f(t)$
15.25	$p^{-\frac{1}{2}}\exp[\frac{1}{4}(a+b)^2/p]$ $\cdot\mathrm{Ei}(-ab/p)$	$\frac{1}{2}(\pi t)^{-\frac{1}{2}}\{\exp[(a+b)t^{\frac{1}{2}}]$ $\cdot[\mathrm{Ei}(-2at^{\frac{1}{2}})+\mathrm{Ei}(-2bt^{\frac{1}{2}})]$ $+\exp[-(a+b)t^{\frac{1}{2}}]$ $\cdot[\overline{\mathrm{Ei}}(2at^{\frac{1}{2}})+\overline{\mathrm{Ei}}(2bt^{\frac{1}{2}})]\}$
15.26	$p^{-\frac{1}{2}}e^{-a/p}\overline{\mathrm{Ei}}(a/p)$	$2(\pi t)^{-\frac{1}{2}}\{\cos[2(at)^{\frac{1}{2}}\mathrm{Ci}[2(at)^{\frac{1}{2}}]$ $+\sin[2(at)^{\frac{1}{2}}]\mathrm{Si}[2(at)^{\frac{1}{2}}]\}$
15.27	$p^{-\frac{1}{2}}\exp[-\frac{1}{4}(a+b)^2/p]$ $\cdot\overline{\mathrm{Ei}}(ab/p)$	$(\pi t)^{-\frac{1}{2}}\{[\mathrm{Ci}(2at^{\frac{1}{2}})+\mathrm{Ci}(2bt^{\frac{1}{2}})]$ $\cdot\cos[t^{\frac{1}{2}}(a+b)]+\sin[t^{\frac{1}{2}}(a+b)]$ $\cdot[\mathrm{Si}(2at^{\frac{1}{2}})+\mathrm{Si}(2bt^{\frac{1}{2}})]\}$
15.28	$p^{-\frac{1}{2}}\exp[-\frac{1}{4}(a-b)^2/p]$ $\cdot\mathrm{Ei}(-ab/p)$	$(\pi t)^{-\frac{1}{2}}\{[\mathrm{Ci}(2at^{\frac{1}{2}})+\mathrm{Ci}(2bt^{\frac{1}{2}})]$ $\cdot\cos[t^{\frac{1}{2}}(a-b)]+\sin[t^{\frac{1}{2}}(a-b)]$ $\cdot[\mathrm{Si}(2at^{\frac{1}{2}})-\mathrm{Si}(2bt^{\frac{1}{2}})]\}$
15.29	$p^{-\frac{1}{2}}\mathrm{Ei}(-ap^{\frac{1}{2}})$	$\frac{1}{2}(\pi t)^{-\frac{1}{2}}\mathrm{Ei}(-\frac{1}{4}a^2/t)$
15.30	$\exp[2(bp)^{\frac{1}{2}}]\mathrm{Ei}(-u)$ $+\exp[-2(bp)^{\frac{1}{2}}]\mathrm{Ei}(-v)$	$(b/\pi)^{\frac{1}{2}}t^{-3/2}e^{-b/t}\mathrm{Ei}(-at)$ $\begin{matrix}u\\v\end{matrix} = 2b^{\frac{1}{2}}[(p+a)^{\frac{1}{2}}\pm p^{\frac{1}{2}}]$

	$g(p) = \int_0^\infty f(t)e^{-pt}dt$	$f(t)$
15.31	$\exp[2(bp)^{\frac{1}{2}}]Ei(-u)$ $+\exp[-2(bp)^{\frac{1}{2}}]\overline{Ei}(v)$	$(b/\pi)^{\frac{1}{2}}t^{-3/2}e^{-bt}\ \overline{Ei}(at)$ $\begin{matrix}u\\v\end{matrix} = 2b^{\frac{1}{2}}[p^{\frac{1}{2}}\pm(p-a)^{\frac{1}{2}}]$
15.32	$p^{-\frac{1}{2}}\{\exp[2(bp)^{\frac{1}{2}}]Ei(-u)$ $+\exp[-2(bp)^{\frac{1}{2}}]Ei(-v)\}$	$\pi^{-\frac{1}{2}}t^{-\frac{1}{2}}e^{-b/t}Ei(-a/t)$ $\begin{matrix}u\\v\end{matrix} = 2p^{\frac{1}{2}}[(b+a)^{\frac{1}{2}}\pm b^{\frac{1}{2}}]$
15.33	$p^{-\frac{1}{2}}\{\exp[2(bp)^{\frac{1}{2}}]Ei(-u)$ $+\exp[-2(bp)^{\frac{1}{2}}]\overline{Ei}(v)\}$	$\pi^{-\frac{1}{2}}t^{-\frac{1}{2}}e^{-b/t}\ \overline{Ei}(-a/t)$ $\begin{matrix}u\\v\end{matrix} = 2p^{\frac{1}{2}}[b^{\frac{1}{2}}\pm(b-a)^{\frac{1}{2}}]$
15.34	$r^{-1}\{\exp[-b(r-p)$ $\cdot\overline{Ei}[b(r-p)$ $-\exp[b(r+p)]$ $\cdot Ei[-b(r+p)]\}$	$\pi Y_o[a(t^2+2bt)^{\frac{1}{2}}]$ $r = (p^2+a^2)^{\frac{1}{2}}$
15.35	$s^{-1}\{\exp[b(p-s)]$ $\cdot Ei[-b(p-s)]$ $-\exp[b(p+s)]$ $\cdot Ei[-b(p+s)]\}$	$-2K_o[a(t^2+2bt)^{\frac{1}{2}}]$ $s = (p^2-a^2)^{\frac{1}{2}}$

	$g(p) = \int_0^\infty f(t) e^{-pt} dt$	$f(t)$	
15.36	$r^{-1}[e^{-br}\overline{Ei}(br-bp)$ $-e^{br}Ei(-br-bp)]$ $r = (p^2+a^2)^{\frac{1}{2}}$	0 $\pi Y_o[a(t^2-b^2)^{\frac{1}{2}}]$	$t < b$ $t > b$
15.37	$s^{-1}[e^{-bs}Ei(-bp+bs)$ $-e^{bs}Ei(-bp-bs)]$ $s = (p^2-a^2)^{\frac{1}{2}}$	0 $-2K_o[a(t^2-b^2)^{\frac{1}{2}}]$	$t < b$ $t > b$
15.38	$Ci(ap)\cos(ap)$ $+si(ap)\sin(ap)$	$-t(t^2+a^2)^{-1}$	
15.39	$Ci(ap)\sin(ap)$ $-si(ap)\cos(ap)$	$a(t^2+a^2)^{-1}$	
15.40	$[a \sin(cp)+b/c \cos(cp)]si(cp)$ $+[a \cos(cp)-b/c \sin(cp)]Ci(cp)$	$-(at+b)(t^2+c^2)^{-1}$	
15.41	$p^{-1}[Ci(ap)\cos(ap)$ $+si(ap)\sin(ap)]$	$-\log(1+t^2/a^2)$	

	$g(p) = \int_0^\infty f(t)e^{-pt}dt$	$f(t)$
15.42	$p^{-1}[\mathrm{Ci}(ap)\sin(ap)$ $-\mathrm{si}(ap)\cos(ap)$	$\arctan(p/a)$
15.43	$[\mathrm{Ci}(ap)]^2+[\mathrm{si}(ap)]^2$	$t^{-1}\log(1+t^2/a^2)$
15.44	$\mathrm{Ci}(ap^2)\sin(ap^2)$ $-\mathrm{si}(ap^2)\cos(ap^2)$	$(\frac{1}{2}\pi a)^{-\frac{1}{2}}[\cos(\frac{1}{4}t^2/a)C(\frac{1}{4}t^2/a)$ $+\sin(\frac{1}{4}t^2/a)S(\frac{1}{4}t^2/a)]$
15.45	$\mathrm{Ci}(ap^2)\cos(ap^2)$ $+\mathrm{si}(ap^2)\sin(ap^2)$	$(\frac{1}{2}\pi a)^{-\frac{1}{2}}[\cos(\frac{1}{4}t^2/a)S(\frac{1}{4}t^2/a)$ $-\sin(\frac{1}{4}t^2/a)C(\frac{1}{4}t^2/a)]$
15.46	$p^{-\frac{1}{2}}[\mathrm{Ci}(ap^{\frac{1}{2}})\sin(ap^{\frac{1}{2}})$ $-\mathrm{si}(ap^{\frac{1}{2}})\cos(ap^{\frac{1}{2}})]$	$\frac{1}{2}(t/\pi)^{-\frac{1}{2}}\exp(\frac{1}{4}a^2/t)\mathrm{Erfc}(\frac{1}{2}at^{-\frac{1}{2}})$
15.47	$p^{-\frac{1}{2}}[\mathrm{Ci}(ap^{\frac{1}{2}})\cos(ap^{\frac{1}{2}})$ $+\mathrm{si}(ap^{\frac{1}{2}})\sin(ap^{\frac{1}{2}})]$	$\frac{1}{2}(\pi t)^{-\frac{1}{2}}\exp(\frac{1}{4}a^2/t)\mathrm{Ei}(-\frac{1}{4}a^2/t)$
15.48	$p^{-1}[\cos(a/p)\mathrm{Ci}(a/p)$ $+\sin(a/p)\mathrm{si}(a/p)]$	$-2\ \mathrm{ker}[2(at)^{\frac{1}{2}}]$

	$g(p) = \int_0^\infty f(t)e^{-pt}dt$	$f(t)$
15.49	$p^{-1}[\sin(a/p)\,\mathrm{Ci}(a/p)$ $-\cos(a/p)\,\mathrm{si}(a/p)]$	$-2\,\mathrm{kei}[2(at)^{\frac{1}{2}}]$
15.50	$[\tfrac{1}{2}-C(ap^2)]\cos(ap^2)$ $+[\tfrac{1}{2}-S(ap^2)]\sin(ap^2)$	$(2\pi a)^{-\frac{1}{2}}\sin(\tfrac{1}{4}t^2/a)$
15.51	$[\tfrac{1}{2}-S(ap^2)]\cos(ap^2)$ $-[\tfrac{1}{2}-C(ap^2)]\sin(ap^2)$	$(2\pi a)^{-\frac{1}{2}}\cos(\tfrac{1}{4}t^2/a)$
15.52	$p^{-1}\{\cos(ap^2)[\tfrac{1}{2}-S(ap^2)]$ $-\sin(ap^2)[\tfrac{1}{2}-C(ap^2)]\}$	$C(\tfrac{1}{4}t^2/a)$
15.53	$p^{-1}\{\cos(ap^2)[\tfrac{1}{2}-C(ap^2)]$ $+\sin(ap^2)[\tfrac{1}{2}-S(ap^2)]\}$	$S(\tfrac{1}{4}t^2/a)$
15.54	$[\tfrac{1}{2}-C(ap^2)]^2$ $+[\tfrac{1}{2}-S(ap^2)]^2$	$2\pi^{-1}t^{-1}\sin(\tfrac{1}{4}t^2/a)$
15.55	$p^{-1}\{[\tfrac{1}{2}-C(ap^2)]^2$ $+[\tfrac{1}{2}-S(ap^2)]^2\}$	$\pi^{-1}\mathrm{Si}(\tfrac{1}{4}t^2/a)$

	$g(p) = \int_0^\infty f(t)e^{-pt}dt$	$f(t)$
15.56	$\exp(a^2p^2)\mathrm{Erfc}(ap)$	$\pi^{-\frac{1}{2}}a^{-1}\exp(-\frac{1}{4}t^2/a^2)$
15.57	$\exp(a^2p^2)$ $\cdot\mathrm{Erfc}(ap+b)$	$0 \quad t < 2ab$ $\pi^{-\frac{1}{2}}a^{-1}\exp(-\frac{1}{4}t^2/a^2) \quad t > 2ab$
15.58	$1-\pi^{\frac{1}{2}}ap\,\exp(a^2p^2)$ $\cdot\mathrm{Erfc}(ap)$	$\frac{1}{2}a^{-2}t\,\exp(-\frac{1}{4}t^2/a^2)$
15.59	$p^{-1}\exp(a^2p^2)\mathrm{Erfc}(ap)$	$\mathrm{Erf}(\frac{1}{2}t/a)$
15.60	$p^{-1}[1-\exp(a^2p^2)\mathrm{Erfc}(ap)]$	$\mathrm{Erfc}(\frac{1}{2}t/a)$
15.61	$(p-a)^{-1}e^{p^2}\mathrm{Erfc}(p)$	$\exp[a(t+a)][\mathrm{Erf}(a+\frac{1}{2}t)-\mathrm{Erf}(a)]$
15.62	$p^{-1}e^{p^2}\mathrm{Erfc}(p+a)$	$0 \quad t < 2a$ $\mathrm{Erf}(\frac{1}{2}t)-\mathrm{Erf}(a) \quad t > 2a$
15.63	$p^{-1}\exp(a^2p^2)$ $\cdot[\mathrm{Erf}(ap)-\mathrm{Erf}(ap+b)]$	$\mathrm{Erf}(\frac{1}{2}t/a) \quad t < 2ab$ $\mathrm{Erf}(b) \quad t > 2ab$

	$g(p) = \int_0^\infty f(t)e^{-pt}dt$	f(t)	
15.64	$\mathrm{Erfc}[(ap)^{\frac{1}{2}}]$	0	$t < a$
		$\pi^{-1}a^{\frac{1}{2}}t^{-1}(t-a)^{-1}$	$t > a$
15.65	$p^{-\frac{1}{2}}\mathrm{Erf}[(ap)^{\frac{1}{2}}]$	$(\pi t)^{-\frac{1}{2}}$	$t < a$
		0	$t > a$
15.66	$p^{-\frac{1}{2}}\mathrm{Erfc}[(ap^{\frac{1}{2}}]$	0	$t < a$
		$(\pi t)^{-\frac{1}{2}}$	$t > a$
15.67	$e^{ap}\mathrm{Erfc}[(ap)^{\frac{1}{2}}]$	$\pi^{-1}(t/a)^{-\frac{1}{2}}(t+a)^{-1}$	
15.68	$p^{-\frac{1}{2}}e^{ap}\mathrm{Erfc}[(ap)^{\frac{1}{2}}]$	$\pi^{-\frac{1}{2}}(t+a)^{-\frac{1}{2}}$	
15.69	$p^{-3/2}e^{ap}\mathrm{Erfc}[(ap)^{\frac{1}{2}}]$	$2\pi^{-\frac{1}{2}}[(t^2+a)^{\frac{1}{2}}-a^{\frac{1}{2}}]$	
15.70	$p^{-\frac{1}{2}}e^{a^2/p}\mathrm{Erfc}(ap^{-\frac{1}{2}})$	$(\pi t)^{-\frac{1}{2}}\exp(-2at^{\frac{1}{2}})$	
15.71	$p^{-3/2}e^{a^2/p}\mathrm{Erfc}(ap^{-\frac{1}{2}})$	$\pi^{-\frac{1}{2}}a^{-1}[1-\exp(-2at^{\frac{1}{2}})]$	
15.72	$\mathrm{Erf}(ap^{-\frac{1}{2}})$	$(\pi t)^{-1}\sin(2at^{\frac{1}{2}})$	

	$g(p) = \int_0^\infty f(t)e^{-pt}dt$	$f(t)$
15.73	$p^{-3/2}\exp(-a^2/p)$ $\cdot\mathrm{Erf}(iap^{-1/2})$	$i\pi^{-1/2}a^{-1}[1-\cos(2at^{1/2})]$
15.74	$p^{-1/2}\exp(-a^2/p)$ $\cdot\mathrm{Erf}(iap^{-1/2})$	$i(\pi t)^{-1/2}\sin(2at^{1/2})$
15.75	$e^{bp}\{e^{-2ab^{1/2}}\mathrm{Erfc}[(bp)^{1/2}-a/p]$ $+e^{2ab^{1/2}}\mathrm{Erfc}[(bp)^{1/2}+a/p]\}$	$2\pi^{-1}b^{1/2}t^{-1/2}(b+t)^{-1}\cos(2at^{1/2})$
15.76	$p^{-3/2}\exp(a^2/p)$ $\cdot\mathrm{Erf}(ap^{-1/2})$	$a^{-1}\pi^{-1/2}[\cosh(2at^{1/2})-1]$
15.77	$p^{-1/2}\exp(a^2/p)$ $\cdot\mathrm{Erf}(ap^{-1/2})$	$(\pi t)^{-1/2}\sinh(2at^{1/2})$
15.78	$\mathrm{Erf}(iap^{-1/2})$	$(\pi t)^{-1}\sinh(2at^{1/2})$
15.79	$p^{-\nu}\exp(a^2/p)\mathrm{Erf}(ap^{-1/2})$ $\mathrm{Re}\,\nu > -\frac{1}{2}$	$a^{1-\nu}t^{\frac{1}{2}\nu-\frac{1}{2}}\mathbf{L}_{\nu-1}(2at^{1/2})$

	$g(p) = \int_0^\infty f(t)e^{-pt}dt$	$f(t)$
15.80	$p^{-\nu}\exp(-a^2/p)\mathrm{Erf}(iap^{-\frac{1}{2}})$ $\mathrm{Re}\ \nu > -\frac{1}{2}$	$ia^{1-\nu}t^{\frac{1}{2}\nu-\frac{1}{2}}\mathbf{H}_{\nu-1}(2at^{\frac{1}{2}})$
15.81	$p^{-\nu}\exp(a^2/p)\mathrm{Erfc}(ap^{-\frac{1}{2}})$ $\mathrm{Re}\ \nu > 0$	$a^{1-\nu}t^{\frac{1}{2}\nu-\frac{1}{2}}[I_{\nu-1}(2at^{\frac{1}{2}})-\mathbf{L}_{\nu-1}(2at^{\frac{1}{2}})]$
15.82	$p^{-\nu}e^{ap}\mathrm{Erfc}[(ap)^{\frac{1}{2}}]$ $\mathrm{Re}\ \nu > -\frac{1}{2}$	$(\pi a)^{-\frac{1}{2}}[\Gamma(\frac{1}{2}+\nu)]^{-1}t^{\nu-\frac{1}{2}}$ $\cdot {}_2F_1(1,\frac{1}{2};\frac{1}{2}+\nu;-t/a)$
15.83	$\mathrm{Erfc}(u^{\frac{1}{2}})\mathrm{Erfc}(v^{\frac{1}{2}})$ ${u \atop v} = a[b\pm(b^2-p^2)^{\frac{1}{2}}]$	$\pi^{-1}(2a)^{\frac{1}{2}}e^{-ab}(a^2+t^2)^{-\frac{1}{2}}$ $\cdot[(a^2+t^2)^{\frac{1}{2}}+a]^{-\frac{1}{2}}\exp[-b(a^2+t^2)^{\frac{1}{2}}]$
15.84	$p(b^2-p^2)^{-\frac{1}{2}}$ $\cdot\{v^{-\frac{1}{2}}\exp[-a(b^2-p^2)^{\frac{1}{2}}]$ $\cdot\mathrm{Erfc}(v^{\frac{1}{2}})-u^{-\frac{1}{2}}$ $\cdot\exp[a(b^2-p^2)^{\frac{1}{2}}]\mathrm{Erfc}(u^{\frac{1}{2}})\}$	$(\frac{1}{2}\pi)^{-\frac{1}{2}}(a^2+t^2)^{-\frac{1}{2}}[(a^2+t^2)^{\frac{1}{2}}+a]^{\frac{1}{2}}$ $\cdot\exp[-b(a^2+t^2)^{\frac{1}{2}}]$ ${u \atop v} = a[b\pm(b^2-p^2)^{\frac{1}{2}}]$
15.85	$e^{ap}\mathrm{Erfc}(u^{\frac{1}{2}})\mathrm{Erfc}(v^{\frac{1}{2}})$ ${u \atop v} = a[p\pm(p^2-b^2)^{\frac{1}{2}}]$	$0 \quad t < a$ $\pi^{-1}(2a)^{\frac{1}{2}}(t-a)^{-\frac{1}{2}}(t+a)^{-1}\exp[-b(t^2-a^2)^{\frac{1}{2}}]$ $t > a$

	$g(p) = \int_0^\infty f(t)e^{-pt}dt$	$f(t)$
15.86	$e^{ap}\mathrm{Erfc}(u)\mathrm{Erf}(iv)$ $\begin{matrix}u\\v\end{matrix} = (\frac{1}{2}a)^{\frac{1}{2}}[(p^2+b^2)^{\frac{1}{2}}\pm p]^{\frac{1}{2}}$	$i\pi^{-1}a^{\frac{1}{2}}(a+t)^{-\frac{1}{2}}(t^2+at)^{-\frac{1}{2}}$ $\cdot\ \sin[b(t^2+at)^{\frac{1}{2}}]$
15.87	$e^{2ap}\mathrm{Erfc}\{a^{\frac{1}{2}}[(p^2+b^2)^{\frac{1}{2}}+p]^{\frac{1}{2}}\}$	$\pi^{-1}(2a)^{\frac{1}{2}}(t+2a)^{-\frac{1}{2}}(t^2+2at)^{-\frac{1}{2}}$ $\cdot\cos[b(t^2+2at)^{\frac{1}{2}}]$
15.88	$e^{ap}\mathrm{Erfc}(u)\mathrm{Erf}(iv)$ $\begin{matrix}u\\v\end{matrix} = a^{\frac{1}{2}}[(p^2+b^2)^{\frac{1}{2}}\pm p]^{\frac{1}{2}}$	$0 \qquad t < a$ $i\pi^{-1}(2a)^{\frac{1}{2}}(t+a)^{-\frac{1}{2}}(t^2-a^2)^{-\frac{1}{2}}$ $\cdot\ \sin[b(t^2-a^2)^{\frac{1}{2}}] \qquad t > a$
15.89	$e^{ap}\mathrm{Erfc}\{a^{\frac{1}{2}}[(p^2+b^2)^{\frac{1}{2}}+p]^{\frac{1}{2}}\}$	$0 \qquad t < a$ $(t+a)^{-\frac{1}{2}}(t^2-a^2)^{-\frac{1}{2}}\cos[b(t^2-a^2)^{\frac{1}{2}}]$ $t > a$
15.90	$e^{ap}\mathrm{Erfc}(u)\mathrm{Erf}(v)$ $\begin{matrix}u\\v\end{matrix} = (\frac{1}{2}a)^{\frac{1}{2}}[p+(p^2-b^2)^{\frac{1}{2}}]^{\frac{1}{2}}$	$\pi^{-1}a^{\frac{1}{2}}(a+t)^{-\frac{1}{2}}(t^2+at)^{-\frac{1}{2}}$ $\cdot\sinh[b(t^2+at)^{\frac{1}{2}}]$
15.91	$e^{2ap}\mathrm{Erfc}\{a^{\frac{1}{2}}[p+(p^2-b^2)^{\frac{1}{2}}]^{\frac{1}{2}}\}$	$\pi^{-1}(2a)^{\frac{1}{2}}(t+2a)^{-\frac{1}{2}}(t^2+2at)^{-\frac{1}{2}}$ $\cdot\cosh[b(t^2+2at)^{\frac{1}{2}}]$

	$g(p) = \int_0^\infty f(t)e^{-pt}dt$	$f(t)$
15.92	$e^{ap}\mathrm{Erfc}(u)\,\mathrm{Erf}(v)$ $\begin{matrix}u\\v\end{matrix} = a^{\frac{1}{2}}[p\pm(p^2-b^2)^{\frac{1}{2}}]^{\frac{1}{2}}$	$0 \qquad t < a$ $\pi^{-1}(2a)^{\frac{1}{2}}(t+a)^{-\frac{1}{2}}(t^2-a^2)^{-\frac{1}{2}}$ $\cdot\sinh[b(t^2-a^2)^{\frac{1}{2}}] \qquad t > a$
15.93	$e^{ap}\mathrm{Erfc}\{a^{\frac{1}{2}}[p+(p^2-b^2)^{\frac{1}{2}}]^{\frac{1}{2}}\}$	$0 \qquad t < a$ $(t+a)^{-\frac{1}{2}}(t^2-a^2)^{-\frac{1}{2}}\cosh[b(t^2-a^2)^{\frac{1}{2}}]$ $t > a$

2.16 Parabolic Cylinder Functions and Whittaker Functions

	$g(p) = \int_0^\infty f(t)e^{-pt}dt$	$f(t)$
16.1	$p^{-1}\exp(\frac{1}{4}a^2p^2)$ $\cdot D_\nu(ap)$, $\mathrm{Re}\,\nu < 0$	$2^{1-\frac{1}{2}\nu}[\Gamma(-\nu)]^{-1}\gamma(-\frac{1}{2}\nu,\frac{1}{2}t^2/a^2)$
16.2	$\exp(\frac{1}{4}a^2p^2)D_\nu(ap)$ $\mathrm{Re}\,\nu < 0$	$a^\nu[\Gamma(-\nu)]^{-1}t^{-\nu-1}\exp(-\frac{1}{2}t^2/a^2)$
16.3	$p^n e^{p^2/4}D_{-n-1}(p)$	$\frac{d^n}{dt^n}(e^{-\frac{1}{2}t^2}\frac{t^n}{n!})$ $n = 0,1,2,\cdots$
16.4	$D_\nu(ape^{i\pi/4})D_\nu(ape^{-i\pi/4})$ $\mathrm{Re}\,\nu < 0$	$\pi^{\frac{1}{2}}[a\Gamma(\nu)]^{-1}J_{-\nu-\frac{1}{2}}(\frac{1}{2}t^2/a^2)$
16.5	$\exp[\frac{1}{2}a^2p(\frac{1}{2}p+ib)]$ $\cdot D_\nu[a(p+ib)]$ $-\exp[\frac{1}{2}a^2p(\frac{1}{2}p-ib)]$ $\cdot D_\nu[a(p-ib)]$	$-2ia^\nu[\Gamma(-\nu)]^{-1}t^{-\nu-1}\sin(bt)$ $\cdot\exp(-\frac{1}{2}t^2/a^2+\frac{1}{4}a^2b^2)$ $\mathrm{Re}\,\nu < 1$
16.6	$\exp[\frac{1}{2}a^2p(\frac{1}{2}p+ib)]$ $\cdot D_\nu[a(p+ib)]$ $+\exp[\frac{1}{2}a^2p(\frac{1}{2}p-ib)]$ $\cdot D_\nu[a(p-ib)]$	$2a^\nu[\Gamma(-\nu)]^{-1}t^{-\nu-1}\cos(bt)$ $\cdot\exp(-\frac{1}{2}t^2/a^2+\frac{1}{4}a^2b^2)$ $\mathrm{Re}\,\nu < 0$

	$g(p) = \int_0^\infty f(t)e^{-pt}dt$	$f(t)$
16.7	$\exp(\frac{1}{4}a^2p^2)$ $D_\mu(ap)D_\nu(bp)$ $\mathrm{Re}(\mu+\nu) < 0$	$a^{\mu+\nu}[\Gamma(-\mu-\nu)]^{-1}t^{-\mu-\nu-1}\exp(-\frac{1}{2}t^2/a^2)$ $\cdot {}_2F_2(-\mu,-\nu;-\frac{1}{2}\mu-\frac{1}{2}\nu,\frac{1}{2}-\frac{1}{2}\nu-\frac{1}{2}\mu;\frac{1}{4}t^2/a^2)$
16.8	$\exp[-i(\frac{1}{4}\pi\nu+\frac{1}{4}p^2a^2)]$ $\cdot D_{-\nu}(ape^{-i\pi/4})$ $-\exp[i(\frac{1}{4}\pi\nu+\frac{1}{4}p^2a^2)]$ $\cdot D_{-\nu}(ape^{i\pi/4})$	$-2i[\Gamma(\nu)]^{-1}a^{-\nu}t^{\nu-1}\sin(\frac{1}{2}t^2/a^2)$ $\mathrm{Re}\ \nu > -2$
16.9	$\exp[-i(\frac{1}{4}\pi\nu+\frac{1}{4}p^2a^2)]$ $\cdot D_{-\nu}(ape^{-i\pi/4})$ $+\exp[i(\frac{1}{4}\pi\nu+\frac{1}{4}p^2a^2)]$ $\cdot D_{-\nu}(ape^{i\pi/4})$	$2[\Gamma(\nu)]^{-1}a^{-\nu}t^{\nu-1}\cos(\frac{1}{2}t^2/a^2)$ $\mathrm{Re}\ \nu > -1$
16.10	$D_\nu(ap^{\frac{1}{2}})$ $\mathrm{Re}\ \nu < 0$	$0 \qquad t < \frac{1}{4}a^2$ $2^{\frac{1}{2}\nu-\frac{1}{2}}a[\Gamma(-\frac{1}{2}\nu)]^{-1}$ $\cdot(t-\frac{1}{4}a^2)^{-\frac{1}{2}\nu-1}(t+\frac{1}{4}a^2)^{\frac{1}{2}\nu-\frac{1}{2}}$ $t > \frac{1}{4}a^2$

	$g(p) = \int_0^\infty f(t)e^{-pt}dt$	$f(t)$
16.11	$p^{-\frac{1}{2}}D_\nu(ap^{\frac{1}{2}})$ $\operatorname{Re}\nu < 1$	$0 \qquad t < \frac{1}{4}a^2$ $2^{\frac{1}{2}\nu}[\Gamma(\frac{1}{2}-\frac{1}{2}\nu)]^{-1}(t-\frac{1}{4}a^2)^{-\frac{1}{2}\nu-\frac{1}{2}}$ $\cdot(t+\frac{1}{4}a^2)^{\frac{1}{2}\nu} \qquad t > \frac{1}{4}a^2$
16.12	$e^{\frac{1}{4}a^2p}D_\nu(ap^{\frac{1}{2}})$ $\operatorname{Re}\nu < 0$	$a[\Gamma(-\frac{1}{2}\nu)]^{-1}t^{-\frac{1}{2}\nu-1}(2t+a^2)^{\frac{1}{2}\nu-\frac{1}{2}}$
16.13	$p^{-\frac{1}{2}}e^{\frac{1}{4}a^2p}D_\nu(ap^{\frac{1}{2}})$ $\operatorname{Re}\nu < 1$	$[\Gamma(\frac{1}{2}-\frac{1}{2}\nu)]^{-1}t^{-\frac{1}{2}\nu-\frac{1}{2}}(2t+a^2)^{\frac{1}{2}\nu}$
16.14	$p^{\frac{1}{2}\nu}e^{\frac{1}{2}ap}$ $\cdot D_\nu[(2ap)^{\frac{1}{2}}]$, $\operatorname{Re}\nu < 0$	$2^{-1-\frac{1}{2}\nu}[\Gamma(-\nu)]^{-1}(a+t)^{-\frac{1}{2}}$ $\cdot[(a+t)^{\frac{1}{2}}-a^{\frac{1}{2}}]^{-\nu-1}$
16.15	$p^{\frac{1}{2}\nu-1}e^{\frac{1}{2}ap}$ $\cdot D_\nu[(2ap)^{\frac{1}{2}}]$, $\operatorname{Re}\nu < 1$	$2^{-\frac{1}{2}\nu}[\Gamma(1-\nu)]^{-1}[(a+t)^{\frac{1}{2}}-a^{\frac{1}{2}}]^{-\nu}$
16.16	$p^{-\nu}\exp(-\frac{1}{4}a^2p^{-1})$ $\cdot D_{2\nu-1}(ap^{-\frac{1}{2}})$	$\pi^{-\frac{1}{2}}2^{\frac{1}{2}+\nu}t^{\nu-1}\sin[\pi\nu-a(2t)^{\frac{1}{2}}]$ $\operatorname{Re}\nu > 0$

	$g(p) = \int_0^\infty f(t)e^{-pt}dt$	$f(t)$
16.17	$p^{\nu}\exp(\tfrac{1}{4}a^2p^{-1})$ $\cdot D_{2\nu}(ap^{-\frac{1}{2}})$	$[\Gamma(-2\nu)]^{-1}(2t)^{-\nu-1}\exp[-a(2t)^{\frac{1}{2}}]$ $\mathrm{Re}\ \nu < 0$
16.18	$p^{-\frac{1}{2}-\frac{1}{2}\nu}\exp(-\tfrac{1}{4}a^2/p)$ $\cdot D_{\nu}(ap^{-\frac{1}{2}})$, $\mathrm{Re}\ \nu > -1$	$\pi^{-\frac{1}{2}}2^{\frac{1}{2}\nu}t^{\frac{1}{2}\nu-\frac{1}{2}}\cos[\tfrac{1}{2}\pi\nu-a(2t)^{\frac{1}{2}}]$
16.19	$p^{\nu}\exp(\tfrac{1}{4}a^2/p)$ $\cdot[D_{2\nu}(-ap^{-\frac{1}{2}})+D_{2\nu}(ap^{-\frac{1}{2}})]$	$[\Gamma(-2\nu)]^{-1}2^{-\nu}t^{-\nu-1}\cosh[a(2t)^{\frac{1}{2}}]$ $\mathrm{Re}\ \nu < 0$
16.20	$p^{\nu}\exp(\tfrac{1}{4}a^2/p)$ $\cdot[D_{2\nu}(-ap^{\frac{1}{2}})-D_{2\nu}(ap^{-\frac{1}{2}})]$	$[\Gamma(-2\nu)]^{-1}2^{-\nu}t^{-\nu-1}\sinh[a(2t)^{\frac{1}{2}}]$ $\mathrm{Re}\ \nu < \tfrac{1}{2}$
16.21	$p^{-\nu}\exp(-\tfrac{1}{4}a^2/p)$ $\cdot[D_{2\nu-1}(ap^{-\frac{1}{2}})+D_{2\nu-1}(-ap^{-\frac{1}{2}})]$	$\pi^{-\frac{1}{2}}2^{\frac{1}{2}+\nu}\sin(\pi\nu)t^{\nu-1}\cos[a(2t)^{\frac{1}{2}}]$ $\mathrm{Re}\ \nu > 0$
16.22	$p^{-\nu}\exp(-\tfrac{1}{4}a^2/p)$ $\cdot[D_{2\nu-1}(-ap^{-\frac{1}{2}})-D_{2\nu-1}(ap^{-\frac{1}{2}})]$	$\pi^{-\frac{1}{2}}2^{\frac{1}{2}+\nu}\cos(\pi\nu)t^{\nu-1}\sin[a(2t)^{\frac{1}{2}}]$ $\mathrm{Re}\ \nu < \tfrac{1}{2}$

	$g(p) = \int_0^\infty f(t)e^{-pt}dt$	$f(t)$
16.23	$p^{\frac{1}{2}\nu}\exp(-\frac{1}{4}p^{-1})$ $\cdot[D_\nu(ip^{-\frac{1}{2}})+D_\nu(-ip^{-\frac{1}{2}})]$	$[t\Gamma(-\nu)]^{-1}(2t)^{-\frac{1}{2}\nu}\cos[(2t)^{\frac{1}{2}}]$ Re $\nu < 0$
16.24	$e^{\frac{1}{2}p}D_{\nu-\frac{1}{2}}(p^{\frac{1}{2}})D_{-\nu-\frac{1}{2}}(p^{\frac{1}{2}})$	$[\pi t(t+1)(2t+1)]^{-\frac{1}{2}}$ $\cdot\cos[\nu \arccos(1+2t)^{-1}]$
16.25	$D_{n-\frac{1}{2}}[(2ap)^{\frac{1}{2}}]$ $\cdot D_{-n-\frac{1}{2}}[(2ap)^{\frac{1}{2}}]$ $n = 0,1,2,\cdots$	0 $\quad t < a$ $(\pi/a)^{-\frac{1}{2}}t^{-\frac{1}{2}}(t^2-a^2)^{-\frac{1}{2}}T_n(a/t)$ $\quad t > a$
16.26	$p^{-\frac{1}{2}}\exp[\frac{1}{4}p(a^2+b^2)]$ $\cdot D_\mu(ap^{\frac{1}{2}})D_\nu(bp^{\frac{1}{2}})$ $\mathrm{Re}(\mu+\nu) < 2$	$2^{\frac{1}{2}\mu+\frac{1}{2}\nu}[\Gamma(\frac{1}{2}-\frac{1}{2}\mu-\frac{1}{2}\nu)]^{-1}t^{-\frac{1}{2}}$ $\cdot(1+\frac{1}{2}a^2/t)^{\frac{1}{2}\mu}(1+\frac{1}{2}b^2/t)^{\frac{1}{2}\nu}$ $\cdot {}_2F_1[-\frac{1}{2}\mu,-\frac{1}{2}\nu;\frac{1}{2}-\frac{1}{2}\mu-\frac{1}{2}\nu;$ $2t(a^2+b^2+2t)(a^2+2t)^{-1}(b^2+2t)^{-1}]$
16.27	$p^{\nu+\frac{1}{2}\alpha}e^{ap}$ $\cdot D_\alpha[2(ap)^{\frac{1}{2}}]$ $\mathrm{Re}(\nu+\alpha) < 0$	$2^{-1-\nu-\frac{1}{2}\alpha}t^{-\frac{1}{2}-\frac{1}{2}\nu-\frac{1}{2}\alpha}(2a+t)^{-\frac{1}{2}-\frac{1}{2}\nu}$ $\cdot P_\nu^{\nu+1+\alpha}[(2a)^{\frac{1}{2}}(2a+t)^{-\frac{1}{2}}]$

	$g(p) = \int_0^\infty f(t)e^{-pt}dt$	$f(t)$
16.28	$p^{-\frac{1}{2}}\exp[(a+b)p]$ $\cdot D_\nu[2(ap)^{\frac{1}{2}}]D_\mu[2(bp)^{\frac{1}{2}}]$	$2^{-\frac{1}{2}}t^{-\frac{1}{4}(\nu+\mu+\frac{1}{2})}(2a+2b+t)^{\frac{1}{4}(\nu+\mu+1)}$ $\cdot(2a+t)^{\frac{1}{4}(\nu-\mu-\frac{1}{2})}(2b+t)^{-\frac{1}{4}(\nu-\mu+1)}$ $P^{\frac{1}{2}(\nu+\mu+1)}_{\frac{1}{2}(\nu-\mu-1)}[2(ab)^{\frac{1}{2}}(2a+t)^{-\frac{1}{2}}(2b+t)^{-\frac{1}{2}}]$
16.29	$D_\nu[(2aip)^{\frac{1}{2}}]D_\nu[(-2aip)^{\frac{1}{2}}]$ $\mathrm{Re}\,\nu < 0$	$(\frac{1}{2}a)^{\frac{1}{2}}[\Gamma(-\nu)]^{-1}(t^2+a^2)^{-\frac{1}{2}}t^{-\nu-1}$ $\cdot[a+(t^2+a^2)^{\frac{1}{2}}]^{\nu+\frac{1}{2}}$
16.30	$2^p\Gamma(\nu+p)D_{-2p}(a)$	$2^{-\nu}e^{\frac{1}{2}t}(e^t-1)^{-\frac{1}{2}-\nu}\exp[-\frac{1}{4}a^2e^{-t}(1-e^{-t})^{-1}]$ $\cdot D_{2\nu}[a(1-e^{-t})^{-\frac{1}{2}}]$
16.31	$p^{-\nu}\exp[\frac{1}{4}(b^2-a^2)/p]$ $\cdot\{\exp(i\frac{1}{2}ab/p)$ $\cdot D_{-2\nu}[p^{-\frac{1}{2}}(b+ia)]$ $-\exp(-i\frac{1}{2}ab/p)$ $D_{-2\nu}[p^{-\frac{1}{2}}(b-ia)]\}$ $\mathrm{Re}\,\nu > -\frac{1}{2}$	$-i2^\nu[\Gamma(2\nu)]^{-1}t^{\nu-1}$ $e^{-b(2t)^{\frac{1}{2}}}\sin[a(2t)^{\frac{1}{2}}]$

	$g(p) = \int_0^\infty f(t)e^{-pt}dt$	$f(t)$
16.32	$p^{-\nu}\exp[\frac{1}{4}(b^2-a^2)/p]$ $\cdot\{\exp(\frac{1}{2}iab/p)D_{-2\nu}[p^{-\frac{1}{2}}(b+ia)]$ $+\exp(-\frac{1}{2}iab/p)D_{-2\nu}[p^{-\frac{1}{2}}(b-ia)]\}$	$2^{\nu}[\Gamma(2\nu)]^{-1}t^{\nu-1}$ $\cdot e^{-b(2t)^{\frac{1}{2}}}\cos[a(2t)^{\frac{1}{2}}]$ $\mathrm{Re}\,\nu > 0$
16.33	$D_{\nu}(z_1^{\frac{1}{2}})D_{\nu}(z_2^{\frac{1}{2}})$ $z_{\frac{1}{2}} = 2a[b\pm(b^2-p^2)^{\frac{1}{2}}]$ $\mathrm{Re}\,\nu < 0$	$(\frac{1}{2}a)^{\frac{1}{2}}[\Gamma(-\nu)]^{-1}t^{-\nu-1}(a^2+t^2)^{-\frac{1}{2}}$ $\cdot[(a^2+t^2)^{\frac{1}{2}}+a]^{\frac{1}{2}+\nu}\exp[-b(a^2+t^2)^{\frac{1}{2}}]$
16.34	$D_{\nu}(z_1^{\frac{1}{2}})D_{\nu}(z_2^{\frac{1}{2}})$ $z_{\frac{1}{2}} = 2a[p\pm(p^2-b^2)^{\frac{1}{2}}]$ $\mathrm{Re}\,\nu < 0$	$0 \quad t < a$ $(\frac{1}{2}a)^{\frac{1}{2}}[\Gamma(-\nu)]^{-1}(t^2-a^2)^{-\frac{1}{2}\nu-1}$ $\cdot(t+a)^{\frac{1}{2}+\nu}\exp[-b(t^2-a^2)^{\frac{1}{2}}] \quad t > a$
16.35	$2^{p}\Gamma(p+\frac{1}{2}+\nu)$ $\cdot D_{-2p-1}(a)$	$2^{-\nu-\frac{1}{2}}(e^{t}-1)^{-\frac{1}{2}-\nu}\exp[\frac{1}{4}a^2(e^{t}-1)^{-1}]$ $\cdot D_{2\nu}[a(1-e^{-t})^{-\frac{1}{2}}]$
16.36	$p^{-\mu-\frac{1}{2}}e^{\frac{1}{2}ap}$ $\cdot W_{\nu,\mu}(ap)$	$a^{\frac{1}{2}-\mu}[\Gamma(\frac{1}{2}+\mu-\nu)]^{-1}t^{\mu-\nu-\frac{1}{2}}(t+a)^{\nu+\mu-\frac{1}{2}}$ $\mathrm{Re}(\mu-\nu) > -\frac{1}{2}$

	$g(p) = \int_0^\infty f(t)e^{-pt}dt$	$f(t)$
16.37	$p^{-\mu-\frac{1}{2}}W_{\nu,\mu}(ap)$ $\mathrm{Re}(\mu-\nu) > -\frac{1}{2}$	$0 \qquad t < \frac{1}{2}a$ $a^{\frac{1}{2}-\mu}[\Gamma(\frac{1}{2}+\mu-\nu)]^{-1}$ $\cdot(t-\frac{1}{2}a)^{\mu-\nu-\frac{1}{2}}(t+\frac{1}{2}a)^{\nu+\mu-\frac{1}{2}} \qquad t > \frac{1}{2}a$
16.38	$p^{-1}W_{\nu,\mu}(ap)$ $\mathrm{Re}\ \nu > 1$	$0 \qquad t < \frac{1}{2}a$ $[(2t+a)/(2t-a)]^{\frac{1}{2}\nu}P^{\nu}_{\mu-\frac{1}{2}}(2t/a) \qquad t > \frac{1}{2}a$
16.39	$p^{-1}e^{\frac{1}{2}ap}W_{\nu,\mu}(ap)$	$(1+a/t)^{\frac{1}{2}\nu}P^{\nu}_{\mu-\frac{1}{2}}(1+2t/a)$
16.40	$p^{-\mu-\frac{1}{2}}\exp[\frac{1}{2}(a-b)p]$ $\cdot W_{\nu,\mu}(ap+bp)$	$0 \qquad t < b$ $(a+b)^{\frac{1}{2}-\mu}[\Gamma(\frac{1}{2}+\mu-\nu)]^{-1}$ $\cdot(t+a)^{\nu+\mu-\frac{1}{2}}(t-b)^{\mu-\nu-\frac{1}{2}}, \qquad t > b$
16.41	$p^{\nu-\frac{1}{2}}e^{\frac{1}{2}ap}$ $\cdot W_{\nu,\mu}(ap)$ $\mathrm{Re}\ \nu < \frac{1}{4}$	$2^{-2\nu-\frac{1}{2}}a^{\frac{1}{4}}t^{-\nu-\frac{1}{4}}(a+t)^{-\frac{1}{2}}$ $\cdot P^{2\nu+\frac{1}{2}}_{2\mu-\frac{1}{2}}[(1+t/a)^{\frac{1}{2}}]$ $= \pi^{-\frac{1}{2}}2^{-2\nu}[\Gamma(-2\mu-2\nu)]^{-1}(1+t/a)^{-\frac{1}{2}}$ $\cdot t^{-\nu-\frac{1}{2}}e^{i2\pi\mu}Q^{-2\mu}_{-2\nu-1}[(1+a/t)^{\frac{1}{2}}]$

	$g(p) = \int_0^\infty f(t) e^{-pt} dt$	$f(t)$
16.42	$p^{\nu-3/2} e^{\frac{1}{2}ap}$ $\cdot W_{\nu,\mu}(ap)$ $\mathrm{Re}\, \nu < 3/4$	$2^{\frac{1}{2}-2\nu} a^{\frac{1}{4}} t^{\frac{1}{4}-\nu} P_{2\mu-\frac{1}{2}}^{2\nu-\frac{1}{2}}[(1+t/a)^{\frac{1}{2}}]$ $= \pi^{-\frac{1}{2}} 2^{1-2\nu} a^{\frac{1}{2}} [\Gamma(1-2\mu-2\nu)]^{-1}$ $\cdot t^{-\nu} e^{i2\pi\mu} Q_{-2\nu}^{-2\mu}[(1+a/t)^{\frac{1}{2}}]$
16.43	$p^{\frac{1}{2}-\mu} \exp[-\frac{1}{2}(a+b)p]$ $\cdot M_{\nu,\mu}[p(b-a)]$ $\mathrm{Re}(\mu \pm \nu) > -\frac{1}{2}$	$0 \qquad t < a$ $[B(\frac{1}{2}+\mu-\nu, \frac{1}{2}+\mu+\nu)]^{-1} (b-a)^{\frac{1}{2}-\mu}$ $\cdot (t-a)^{\mu+\nu-\frac{1}{2}} (b-t)^{\mu-\nu-\frac{1}{2}} \qquad a < t < b$ $0 \qquad t > b$
16.44	$p^{-1} e^{ap}$ $\cdot W_{\nu,0}(ap) W_{-\nu,0}(ap)$	$(1+t/a)^{-1} P_\nu[2(1+t/a)^{-2}-1]$
16.45	$p^{\frac{1}{2}} W_{\frac{1}{4},\nu}(ia/p)$ $\cdot W_{\frac{1}{4},\nu}(-ia/p)$ $-\frac{1}{4} < \mathrm{Re}\, \nu < \frac{1}{4}$	$-4\pi^{\frac{1}{2}} a [\Gamma(\frac{1}{4}+\nu)\Gamma(\frac{1}{4}-\nu)]^{-1} (2t)^{-\frac{1}{2}} K_{2\nu}[2(at)^{\frac{1}{2}}]$ $\cdot \{\sin[(\nu-\frac{1}{4})\pi] J_{2\nu}[2(at)^{\frac{1}{2}}]$ $+ \cos[(\nu-\frac{1}{4})\pi] Y_{2\nu}[2(at)^{\frac{1}{2}}]\}$
16.46	$p^{-1} W_{\nu,0}(iap) W_{-\nu,0}(-iap)$ $\mathrm{Re}\, \nu > -1$	$2a[\Gamma(1+\nu)]^{-2}$ $\cdot t^{-1} Q_{-\nu-\frac{1}{2}}(1+2a^2/t^2)$

	$g(p) = \int_0^\infty f(t)e^{-pt}dt$	$f(t)$
16.47	$p^{-\alpha}e^{ap}W_{\nu,\mu}(ap)$ $\mathrm{Re}(\alpha-\nu) > 0$	$a^{\nu}[\Gamma(\alpha-\nu)]^{-1}t^{\alpha-\nu-1}$ $\cdot {}_2F_1(\tfrac{1}{2}-\nu+\mu, \tfrac{1}{2}-\nu-\mu; \alpha-\nu; t/a)$
16.48	$p^{-1}\exp[\tfrac{1}{2}p(a+b)]$ $\cdot W_{\alpha,\beta}(ap)W_{\gamma,\beta}(bp)$ $\mathrm{Re}(\gamma+\alpha) < 1$	$(ab)^{\frac{1}{2}+\beta}[\Gamma(1-\gamma-\alpha)]^{-1}$ $\cdot t^{-\gamma-\alpha}(a+t)^{\alpha-\beta-\frac{1}{2}}(b+t)^{\gamma-\beta-\frac{1}{2}}$ $\cdot {}_2F_1(\tfrac{1}{2}+\beta-\alpha, \tfrac{1}{2}+\beta-\gamma; 1-\gamma-\alpha; z)$ $z = [(a+b)(b+t)]^{-1}t(a+b+t)$
16.49	$p^{-\alpha}W_{\nu,\mu}(iap)$ $\cdot W_{\nu,\mu}(-iap)$ $\mathrm{Re}(\alpha-2\nu) > 0$	$a^{2\nu}[\Gamma(\alpha-2\nu)]^{-1}t^{\alpha-2\nu-1}\,{}_4F_3(\tfrac{1}{2}-\nu+\mu, \tfrac{1}{2}-\nu-\mu,$ $\tfrac{1}{2}-\nu, 1-\nu; 1-2\nu, \tfrac{1}{2}\alpha-\nu, \tfrac{1}{2}+\tfrac{1}{2}\alpha-\nu; -t^2/a^2)$
16.50	$p^{-\nu}\exp(-\tfrac{1}{2}a/p)$ $\cdot W_{\nu,\mu}(a/p)$ $\mathrm{Re}(\nu\pm\mu) > -\tfrac{1}{2}$	$-a^{\frac{1}{2}}\csc(2\pi\mu)t^{\nu-\frac{1}{2}}\{\cos(\pi\nu+\pi\mu)J_{2\mu}[2(at)^{\frac{1}{2}}]$ $-J_{-2\mu}[2(at)^{\frac{1}{2}}]\cos(\pi\nu-\pi\mu)\}$
16.51	$p^{-\nu}\exp(-\tfrac{1}{2}a/p)$ $\cdot M_{\nu,\mu}(a/p)$	$a^{\frac{1}{2}}\Gamma(1+2\mu)[\Gamma(\tfrac{1}{2}+\nu+\mu)]^{-1}t^{\nu-\frac{1}{2}}J_{2\mu}[2(at)^{\frac{1}{2}}]$ $\mathrm{Re}(\nu+\mu) > -\tfrac{1}{2}$

	$g(p) = \int_0^\infty f(t)e^{-pt}dt$	$f(t)$
16.52	$p^{\nu}\exp(\tfrac{1}{2}a/p)$ $\cdot M_{\nu,\mu}(a/p)$ $\mathrm{Re}(\mu-\nu) > -\tfrac{1}{2}$	$a^{\frac{1}{2}}\Gamma(1+2\mu)[\Gamma(\tfrac{1}{2}+\mu-\nu)]^{-1}$ $\cdot t^{-\nu-\frac{1}{2}}I_{2\mu}[2(at)^{\frac{1}{2}}]$
16.53	$p^{\nu}\exp(\tfrac{1}{2}a/p)$ $\cdot W_{\nu,\mu}(a/p)$ $\mathrm{Re}(\nu\pm\mu) < \tfrac{1}{2}$	$2a^{\frac{1}{2}}[\Gamma(\tfrac{1}{2}+\mu-\nu)\Gamma(\tfrac{1}{2}-\mu-\nu)]^{-1}$ $\cdot t^{-\nu-\frac{1}{2}}K_{2\mu}[2(at)^{\frac{1}{2}}]$
16.54	$p^{-3\nu-\frac{1}{2}}\exp(\tfrac{1}{2}a/p)$ $\cdot W_{-\nu,\nu}(a/p)$ $\mathrm{Re}\,\nu > -\tfrac{1}{4}$	$2a^{1-2\nu}[\Gamma(\tfrac{1}{2}+2\nu)]^{-1}$ $\cdot t^{2\nu}I_{2\nu}(at^{\frac{1}{2}})K_{2\nu}(at^{\frac{1}{2}})$
16.55	$p^{-\nu}\exp(-a/p)$ $\cdot W_{-\nu,\mu}(ae^{i\pi}/p)$ $\mathrm{Re}(\nu\pm\mu) > -\tfrac{1}{2}$	$\pi a^{\frac{1}{2}}e^{-i\pi\mu}[\Gamma(\tfrac{1}{2}+\nu+\mu)\Gamma(\tfrac{1}{2}+\nu-\mu)]^{-1}$ $\cdot t^{\nu-\frac{1}{2}}H_{2\mu}^{(2)}[2(at)^{\frac{1}{2}}]$
16.56	$p^{-\nu}\exp(-a/p)$ $\cdot W_{-\nu,\mu}(ae^{-i\pi}/p)$ $\mathrm{Re}(\nu\pm\mu) > -\tfrac{1}{2}$	$\pi a^{\frac{1}{2}}e^{i\pi\mu}[\Gamma(\tfrac{1}{2}+\nu+\mu)\Gamma(\tfrac{1}{2}+\nu-\mu)]^{-1}$ $\cdot t^{\nu-\frac{1}{2}}H_{2\mu}^{(1)}[2(at)^{\frac{1}{2}}]$

	$g(p) = \int_0^\infty f(t)e^{-pt}dt$	$f(t)$
16.57	$p^{-\alpha}\exp(\frac{1}{2}a/p)$ $\cdot W_{\nu,\mu}(a/p)$ $\mathrm{Re}(\alpha\pm\mu) > -\frac{1}{2}$	$\Gamma(-2\mu)[\Gamma(\frac{1}{2}-\nu-\mu)\Gamma(\frac{1}{2}+\mu+\alpha)]^{-1}t^{\mu+\alpha-\frac{1}{2}}$ $\cdot {}_1F_2(\frac{1}{2}-\nu+\mu;1+2\mu,\frac{1}{2}+\alpha+\mu;at)$ $+\Gamma(2\mu)[\Gamma(\frac{1}{2}-\nu+\mu)\Gamma(\frac{1}{2}+\alpha-\mu)]^{-1}t^{\alpha-\mu-\frac{1}{2}}$ $\cdot {}_1F_2(\frac{1}{2}-\nu-\mu;1-2\mu,\frac{1}{2}+\alpha-\mu;at)$
16.58	$e^{ap}M_{\nu,\frac{1}{4}}(z_2)W_{-\nu,\frac{1}{4}}(z_1)$ $z_{\frac{1}{2}} = a[(p^2+b^2)^{\frac{1}{2}}\pm p]$ $\mathrm{Re}\,\nu > -\frac{3}{4}$	$a(\frac{1}{2}b)^{\frac{1}{2}}[\Gamma(\frac{3}{4}+\nu)]^{-1}(2a+t)^{-2\nu}$ $\cdot(t^2+2at)^{\nu-\frac{3}{4}}\sin[b(t^2+2at)^{\frac{1}{2}}]$
16.59	$e^{ap}M_{\nu,-\frac{1}{4}}(z_2)W_{-\nu,\frac{1}{4}}(z_1)$ $z_{\frac{1}{2}} = a[(p^2+b^2)^{\frac{1}{2}}\pm p]$ $\mathrm{Re}\,\nu > -\frac{1}{4}$	$a(2b)^{\frac{1}{2}}[\Gamma(\frac{1}{4}+\nu)]^{-1}(2a+t)^{-2\nu}$ $\cdot(t^2+2at)^{\nu-\frac{3}{4}}\cos[b(t^2+2at)^{\frac{1}{2}}]$
16.60	$M_{\nu,\frac{1}{4}}(z_2)W_{-\nu,\frac{1}{4}}(z_1)$ $z_{\frac{1}{2}} = a[(p^2+b^2)^{\frac{1}{2}}\pm p]$ $\mathrm{Re}\,\nu > -\frac{3}{4}$	$0 \quad t < a$ $a(\frac{1}{2}b)^{\frac{1}{2}}[\Gamma(\frac{3}{4}+\nu)]^{-1}(a+t)^{-2\nu}$ $\cdot(t^2-a^2)^{\nu-\frac{3}{4}}\sin[b(t^2-a^2)^{\frac{1}{2}}] \quad t > a$

	$g(p) = \int_0^\infty f(t)e^{-pt}dt$	$f(t)$
16.61	$M_{\nu,-\frac{1}{4}}(z_1)W_{-\nu,\frac{1}{4}}(z_2)$ $z_{\frac{1}{2}} = a[(p^2+b^2)^{\frac{1}{2}}\pm p]$ $\mathrm{Re}\,\nu > \frac{1}{4}$	$0 \qquad t < a$ $a(2b)^{\frac{1}{2}}[\Gamma(\frac{1}{4}+\nu)]^{-1}(a+t)^{-2\nu}$ $\cdot(t^2-a^2)^{\nu-3/4}\cos[b(t^2-a^2)^{\frac{1}{2}}]$ $t > a$
16.62	$e^{ap}M_{\nu,\mu}(z_2)W_{-\nu,\mu}(z_1)$ $z_{\frac{1}{2}} = a[(p^2+b^2)^{\frac{1}{2}}\pm p]$ $\mathrm{Re}(\mu+\nu) > -\frac{1}{2}$	$ab\Gamma(1+2\mu)[\Gamma(\nu+\mu+\frac{1}{2})]^{-1}t^{\nu-\frac{1}{2}}$ $\cdot(t+2a)^{-\nu-\frac{1}{2}}J_{2\mu}[b(t^2+2at)^{\frac{1}{2}}]$
16.63	$e^{ap}M_{\nu,\mu}(z_2)W_{\nu,\mu}(z_1)$ $z_{\frac{1}{2}} = a[p\pm(p^2-b)^{\frac{1}{2}}]$ $\mathrm{Re}(\mu-\nu) > -\frac{1}{2}$	$ab\Gamma(1+2\mu)[\Gamma(\mu-\nu+\frac{1}{2})]^{-1}t^{-\nu-\frac{1}{2}}$ $\cdot(t+2a)^{\nu-\frac{1}{2}}I_{2\mu}[b(t^2+2at)^{\frac{1}{2}}]$
16.64	$e^{ap}W_{\nu,\mu}(z_1)W_{\nu,\mu}(z_2)$ $z_{\frac{1}{2}} = a[p\pm(p^2-b^2)^{\frac{1}{2}}]$ $\mathrm{Re}(\nu\pm\mu) < \frac{1}{2}$	$2ab[\Gamma(\mu-\nu+\frac{1}{2})\Gamma(\frac{1}{2}-\nu-\mu)]^{-1}t^{-\nu-\frac{1}{2}}$ $\cdot(t+2a)^{\nu-\frac{1}{2}}K_{2\mu}[b(t^2+2at)^{\frac{1}{2}}]$

	$g(p) = \int_0^\infty f(t) e^{-pt} dt$	$f(t)$
16.65	$M_{\nu,\mu}(z_2) W_{\nu,\mu}(z_1)$ $z_{\frac{1}{2}} = a[p \pm (p^2-b^2)^{\frac{1}{2}}]$ $\operatorname{Re}(\mu-\nu) > -\frac{1}{2}$	$0 \qquad t < a$ $ab\Gamma(1+2\mu)[\Gamma(\frac{1}{2}+\mu-\nu)]^{-1}$ $\cdot (t-a)^{-\nu-\frac{1}{2}}(t+a)^{\nu-\frac{1}{2}} I_{2\mu}[b(t^2-a^2)^{\frac{1}{2}}]$ $t > a$
16.66	$W_{\nu,\mu}(z_1) W_{\nu,\mu}(z_2)$ $z_{\frac{1}{2}} = a[p \pm (p^2-b^2)^{\frac{1}{2}}]$ $\operatorname{Re}(\nu \pm \mu) < \frac{1}{2}$	$0 \qquad t < a$ $2ab[\Gamma(\frac{1}{2}+\mu-\nu)\Gamma(\frac{1}{2}-\mu-\nu)]^{-1}(t-a)^{-\nu-\frac{1}{2}}$ $\cdot (t+a)^{\nu-\frac{1}{2}} K_{2\nu}[b(t^2-a^2)^{\frac{1}{2}}] \qquad t > a$
16.67	$p^{-2\nu-1} \exp(\frac{1}{2}ap^2)$ $\cdot W_{-3\nu,\nu}(ap^2)$ $\operatorname{Re} \nu > -\frac{1}{8}$	$2^{8\nu}\Gamma(1+2\nu)[\Gamma(1+8\nu)]^{-1}$ $\cdot t^{6\nu} \exp(-\frac{1}{8}t^2/a) I_\nu(\frac{1}{8}t^2/a)$
16.68	$p^{-2\mu} \exp(\frac{1}{2}ap^2)$ $\cdot W_{\nu,\mu}(ap^2)$ $\operatorname{Re}(\mu-\nu) > 0$	$2^{\mu-\nu} a^{\frac{1}{2}\nu+\frac{1}{2}\mu} [\Gamma(2\mu-2\nu)]^{-1} t^{\mu-\nu-1}$ $\cdot \exp(-\frac{1}{8}t^2/a) M_{\alpha,\beta}(\frac{1}{4}t^2/a)$ $\alpha = \frac{1}{2}(\mu-\nu-1), \ \beta = \frac{1}{2}(1-\nu-3\mu)$

	$g(p) = \int_0^\infty f(t)e^{-pt}dt$	$f(t)$
16.69	$p^{-2\mu-1}\exp(\frac{1}{2}ap^2)$ $\cdot W_{\nu,\mu}(ap^2)$ $\mathrm{Re}(\mu-\nu) > -\frac{1}{2}$	$2^{1+\mu-\nu}a^{\frac{1}{2}+\frac{1}{2}\nu+\frac{1}{2}\mu}[\Gamma(1+2\mu-2\nu)]^{-1}$ $\cdot t^{\mu-\nu-1}\exp(-\frac{1}{8}t^2/a)M_{\alpha,\beta}(\frac{1}{4}t^2/a)$ $\alpha = \frac{1}{2}(\mu-\nu),\ \beta = -\frac{1}{2}(\nu+3\mu)$
16.70	$\Gamma(p+\mu)W_{-p,\nu}(a)$	$ae^t(e^t-1)^{-\mu-1}\exp[-\frac{1}{2}a(e^t-1)^{-1}]$ $\cdot W_{\mu,\nu}[a(e^t-1)^{-1}]$
16.71	$\Gamma(\frac{1}{2}+\nu+p)\Gamma(\frac{1}{2}-\nu+p)$ $\cdot[\Gamma(1-\mu+p)]^{-1}W_{-p,\nu}(a)$	$e^{-\frac{1}{2}a}(1-e^{-t})^{-\mu}\exp[-\frac{1}{2}a(e^t-1)^{-1}]$ $\cdot W_{\mu,\nu}[a(e^t-1)^{-1}]$
16.72	$a^pW_{\nu-p,p}(a)$ $\mathrm{Re}\ \nu < \frac{1}{2}$	$\frac{1}{2}a^{\frac{1}{2}}[\Gamma(\frac{1}{2}-\nu)]^{-1}(1-e^{-\frac{1}{2}t})^{\nu-1}\exp(-ae^{\frac{1}{2}t})$
16.73	$\Gamma(\frac{1}{2}+\nu+cp)$ $\cdot M_{cp,\nu}(a)W_{-cp,\nu}(b)$	$\frac{1}{2}c^{-1}(ab)^{\frac{1}{2}}\Gamma(1+2\nu)\mathrm{csch}(\frac{1}{2}t/c)$ $\cdot\exp[(a-b)\coth(\frac{1}{2}t/c)]$ $\cdot J_{2\nu}[(ab)^{\frac{1}{2}}\mathrm{csch}(\frac{1}{2}t/c)]$
16.74	$\Gamma(\frac{1}{2}+\nu+p)\Gamma(\frac{1}{2}-\nu+p)$ $\cdot W_{-p,\nu}(ia)W_{-p,\nu}(-ia)$	$a\ \mathrm{csch}(\frac{1}{2}t)K_{2\nu}[a\ \mathrm{csch}(\frac{1}{2}t)]$

	$g(p) = \int_0^\infty f(t)e^{-pt}dt$	$f(t)$
16.75	$\Gamma(\tfrac{1}{2}+\nu+p)\Gamma(\tfrac{1}{2}-\nu+p)$ $\cdot W_{-p,\nu}(a)W_{-p,\nu}(b)$	$(ab)^{\frac{1}{2}}\exp[\tfrac{1}{2}(a-b)]$ $\cdot \operatorname{csch}(\tfrac{1}{2}t)\exp[-(ae^{t}+b)(e^{t}-1)^{-1}]$ $\cdot K_{2\nu}[(ab)^{\frac{1}{2}}\operatorname{csch}(\tfrac{1}{2}t)]$
16.76	$\Gamma(\tfrac{1}{2}+\nu+p)\Gamma(\tfrac{1}{2}-\nu+p)$ $\cdot W_{-p,\nu}(a)W_{p,\nu}(b)$	$(ab)^{\frac{1}{2}}\operatorname{csch}(\tfrac{1}{2}t)\exp[\tfrac{1}{2}(a+b)\coth(\tfrac{1}{2}t)]$ $\cdot K_{2\nu}[(ab)^{\frac{1}{2}}\operatorname{csch}(\tfrac{1}{2}t)]$
16.77	$\Gamma(\tfrac{1}{2}-\nu-\mu+2p)[\Gamma(1+2p)]^{-1}$ $\cdot W_{\nu-p,\mu-p}(a)$ $\operatorname{Re}\mu > -\tfrac{1}{2}$	$\tfrac{1}{2}[\Gamma(1+2\mu)]^{-1}(e^{\frac{1}{2}t}-1)^{\mu-\frac{1}{2}}$ $\cdot \exp(-\tfrac{1}{2}ae^{\frac{1}{2}t})M_{-\nu,\mu}[a(e^{\frac{1}{2}t}-1)]$
16.78	$\Gamma(\tfrac{1}{2}+\nu+p)\Gamma(\tfrac{1}{2}-\nu+p)$ $\cdot W_{-p,\nu}(ia)W_{-p,\nu}(-ia)$	$a\operatorname{csch}(\tfrac{1}{2}t)K_{2\nu}[a\operatorname{csch}(\tfrac{1}{2}t)]$

2.17 Elliptic Integrals and Elliptic Functions

	$g(p) = \int_0^\infty f(t)e^{-pt}dt$	$f(t)$
17.1	$p^{-1}K(a/p)$	$\frac{1}{2}\pi I_0^2(\frac{1}{2}at)$
17.2	$K(a/p)-\frac{1}{2}\pi$	$\frac{1}{2}\pi a I_0(\frac{1}{2}at) I_1(\frac{1}{2}at)$
17.3	$p[\frac{1}{2}\pi-E(a/p)]$	$\frac{1}{2}\pi a t^{-1} I_0(at) I_1(at)$
17.4	$p[K(a/p)-\frac{1}{2}\pi]$	$\frac{1}{8}\pi a^2[I_0^2(\frac{1}{2}at)+2I_1^2(\frac{1}{2}at)+I_0(\frac{1}{2}at)I_2(\frac{1}{2}at)]$
17.5	$p[K(a/p)-E(a/p)]$	$\frac{1}{4}\pi a^2[I_0^2(\frac{1}{2}at)+I_1^2(\frac{1}{2}at)]$
17.6	$p(p^2-a^2)^{-1}E(a/p)$	$\frac{1}{2}\pi I_0(\frac{1}{2}at)[I_0(\frac{1}{2}at)+at\ I_1(\frac{1}{2}at)]$
17.7	$p(p^2-a^2)^{-1}E(a/p)$ $-K(a/p)$	$\frac{1}{4}\pi a^2 t[I_0^2(\frac{1}{2}at)+I_1^2(\frac{1}{2}at)]$
17.8	$p(p^2-a^2)^{-1}E(a/p)$ $-p^{-1}K(a/p)$	$\frac{1}{2}\pi a t I_0(\frac{1}{2}at) I_1(\frac{1}{2}at)$
17.9	$(2p-a^2/p)K(a/p)$ $-2pE(a/p)$	$\frac{1}{2}\pi a^2 I_1^2(\frac{1}{2}at)$

	$g(p) = \int_0^\infty f(t)e^{-pt}dt$	$f(t)$
17.10	$(p^2+a^2)^{-\frac{1}{2}}$ $\cdot K[a(p^2+a^2)^{-\frac{1}{2}}]$	$\frac{1}{2}\pi J_0^2(\frac{1}{2}at)$
17.11	$(p^2+a^2)^{-\frac{1}{2}}$ $\cdot E[a(p^2+a^2)^{-\frac{1}{2}}]$	$\frac{1}{2}\pi J_0(\frac{1}{2}at)[J_0(\frac{1}{2}at)-at\ J_1(\frac{1}{2}at)$
17.12	$(p-a)^{-\frac{1}{2}}E[2a(p+a)^{\frac{1}{2}}]$	$(\pi t)^{\frac{1}{2}}I_0(at)$
17.13	$(p+a)^{-\frac{1}{2}}K[2a(p+a)^{-\frac{1}{2}}]$	$\frac{1}{2}(\pi/t)^{\frac{1}{2}}I_0(at)$
17.14	$[p^2+(a+b)^2]^{-\frac{1}{2}}$ $\cdot K\{2(ab)^{\frac{1}{2}}[p^2+(a+b)^2]^{-\frac{1}{2}}\}$	$\frac{1}{2}\pi J_0(at)J_0(bt)$
17.15	$[p^2-(a-b)^2]^{-\frac{1}{2}}$ $\cdot K\{2(ab)^{\frac{1}{2}}[p^2-(a-b)^2]^{-\frac{1}{2}}\}$	$\frac{1}{2}\pi I_0(at)I_0(bt)$
17.16	$p^{-1}K[(1-a^2/p^2)^{\frac{1}{2}}]$	$I_0(\frac{1}{2}at)K_0(\frac{1}{2}at)$
17.17	$[p^2-(a-b)^2]^{-\frac{1}{2}}$ $\cdot K\{[\frac{p^2-(a+b)^2}{p^2-(a-b)^2}]\}$	$\frac{1}{2}[I_0(at)K_0(at)+I_0(bt)K_0(at)]$

	$g(p) = \int_0^\infty f(t)e^{-pt}dt$	$f(t)$
17.18	$(p^2+a^2)^{-\frac{1}{2}}K[p(p^2+a^2)^{-\frac{1}{2}}]$	$-\frac{1}{2}\pi J_0(\frac{1}{2}at)Y_0(\frac{1}{2}at)$
17.19	$[p^2+(a+b)^2]^{-\frac{1}{2}}$ $\cdot K\{[\frac{p^2+(a-b)^2}{p^2+(a+b)^2}]\}$	$-\frac{1}{4}\pi[J_0(at)Y_0(bt)+Y_0(at)J_0(bt)]$
17.20	$(p^2+a^2)^{-\frac{1}{4}}$ $\cdot K\{[\frac{1}{2}-\frac{1}{2}p(p^2+a^2)^{-\frac{1}{2}}]^{\frac{1}{2}}\}$	$\frac{1}{2}\pi^{\frac{1}{2}}t^{-\frac{1}{2}}J_0(at)$
17.21	$a^{-1}-2(\pi a^{-1})(p^2+a^2)^{-\frac{1}{2}}$ $\cdot pK[a(p^2+a^2)^{-\frac{1}{2}}]$	$J_0(\frac{1}{2}at)J_1(\frac{1}{2}at)$
17.22	$(p^2+a^2)^{\frac{1}{2}}E[a(p^2+a^2)^{-\frac{1}{2}}]$ $-\frac{1}{2}\pi p$	$\pi at^{-1}J_0(at)J_1(at)$
17.23	$r^{-3/2}\{2E[(\frac{1}{2}-\frac{1}{2}p/r)^{\frac{1}{2}}]$ $-K[(\frac{1}{2}-\frac{1}{2}p/r)^{\frac{1}{2}}]\}$	$(\pi t)^{\frac{1}{2}}J_0(at)$ $r = (p^2+a^2)^{\frac{1}{2}}$
17.24	$s^{-2}\{p(p+s)^{-\frac{1}{2}}$ $\cdot K[(2s)^{\frac{1}{2}}(p+s)^{-\frac{1}{2}}]$ $-(p+s)^{\frac{1}{2}}E[(2s)^{\frac{1}{2}}(p+s)^{-\frac{1}{2}}]\}$	$(\frac{1}{2}\pi)^{-\frac{1}{2}}t^{\frac{1}{2}}K_0(at)$ $s = (p^2-a^2)^{\frac{1}{2}}$

	$g(p) = \int_0^\infty f(t)e^{-pt}dt$	$f(t)$
17.25	$(p^2+a^2)^{-\frac{1}{2}}\{K[a(p^2+a^2)^{-\frac{1}{2}}]$ $-E[a(p^2+a^2)^{-\frac{1}{2}}]\}$	$\frac{1}{2}\pi at J_0(\frac{1}{2}at)J_1(\frac{1}{2}at)$
17.26	$r^{-1}(2p^2+a^2)K(a/r)$ $-2rE(a/r)$	$\frac{1}{2}\pi a^2 J_1^2(\frac{1}{2}at)$ $r = (p^2+a^2)^{\frac{1}{2}}$
17.27	$r^{-1}\{p^2[K(a/r)-E(a/r)]$ $-a^2E(a/r)\}$	$\frac{1}{2}\pi a^2[J_1^2(\frac{1}{2}at)-J_0^2(\frac{1}{2}at)]$
17.28	$z^{-1}(p^2+a^2+b^2)K[2z(ab)^{\frac{1}{2}}]$ $-zE[2z(ab)^{\frac{1}{2}}]$	$\pi ab\ J_1(at)\ J_1(bt)$ $z = [p^2+(a+b)^2]^{\frac{1}{2}}$
17.29	$z^{-1}(p^2-a^2-b^2)K[2z(ab)^{\frac{1}{2}}]$ $-zE[2z(ab)^{\frac{1}{2}}]$	$\pi ab\ I_1(at)\ I_1(bt)$ $z = [p^2-(a-b)^2]^{\frac{1}{2}}$
17.30	$p^{-1}\ \theta_2(0\|p)$ $n = 0,1,2,\cdots$	$0 \quad 0 < t < \pi^2/4$ $2n+2 \quad \pi^2(n+\frac{1}{2})^2<t<\pi^2(n+\frac{3}{2})^2$
17.31	$p^{-1}\theta_3(0\|p)$	$2n+1 \quad n^2\pi^2<t<(n+1)^2\pi^2$ $n = 0,1,2,\cdots$

	$g(p) = \int_0^\infty f(t)e^{-pt}dt$	$f(t)$
17.32	$p^{-1}\theta_4(0\|p)$ $n = 0,1,2,\cdots$	$1 \qquad (2n)^2\pi^2 < t < (2n+1)^2\pi^2$ $-1 \qquad (2n+1)^2\pi^2 < t < (2n+2)^2\pi^2$
17.33	$p^{-1}\theta_1(a\|p)$	$\pi^{-1}\sum_{n=-\infty}^{\infty}(-1)^n(a-\frac{1}{2}+n)^{-1}\sin[2t^{\frac{1}{2}}(a-\frac{1}{2}+n)]$ $= 2\sum_{n=0}^{k}(-1)^n\sin[(2n+1)\pi a]$ $k = [-\frac{1}{2}+t^{\frac{1}{2}}/\pi]$ if $-\frac{1}{2}+t^{\frac{1}{2}}/\pi < 0; \quad f(t) = 0$
17.34	$p^{-1}\theta_2(a\|p)$	$\pi^{-1}\sum_{n=-\infty}^{\infty}(-1)^n(a+n)^{-1}\sin[2t^{\frac{1}{2}}(a+n)]$ $= 2\sum_{n=0}^{k}\cos[(2n+1)a]$ k as in (33)
17.35	$p^{-1}\theta_3(a/p)$	$\pi^{-1}\sum_{n=-\infty}^{\infty}(a+n)^{-1}\sin[2t^{\frac{1}{2}}(a+n)]$ $= \sum_{n=0}^{k}\varepsilon_n\cos(2\pi na)$ $k = [t^{\frac{1}{2}}/\pi]$

	$g(p) = \int_0^\infty f(t)e^{-pt}dt$	$f(t)$
17.36	$p^{-1}\theta_4(a\|p)$	$\pi^{-1}\sum_{n=-\infty}^{\infty}(a+\frac{1}{2}+n)^{-1}\sin[2t^{\frac{1}{2}}(a+\frac{1}{2}+n)]$ $=\sum_{n=0}^{k}(-1)^n\varepsilon_n\cos(2\pi na)$ k as in (35)
17.37	$p^{-\nu}\theta_1(a\|p)$ $\mathrm{Re}\,\nu > 0$	$\pi^{-\frac{1}{2}}t^{\frac{1}{2}\nu-\frac{1}{4}}\sum_{n=-\infty}^{\infty}(-1)^n(a-\frac{1}{2}+n)^{\frac{1}{2}-\nu}$ $\cdot J_{\nu-\frac{1}{2}}[2t^{\frac{1}{2}}(a-\frac{1}{2}+n)]$ $=2[\Gamma(\nu)]^{-1}\sum_{n=0}^{k}(-1)^n\sin[(2n+1)\pi a]$ $\cdot[t-(n+\frac{1}{2})^2\pi^2]^{\nu-1}$ k as in (33)
17.38	$p^{-\nu}\theta_2(a\|p)$ $\mathrm{Re}\,\nu > 0$	$\pi^{-\frac{1}{2}}t^{\frac{1}{2}\nu-\frac{1}{4}}\sum_{n=-\infty}^{\infty}(-1)^n(a+n)^{\frac{1}{2}-\nu}J_{\nu-\frac{1}{2}}[2t^{\frac{1}{2}}(a+n)]$ $=2[\Gamma(\nu)]^{-1}\sum_{n=0}^{k}\cos[(2n+1)\pi a]$ $\cdot[t-(n+\frac{1}{2})^2\pi^2]^{\nu-1}$ k as in (33)

	$g(p) = \int_0^\infty f(t)e^{-pt}dt$	$f(t)$
17.39	$p^{-\nu}\theta_3(a\|p)$ $\text{Re } \nu > 0$	$\pi^{-\frac{1}{2}}t^{\frac{1}{2}\nu-\frac{1}{4}}\sum_{n=-\infty}^{\infty}(a+n)^{\frac{1}{2}-\nu}J_{\nu-\frac{1}{2}}[2t^{\frac{1}{2}}(a+n)]$ $=[\Gamma(\nu)]^{-1}\sum_{n=0}^{k}\varepsilon_n\cos(2\pi na)(t-n^2\pi^2)^{\nu-1}$ k as in (35)
17.40	$p^{-\nu}\theta_4(a\|p)$ $\text{Re } \nu > 0$	$\pi^{-\frac{1}{2}}t^{\frac{1}{2}\nu-\frac{1}{4}}\sum_{n=-\infty}^{\infty}(a+\frac{1}{2}+n)^{\frac{1}{2}-\nu}J_{\nu-\frac{1}{2}}[2t^{\frac{1}{2}}(a+\frac{1}{2}+n)]$ $=[\Gamma(\nu)]^{-1}\sum_{n=0}^{k}(-1)^n\varepsilon_n\cos(2\pi na)(t-n^2\pi^2)^{\nu-1}$ k as in (35)

2.18 Gauss' Hypergeometric Functions

	$g(p) = \int_0^\infty f(t)e^{-pt}dt$	$f(t)$
18.1	${}_2F_1(a,b;c;-p/d)$ $\mathrm{Re}(a,b) > 0$	$d^{\frac{1}{2}(a+b-5)} t^{\frac{1}{2}(a+b-3)} e^{-\frac{1}{2}dt}$ $\cdot W_{\frac{1}{2}(a+b+1)-c,\frac{1}{2}a-\frac{1}{2}b}(dt)$
18.2	${}_2F_1(a,b;c;\tfrac{1}{2}-p/d)$ $\mathrm{Re}(a,b) > 0$	$d[\Gamma(a)\Gamma(b)]^{-1}\Gamma(c)(dt)^{\frac{1}{2}(a+b-3)}$ $\cdot W_{\frac{1}{2}(a+b+1)-c,\frac{1}{2}(a-b)}(dt)$
18.3	$p^{-\nu}{}_2F_1(\nu,b;c;a/p)$ $\mathrm{Re}\ \nu > 0$	$a^{-\frac{1}{2}c} t^{\nu-1-\frac{1}{2}c} M_{\frac{1}{2}c-b,\frac{1}{2}c-\frac{1}{2}}(at)$
18.4	$p^{-\nu}{}_2F_1(\nu,b;2\nu-2;a/p)$ $\mathrm{Re}\ \nu > 0$	$[\Gamma(\nu)]^{-1} a^{1-\nu} e^{\frac{1}{2}at}$ $\cdot M_{\nu-b-1,\nu-3/2}(at)$
18.5	$p^{-a-\nu-1}$ $\cdot {}_2F_1(a+\nu+1,\mu+\nu;2\nu;2a/p)$ $\mathrm{Re}(a+\nu) > -1$	$(2a)^{-\nu}[\Gamma(a+\nu+1)]^{-1} t^a e^{at}$ $\cdot M_{-\mu,\nu-\frac{1}{2}}(2at)$
18.6	$p^{-2\nu-1}{}_2F_1(\nu,\nu+1;2\nu;-a^2p^{-2})$ $\mathrm{Re}\ \nu > -\frac{1}{2}$	$[2\nu B(\nu,\nu)]^{-1}\pi a^{1-2\nu}$ $\cdot t\ J^2_{\nu-\frac{1}{2}}(\tfrac{1}{2}at)$

	$g(p) = \int_0^\infty f(t) e^{-pt} dt$	$f(t)$
18.7	$p^{1-4\nu} {}_2F_1(\nu, 2\nu-\frac{1}{2}; \nu+\frac{1}{2}; -a^2/p^2)$ $\mathrm{Re}\,\nu > \frac{1}{4}$	$\pi\Gamma(\frac{1}{2}+\nu)[\Gamma(\nu)\Gamma(2\nu-\frac{1}{2})]^{-1}$ $\cdot(\frac{1}{2}t/a)^{2\nu-1} J^2_{\nu-\frac{1}{2}}(\frac{1}{2}at)$
18.8	$p^{-2\nu} {}_2F_1(\nu, \frac{1}{2}+\nu; \mu; a^2/p^2)$ $\mathrm{Re}(\nu,\mu) > 0$	$\Gamma(\mu)[\Gamma(2\nu)]^{-1} a^{1-\mu}(2t)^{2\nu-\mu} I_{\mu-1}(at)$
18.9	$p^{-2\nu} {}_2F_1(\nu, \frac{1}{2}+\nu; \mu; -a^2/p^2)$ $\mathrm{Re}(\nu,\mu) > 0$	$\Gamma(\mu)[\Gamma(2\nu)]^{-1} a^{1-\mu}(2t)^{2\nu-\mu} J_{\mu-1}(at)$
18.10	$(p^2+c^2)^{-\nu} {}_2F_1[\nu, b; \frac{1}{2}+\nu+b;$ $c^2(p^2+c^2)^{-1}]$ $\mathrm{Re}\,\nu > 0$	$(\frac{1}{2}c)^{-2b}[\Gamma(2\nu)]^{-1}\Gamma(\frac{1}{2}+\nu+b)$ $\cdot(\frac{1}{2}c/t)^{\frac{1}{2}-\nu+b} J_{\nu+b-\frac{1}{2}}(ct)$
18.11	$(p+h)^{-\nu} {}_2F_1(\nu, b; c; \frac{p-h}{p+h})$ $\mathrm{Re}\,\nu>0,\ \mathrm{Re}(c-b)>0$	$\Gamma(c)[\Gamma(\nu)\Gamma(c-b)]^{-1}(2h)^{\frac{1}{2}(c-1-\nu-b)}$ $\cdot t^{\frac{1}{2}(\nu-b+c-3)} W_{\frac{1}{2}(\nu-b-c+1),\frac{1}{2}(\nu+b-c)}(2ht)$
18.12	$(p+h)^{-\nu} {}_2F_1(\nu, b; c; \frac{2h}{p+h})$ $\mathrm{Re}\,\nu > 0$	$(2h)^{-\frac{1}{2}c}[\Gamma(\nu)]^{-1} t^{\nu-1-\frac{1}{2}c}$ $\cdot M_{\frac{1}{2}c-b,\frac{1}{2}c-\frac{1}{2}}(2ht)$

	$g(p) = \int_0^\infty f(t)e^{-pt}dt$	$f(t)$
18.13	$B(p,\nu)\,{}_2F_1(a,b;p+\nu;h)$ $\mathrm{Re}\ \nu > 0$	$(1-e^{-t})^{\nu-1}\,{}_2F_1[a,b;\nu;h(1-e^{-t})]$
18.14	$B(p,\nu)\,{}_2F_1(\mu,p;p+\nu;h)$ $\mathrm{Re}\ \nu > 0$	$(1-he^{-t})^{-\mu}(1-e^{-t})^{\nu-1}$
18.15	$B(\nu,p-\nu)\,{}_2F_1(a,\nu;p;c)$ $\mathrm{Re}\ \nu > 0$	$[1-c(1-e^{-t})]^{-a}e^{t}(e^{t}-1)^{\nu-1}$
18.16	$p^{-m-n-\frac{1}{2}}(p-a)^{n}(p-b)^{m}$ $\cdot\,{}_2F_1(-m,-n;\tfrac{1}{2}-m-n;z)$ $z = \dfrac{p(p-a-b)}{(p-a)(p-b)}$	$\pi^{-\frac{1}{2}}[(2m+2n)!]^{-1}(-2)^{m+n}(m+n)!$ $\cdot t^{-\frac{1}{2}}He_{2n}[(2at)^{\frac{1}{2}}]He_{2m}[(2bt)^{\frac{1}{2}}]$ $m,\ n = 0,1,2,\cdots$
18.17	$p^{-m-n-3/2}(p-a)^{n}(p-b)^{m}$ $\cdot\,{}_2F_1(-m,-n;-\tfrac{1}{2}-m-n;z)$ $z = \dfrac{p(p-a-b)}{(p-a)(p-b)}$	$-(ab\pi)^{-\frac{1}{2}}[(2m+2n+2)]^{-1}(m+n+1)!$ $\cdot(-2)^{m+n+1}t^{-\frac{1}{2}}He_{2n+1}[(2at)^{\frac{1}{2}}]$ $\cdot He_{2m+1}[(2bt)^{\frac{1}{2}}]$ $m,\ n = 0,1,2,\cdots$

	$g(p) = \int_0^\infty f(t)e^{-pt}dt$	$f(t)$
18.18	$p^{-m-n-a-1}(p-b)^n(p-c)^m$ $\cdot {}_2F_1(-m,-n;-m-n-a;z)$ $z = \frac{p(p-b-c)}{(p-b)(p-c)}$	$m!n![\Gamma(m+n+a+1)]^{-1}$ $\cdot t^a L_n^a(bt) L_m^a(ct)$ Re $a > -1$, $m, n=0,1,2,\cdots$
18.19	$p^{-\nu}{}_2F_1[-n,\frac{1}{2}\nu;\frac{1}{2}-n;(1-1/p)^2]$ Re $\nu>0$, $n = 0,1,2,\cdots$	$2^{1-\nu}[\Gamma(\frac{1}{2}\nu)\Gamma(\frac{1}{2}+n)]^{-1}\pi(n!)^2$ $\cdot t^{\nu-1}[L_n^{\frac{1}{2}\nu-\frac{1}{2}}(\frac{1}{2}t)]^2$
18.20	$p^{-n-\nu}(a-p)^n$ $\cdot {}_2F_1(n+\nu,\frac{1}{2}-\nu;n+1;1-a/p)$ Re $\nu > 0$	$\pi^{-\frac{1}{2}}[\Gamma(2n+2\nu)]^{-1}n!2^{2n+\nu-\frac{1}{2}}$ $\cdot t^{\nu-1}e^{\frac{1}{2}at}D_{2\nu+2n-1}[(2at)^{\frac{1}{2}}]$ $n = 0,1,2,\cdots$
18.21	$p^{-\nu}(p-1)^n$ $\cdot {}_2F_1(-n,a;1-\nu;\frac{p}{p-1})$ $n = 0,1,2,\cdots$	$n![\Gamma(\nu)]^{-1}t^{\nu-1-n}L_n^{\nu+a-n-1}(t)$ Re $\nu >$ Max$(n,n-$Re $a)$
18.22	$p^{-\nu}{}_2F_1(-n,b+n;c;a/p)$ Re $\nu > -1$, $n=0,1,2,\cdots$	$[\Gamma(\nu)]^{-1}t^{\nu-1}{}_2F_2(-n,b+n;c;\nu;at)$
18.23	$p^{-\nu}{}_2F_1(-\mu,\nu;c;a/p)$ Re $\nu > 0$, $n = 0,1,2,\cdots$	$n!\Gamma(c)[\Gamma(\nu)\Gamma(n+c)]^{-1}t^{\nu-1}L_n^{c-1}(at)$

2.19 Generalized Hypergeometric Functions

	$g(p) = \int_0^\infty f(t)e^{-pt}dt$	$f(t)$
19.1	$p^{-\nu}\Phi_2(b,\nu,c;x,y/p)$ $\mathrm{Re}\ \nu > 0$	$[\Gamma(\nu)]^{-1}t^{\nu-1}\Phi_3(b,c;x,yt)$
19.2	$p^{-\nu}Y_1(a,\nu,c,d;x/p,y)$ $\mathrm{Re}\ \nu > 0$	$[\Gamma(\nu)]^{-1}t^{\nu-1}Y_2(a,c,d;xt,y)$
19.3	$p^{-\nu}Y_3(a,\nu,c,d;x,y/p)$ $\mathrm{Re}\ \nu > 0$	$[\Gamma(\nu)]^{-1}t^{\nu-1}Y_4(a,c,d;x,yt)$
19.4	$p^{-\nu}Y_3(a,b,\nu,c;x/p,y)$ $\mathrm{Re}\ \nu > 0$	$[\Gamma(\nu)]^{-1}t^{\nu-1}\Phi_2(a,b,c;xt,y)$
19.5	$p^{-\nu}Y_4(\nu,b,c;x/p,y)$ $\mathrm{Re}\ \nu > 0$	$[\Gamma(\nu)]^{-1}t^{\nu-1}\Phi_3(b,c;xt,y)$
19.6	$p^{-2\nu}Y_3(\nu,b,\tfrac{1}{2}+\nu,c;4yp^{-2},x)$ $\mathrm{Re}\ \nu > 0$	$[\Gamma(2\nu)]^{-1}t^{2\nu-1}\Phi_3(b,c;x,yt^2)$

	$g(p) = \int_0^\infty f(t)e^{-pt}dt$	$f(t)$
19.7	$\Gamma(p)[\Gamma(\nu+p)]^{-1}$ $\cdot\Phi_1(p,\mu,\nu;c,a)$ Re $\nu > 0$	$[\Gamma(\nu)]^{-1}(1-ce^{-t})^{-\mu}$ $\cdot(1-e^{-t})^{\nu-1}\exp(ae^{-t})$
19.8	$p^{-\nu}F_1(a,b,\nu,c;x,y/p)$ Re $\nu > 0$	$[\Gamma(\nu)]^{-1}t^{\nu-1}\Phi_1(a,b,c;x,yt)$
19.9	$p^{-\nu}F_1(\nu,a,b,c;x/p,y/p)$ Re $\nu > 0$	$[\Gamma(\nu)]^{-1}t^{\nu-1}\Phi_2(a,b,c;xt,yt)$
19.10	$p^{-\nu}F_2(a,b,\nu,c,d;x,y/p)$ Re $\nu > 0$	$[\Gamma(\nu)]^{-1}t^{\nu-1}\Psi_1(a,b,c,d;x,yt)$
19.11	$p^{-\nu}F_3(a,b,c,\nu;k;x,y/p)$ Re $\nu > 0$	$[\Gamma(\nu)]^{-1}t^{\nu-1}\Psi_3(a,b,c,k;x,yt)$
19.12	$p^{-\nu}F_4(\nu,b,c,d;x/p,y/p)$ Re $\nu > 0$	$[\Gamma(\nu)]^{-1}t^{\nu-1}\Psi_2(b,c,d;xt,yt)$

	$g(p) = \int_0^\infty f(t)e^{-pt}dt$	$f(t)$
19.13	$p^{-\nu}F_3(a,\nu,b,\frac{1}{2}+\frac{1}{2}\nu,c;x,4yp^{-2})$ $\mathrm{Re}\ \nu > 0$	$[\Gamma(\nu)]^{-1}t^{\nu-1}Y_4(a,b,c;x,yt^2)$
19.14	$p^{-1}{}_rF_s(b_1,\cdots,b_r;$ $c_1,\cdots,c_s;a/p)$ $r \geqq s-1$	${}_rF_{s+1}(b_1,\cdots,br;$ $c_1,\cdots,c_s,1;at)$
19.15	$p^{-\nu}{}_rF_s(a_1,\cdots,a_r;$ $b_1,\cdots,b_s;k/p)$ $r \leqq s+1,\ \mathrm{Re}\ \nu > 0$	$[\Gamma(\nu)]^{-1}t^{\nu-1}{}_rF_{s+1}(a_1,\cdots,a_r;$ $b_1,\cdots,b_s,\nu;kt)$
19.16	$p^{-\nu}{}_rF_s(a_1,\cdots,a_r;$ $b_1,\cdots,b_s;4k^2/p^2)$ $r \leqq s+1,\ \mathrm{Re}\ \nu > 0$	$[\Gamma(\nu)]^{-1}t^{\nu-1}{}_rF_{s+2}(a_1,\cdots,a_r;$ $b_1,\cdots,b_s,\frac{1}{2}\nu+\frac{1}{2},\frac{1}{2}\nu;k^2t^2)$
19.17	$p^{-2}{}_{r+2}F_s(a_1,\cdots,a_r,\frac{1}{2}\nu,\frac{1}{2}+\frac{1}{2}\nu;$ $b_1,\cdots,b_s;4k^2/p^2)$ $r < s,\ \mathrm{Re}\ \nu > 0$	$[\Gamma(2\nu)]^{-1}t^{2\nu-1}{}_rF_s(a_1,\cdots,a_r;$ $b_1,\cdots,b_s;k^2t^2)$

	$g(p) = \int_0^\infty f(t)e^{-pt}dt$	$f(t)$
19.18	$B(p,\nu)\,{}_{r+1}F_{s+1}(a_1,\cdots,a_r,\nu;$ $b_1,\cdots,b_s,s+\nu;k)$ $\text{Re}\,\nu > 0, \quad r \leqq s$	$(1-e^{-t})^{\nu-1}\,{}_rF_s[a_1,\cdots,a_r;$ $b_1,\cdots,b_s;k(1-e^{-t})]$
19.19	$B(p,\nu)\,{}_{r+1}F_{s+1}(a_1,\cdots,a_r,p;$ $b_1,\cdots,b_s,p+\nu;k)$ $\text{Re}\,\nu > 0, \quad r \leqq s$	$(1-e^{-t})^{\nu-1}\,{}_rF_s(a_1,\cdots,a_r;$ $b_1,\cdots,b_s;ke^{-t})$

2.20 Miscellaneous Functions

	$g(p) = \int_0^\infty f(t)e^{-pt}dt$	$f(t)$
20.1	$p^{-\frac{1}{2}}\nu(a/p)$	$\frac{1}{2}(\pi t)^{-\frac{1}{2}}\nu[2(at)^{\frac{1}{2}}]$
20.2	$p^{-\frac{1}{2}}\nu(a/p,b)$	$\frac{1}{2}(\pi t)^{-\frac{1}{2}}\nu[2(at)^{\frac{1}{2}},2b]$
20.3	$p^{-3/2}\nu(a/p,b)$	$2(\pi a)^{-\frac{1}{2}}\nu[2(at)^{\frac{1}{2}},1+2b]$
20.4	$\Gamma(ap)\nu(1,ap)$	$a^{-1}\nu(1-e^{-t/a})$
20.5	$p^{-\frac{1}{2}}\mu(a/p,b)$	$2^{-b-1}(\pi t)^{-\frac{1}{2}}\mu[2(at)^{\frac{1}{2}},b]$

Appendix. List of Notations and Definitions

Abbreviations: ε_n = Neumann's number

$\varepsilon_0 = 1, \quad \varepsilon_n = 2, \quad n = 1, 2, 3, \cdots$

$\gamma = 0.57721\cdots$ Euler's constant

$$(\alpha)_n = \alpha(\alpha+1)\cdots(\alpha+n-1) = \frac{\Gamma(\alpha+n)}{\Gamma(\alpha)}; \quad (\alpha)_0 = 1$$

$$\binom{\alpha}{n} = \alpha(\alpha-1)\cdots(\alpha-n+1)/n! = \frac{(-1)^n}{n!}\Gamma(n-\alpha)/\Gamma(-\alpha)$$

$$= \Gamma(1+\alpha)[n!\Gamma(1+\alpha-n)]^{-1}$$

1. Elementary functions

Trigonometric and inverse trigonometric functions:

$$\sin x, \quad \cos x, \quad \tan x = \frac{\sin x}{\cos x}, \quad \cot x = \frac{\cos x}{\sin x}$$

$$\sec x = \frac{1}{\cos x}, \quad \csc x = \frac{1}{\sin x}, \quad \arcsin x, \quad \arccos x,$$

$$\arctan x, \quad \operatorname{arccot} x.$$

Hyperbolic and inverse hyperbolic functions:

$$\sinh x = \tfrac{1}{2}(e^{x}-e^{x}), \qquad \cosh x = \tfrac{1}{2}(e^{x}+e^{-x}), \quad \sinh^{-1}x, \quad \cosh^{-1}x,$$

$$\tanh x = \frac{\sinh x}{\cosh x}, \qquad \coth x = \frac{\cosh x}{\sinh x}, \qquad \tanh^{-1}x, \quad \coth^{-1}x,$$

$$\operatorname{sech} x = \frac{1}{\cosh x}, \qquad \operatorname{csch} x = \frac{1}{\sinh x}.$$

2. Orthogonal polynomials

Legendre polynomials $P_n(x)$.

$$P_n(x) = 2^{-n}(n!)^{-1}\frac{d^n}{dx^n}(x^2-1)^n = {}_2F_1(-n, n+1; 1; \tfrac{1}{2}-\tfrac{1}{2}x)$$

Gegenbauer's polynomials $C_n^{\nu}(x)$

$$C_n^{\nu}(x) = [n!\Gamma(2\nu)]^{-1}\Gamma(2\nu+n)\,{}_2F_1(-n,2\nu+n;\tfrac{1}{2}+\nu;\tfrac{1}{2}-\tfrac{1}{2}x)$$
$$= (-2)^{-n}(1-x^2)^{\frac{1}{2}-\nu}(2\nu)_n[n!(\nu+\tfrac{1}{2})_n]^{-1}\frac{d^n}{dx^n}[(1-x^2)^{n+\nu-\frac{1}{2}}]$$

Chebycheff polynomials $T_n(x)$, $U_n(x)$

$$T_n(x) = \cos(n\arccos x) = {}_2F_1(-n,n;\tfrac{1}{2};\tfrac{1}{2}-\tfrac{1}{2}x) = \tfrac{1}{2}n \lim_{\nu=0}\cdot\Gamma(\nu)C_n^{\nu}(x)$$
$$= \tfrac{1}{2}\{[x+i(1-x^2)^{\frac{1}{2}}]^n + [x-i(1-x^2)^{\frac{1}{2}}]^n\}$$
$$U_n(x) = (1-x^2)^{-\frac{1}{2}}\sin[(n+1)\arccos x] = C_n^1(x)$$
$$= x(n+1)\,{}_2F_1(\tfrac{1}{2}-\tfrac{1}{2}n,\tfrac{3}{2}+\tfrac{1}{2}n;\tfrac{3}{2};1-x^2)$$
$$= -\tfrac{1}{2}i(1-x^2)^{-\frac{1}{2}}\{[x+i(1-x^2)^{\frac{1}{2}}]^n-[x-i(1-x^2)^{\frac{1}{2}}]^n\}$$

Jacobi polynomials $P_n^{\alpha,\beta}(x)$

$$P_n^{(\alpha,\beta)}(x) = [n!\Gamma(1+\alpha)]^{-1}\Gamma(1+\alpha+n)\,{}_2F_1(-n,n+\alpha+\beta+1;\alpha+1;\tfrac{1}{2}-\tfrac{1}{2}x)$$
$$= (-1)^n 2^{-n}(n!)^{-1}(1-x)^{-\alpha}(1+x)^{-\beta}\frac{d^n}{dx^n}[(1-x)^{\alpha+n}(1+x)^{\beta+n}]$$

Laguerre polynomials

$$L_n^{\alpha}(x) = (n!)^{-1}x^{-\alpha}e^x\frac{d^n}{dx^n}(e^{-x}x^{n+\alpha})$$
$$= [n!\Gamma(1+\alpha)]^{-1}\Gamma(\alpha+1+n)\,{}_1F_1(-n;\alpha+1;x)$$
$$L_n^0(x) = L_n(x)$$

Hermite polynomials

$$He_n(x) = (-1)^n \exp(\tfrac{1}{2}x^2)\frac{d^n}{dx^n}\exp(-\tfrac{1}{2}x^2)$$

$$H_n(x) = (-1)^n e^{x^2}\frac{d^n}{dx^n}e^{-x^2} = 2^{\frac{1}{2}n}He_n(2^{\frac{1}{2}}x)$$

$$He_{2n}(x) = (-1)^n 2^{-n}(n!)^{-1}(2n)!\,{}_1F_1(-n;\tfrac{1}{2};\tfrac{1}{2}x^2)$$

$$He_{2n+1}(x) = x(-1)^n 2^{-n}(n!)^{-1}(2n+1)!\,{}_1F_1(-n;\tfrac{3}{2};\tfrac{1}{2}x^2)$$

3. The Gamma function and related functions

$$\Gamma(z) = \int_0^\infty e^{-t}\,t^{z-1}\,dt\quad, \qquad \mathrm{Re}\ z > 0$$

ψ-function

$$\psi(z) = \frac{d}{dz}\log z = \frac{\Gamma'(z)}{\Gamma(z)}$$

Beta function $B(x,y)$

$$B(x,y) = \frac{\Gamma(x)\Gamma(y)}{\Gamma(x+y)}$$

4. Legendre functions

(Definition according to Hobson)

$$p_\nu^\mu(z) = [\Gamma(1-\mu)]^{-1}\left(\frac{z+1}{z-1}\right)^{\frac{1}{2}\mu}{}_2F_1(-\nu,\nu+1;1-\mu;\tfrac{1}{2}-\tfrac{1}{2}z)$$

$$e^{-i\pi\mu}q_\nu^\mu(z) = 2^{-\nu-1}[\Gamma(\tfrac{3}{2}+\nu)]^{-1}\pi^{\frac{1}{2}}\Gamma(1+\nu+\mu)\,z^{-\nu-\mu-1}(z^2-1)^{\frac{1}{2}\mu}\,.$$

$$\cdot\,{}_2F_1(\tfrac{1}{2}\nu+\tfrac{1}{2}\mu+\tfrac{1}{2},\tfrac{1}{2}\nu+\tfrac{1}{2}\mu+1;\ \nu+\tfrac{3}{2};\ z^{-2})$$

z is a point in the complex z-plane cut along the real z-axis from $-\infty$ to $+1$

$$(z^2-1)^{\frac{1}{2}\mu} = (z-1)^{\frac{1}{2}\mu}(z+1)^{\frac{1}{2}\mu} \quad -\pi < \arg z < \pi,\ -\pi < \arg(z\pm 1) < \pi$$

$$P_\nu^\mu(x) = [\Gamma(1-\mu)]^{-1}\left(\frac{1+x}{1-x}\right)^{\frac{1}{2}\mu} {}_2F_1(-\nu,\nu+1;1-\mu;\tfrac{1}{2}-\tfrac{1}{2}z),\ -1 < x < 1$$

$$Q_\nu^\mu(x) = \tfrac{1}{2}e^{-i\pi\mu}[e^{-\frac{1}{2}i\pi\mu}q_\nu^\mu(x+i0) + e^{\frac{1}{2}i\pi\mu}q_\nu^\mu(x-i0)],\ -1 < x < 1$$

$$p_\nu^\circ(z) = p_\nu(z); \qquad q_\nu^\circ(z) = q_\nu(z)$$

$$P_\nu^\circ(z) = P_\nu(z); \qquad Q_\nu^\circ(z) = Q_\nu(z)$$

5. Bessel functions

$$J_\nu(z) = (\tfrac{1}{2}z)^\nu \sum_{n=0}^{\infty} \frac{(-1)^n \left(\frac{z}{2}\right)^{2n}}{n!\Gamma(\nu+n+1)}$$

$$J_{-\nu}(z) = J_\nu(z)\cos(\pi\nu) - Y_\nu(z)\sin(\pi\nu);$$

$$Y_{-\nu}(z) = J_\nu(z)\sin(\pi\nu) + Y_\nu(z)\cos(\pi\nu);$$

$$Y_\nu(z) = \cot(\pi\nu)J_\nu(z) - \csc(\pi\nu)J_{-\nu}(z)$$

$$H_\nu^{(1)}(z) = J_\nu(z)+iY_\nu(z); \quad H_\nu^{(2)}(z) = J_\nu(z) - iY_\nu(z)$$

6. Modified Bessel functions

$$I_\nu(z) = e^{-\frac{1}{2}i\pi\nu}J_\nu(ze^{i\frac{1}{2}\pi}) = (\tfrac{1}{2}z)^\nu \sum_{n=0}^{\infty} \frac{(\frac{1}{2}z)^{2n}}{n!\Gamma(\nu+n+1)}$$

$$K_\nu(z) = \tfrac{1}{2}\pi \csc(\pi\nu)[I_{-\nu}(z)-I_\nu(z)]$$

$$= \tfrac{1}{2}i\pi e^{\frac{1}{2}i\pi\nu}H_\nu^{(1)}(ze^{i\frac{1}{2}\pi}) = -\tfrac{1}{2}i\pi e^{-i\frac{1}{2}\pi\nu}H_\nu^{(2)}(ze^{-i\frac{1}{2}\pi\nu})$$

7. Anger-Weber functions

$$\mathbf{J}_\nu(z) = \pi^{-1}\int_0^\pi \cos(z\sin t-\nu t)\,dt$$

$$\mathbf{J}_n(z) = J_n(z), \quad n = 0,1,2,\cdots; \quad \mathbf{E}_0(z) = -\mathbf{H}_0(z)$$

$$\mathbf{J}_{\frac{1}{2}}(z) = (\tfrac{1}{2}\pi z)^{-\frac{1}{2}}\{[C(z)-S(z)]\cos z+[C(z)+S(z)]\sin z\} = \mathbf{E}_{-\frac{1}{2}}(z)$$

$$\mathbf{J}_{-\frac{1}{2}}(z) = (\tfrac{1}{2}\pi z)^{-\frac{1}{2}}\{[C(z)+S(z)]\cos z-[C(z)-S(z)]\sin z\} = \mathbf{E}_{\frac{1}{2}}(z)$$

8. Struve functions

$$\mathbf{H}_\nu(z) = \sum_{n=0}^{\infty}\frac{(-1)^n(\frac{1}{2}z)^{\nu+2n+1}}{\Gamma(n+\frac{3}{2})\Gamma(\nu+n+\frac{3}{2})}$$

$$\mathbf{L}_\nu(z) = -ie^{-\frac{1}{2}i\pi\nu}\mathbf{H}_\nu(ze^{i\frac{1}{2}\nu}) = \sum_{n=0}^{\infty}\frac{(\frac{1}{2}z)^{\nu+2n+1}}{\Gamma(n+\frac{3}{2})\Gamma(\nu+n+\frac{3}{2})}$$

9. Lommel functions

$$s_{\mu,\nu}(z) = \frac{z^{\mu+1}}{(\mu-\nu+1)(\mu+\nu+1)}\;{}_1F_2(1;\tfrac{1}{2}\mu-\tfrac{1}{2}\nu+\tfrac{3}{2},\tfrac{1}{2}\mu+\tfrac{1}{2}\nu+\tfrac{3}{2};-\tfrac{1}{4}z^2)$$

$$\mu\pm\nu \neq -1,\ -2,\ -3,\ \cdots$$

$$S_{\mu,\nu}(z) = s_{\mu,\nu}(z)-2^{\mu-1}\Gamma(\tfrac{1}{2}+\tfrac{1}{2}\mu-\tfrac{1}{2}\nu)\Gamma(\tfrac{1}{2}+\tfrac{1}{2}\mu+\tfrac{1}{2}\nu)\ .$$

$$\cdot\{\sin[\tfrac{1}{2}\pi(\nu-\mu)]J_\nu(z)+\cos[\tfrac{1}{2}\pi(\nu-\mu)]Y_\nu(z)\}$$

$$s_{\nu,\mu}(z) = s_{\nu,-\mu}(z); \qquad S_{\mu,\nu}(z) = S_{\mu,-\nu}(z)$$

Special cases:

$$s_{\nu,\nu}(z) = \Pi^{\frac{1}{2}}2^{\nu-1}\Gamma(\tfrac{1}{2}+\nu)\mathbf{H}_\nu(z)$$

$$S_{\nu,\nu}(z) = \pi^{\frac{1}{2}}2^{\nu-1}\Gamma(\tfrac{1}{2}+\nu)[\mathbf{H}_\nu(z) - Y_\nu(z)]$$

$$s_{\nu+1,\nu}(z) = z^{\nu}-2^{\nu}\Gamma(1+\nu)J_{\nu}(z); \qquad S_{\nu+1,\nu} = z^{\nu}$$

$$\lim_{\mu\to\nu} \frac{s_{\mu-1,\nu}(z)}{\Gamma(\mu-\nu)} = 2^{\nu-1}\Gamma(\nu)J_{\nu}(z)$$

$$s_{o,\nu}(z) = \tfrac{1}{2}\pi\csc(\pi\nu)[\mathbf{J}_{\nu}(z) - \mathbf{J}_{-\nu}(z)]$$

$$S_{o,\nu}(z) = \tfrac{1}{2}\pi\csc(\pi\nu)[\mathbf{J}_{\nu}(z) - \mathbf{J}_{-\nu}(z) - J_{\nu}(z) + J_{-\nu}(z)]$$

$$s_{-1,\nu}(z) = -\tfrac{1}{2}\pi\nu^{-1}\csc(\pi\nu)[\mathbf{J}_{\nu}(z)+\mathbf{J}_{-\nu}(z)]$$

$$S_{-1,\nu}(z) = \tfrac{1}{2}\pi\nu^{-1}\csc(\pi\nu)[J_{\nu}(z)+J_{-\nu}(z)-\mathbf{J}_{\nu}(z)-\mathbf{J}_{-\nu}(z)]$$

$$s_{1,\nu}(z) = 1-\tfrac{1}{2}\pi\nu\csc(\pi\nu)[\mathbf{J}_{\nu}(z)+\mathbf{J}_{-\nu}(z)]$$

$$S_{1,\nu}(z) = 1+\tfrac{1}{2}\pi\nu\csc(\pi\nu)[J_{\nu}(z)+J_{-\nu}(z)-\mathbf{J}_{\nu}(z)-\mathbf{J}_{-\nu}(z)]$$

$$S_{\frac{1}{2},\frac{1}{2}}(z) = z^{-\frac{1}{2}}, \qquad S_{\frac{3}{2},\frac{1}{2}}(z) = z^{\frac{1}{2}}$$

$$S_{-\frac{1}{2},\frac{1}{2}}(z) = z^{-\frac{1}{2}}[\sin z\ \mathrm{Ci}(z)-\cos z\ \mathrm{si}(z)]$$

$$S_{-\frac{3}{2},\frac{1}{2}}(z) = -z^{-\frac{1}{2}}[\sin z\ \mathrm{si}(z)+\cos z\ \mathrm{Ci}(z)]$$

Lommel functions of two variables

$$U_{\nu}(w,z) = \sum_{n=0}^{\infty}(-1)^{n}\left(\frac{w}{z}\right)^{\nu+2n} J_{\nu+2n}(z)$$

$$V_{\nu}(w,z) = \cos(\tfrac{1}{2}w+\tfrac{1}{2}z^{2}w^{-1}+\tfrac{1}{2}\pi\nu)+U_{2-\nu}(w,z)$$

Kelvin's functions

$$J_{\nu}(ze^{i\frac{3}{4}\pi}) = \mathrm{ber}_{\nu}(z)+i\,\mathrm{bei}_{\nu}(z)$$

$$J_{\nu}(ze^{-i\frac{3}{4}\pi}) = \mathrm{ber}_{\nu}(z)-i\,\mathrm{bei}_{\nu}(z)$$

$$K_\nu(ze^{i\frac{\pi}{4}}) = \mathrm{ker}_\nu(z) + i\,\mathrm{kei}_\nu(z)$$

$$K_\nu(ze^{-i\frac{\pi}{4}}) = \mathrm{ker}_\nu(z) - i\,\mathrm{kei}_\nu(z)$$

$$\mathrm{ber}_o(z) = \mathrm{ber}(z), \quad \mathrm{bei}_o(z) = \mathrm{bei}(z),$$

$$\mathrm{ker}_o(z) = \mathrm{ker}(z), \quad \mathrm{kei}_o(z) = \mathrm{kei}(z)$$

Bessel integral functions

$$\mathrm{Ji}_\nu(x) = \int_x^\infty t^{-1} J_\nu(t)\,dt, \qquad \mathrm{Yi}_\nu(x) = \int_x^\infty t^{-1} Y_\nu(t)\,dt$$

$$\mathrm{Ki}_\nu(x) = \int_x^\infty t^{-1} K_\nu(t)\,dt$$

Neumann polynomials

$$O_n(x) = \tfrac{1}{4} \sum_{m=0}^{\leq \frac{1}{2}n} n(n-m-1)!(\tfrac{1}{2}x)^{2m-n-1}/m! \; ; \qquad O_o(x) = x^{-1}$$

Schläfli polynomials

$$S_n(x) = \sum_{m=0}^{\leq \frac{1}{2}n} (n-m-1)!(\tfrac{1}{2}z)^{2m-n}/m! \; ; \qquad S_o(x) = 0$$

10. Gauss' hypergeometric series

$$_2F_1(a,b;c;z) = \frac{\Gamma(c)}{\Gamma(a)\Gamma(b)} \sum_{n=0}^\infty \frac{\Gamma(a+n)\Gamma(b+n)}{\Gamma(c+n)} \frac{z^n}{n!}, \qquad |z| < 1$$

11. Confluent hypergeometric functions

Kummer's functions

$$_1F_1(a;c;z) = \frac{\Gamma(c)}{\Gamma(a)} \sum_{n=0}^\infty \frac{\Gamma(a+n)}{\Gamma(c+n)} \frac{z^n}{n!} = z^{-\frac{1}{2}c} e^{\frac{1}{2}z} M_{\frac{1}{2}c-a,\frac{1}{2}c-\frac{1}{2}}(z)$$

$$_1F_1(a;c;z) = e^z\, {}_1F_1(c-a;c;-z)$$

Whittaker functions

$$M_{k,\mu}(z) = z^{\mu+\frac{1}{2}} e^{-\frac{1}{2}z} \, {}_1F_1(\tfrac{1}{2}+\mu-k; 2\mu+1; z)$$

$$= z^{\mu+\frac{1}{2}} e^{\frac{1}{2}z} \, {}_1F_1(\tfrac{1}{2}+\mu+k; 2\mu+1; -z)$$

$$W_{k,\mu}(z) = \frac{\Gamma(-2\mu)}{\Gamma(\frac{1}{2}-\mu-k)} M_{k,\mu}(z) + \frac{\Gamma(2\mu)}{\Gamma(\frac{1}{2}+\mu-k)} M_{k,-\mu}(z)$$

$$W_{k,-\mu}(z) = W_{k,\mu}(z)$$

Parabolic cylinder function

$$D_\alpha(z) = 2^{\frac{1}{4}+\frac{1}{2}\alpha} z^{-\frac{1}{2}} W_{\frac{1}{4}+\frac{1}{2}\alpha, \frac{1}{4}}(\tfrac{1}{2}z^2)$$

$$D_n(z) = e^{-\frac{1}{4}z^2} He_n(z) , \qquad n = 0, 1, 2, \cdots$$

$$D_{-1}(z) = (\tfrac{1}{2}\pi)^{\frac{1}{2}} e^{\frac{1}{4}z^2} \, \mathrm{Erfc}(2^{-\frac{1}{2}}z)$$

$$D_{-\frac{1}{2}}(z) = (2\pi z^{-1})^{-\frac{1}{2}} K_{\frac{1}{4}}(\tfrac{1}{4}z^2)$$

Error integrals

$$\mathrm{Erf}(x) = 2\pi^{-\frac{1}{2}} \int_0^x e^{-t^2} dt = 2\pi^{-\frac{1}{2}} x \, {}_1F_1(\tfrac{1}{2}; \tfrac{3}{2}; -x^2)$$

$$= 2(\pi x)^{-\frac{1}{2}} e^{-\frac{1}{2}x^2} M_{-\frac{1}{4},\frac{1}{4}}(x^2)$$

$$\mathrm{Erfc}(x) = 1-\mathrm{Erf}(x) = 2\pi^{-\frac{1}{2}} \int_x^\infty e^{-t^2} dt = (\pi x)^{-\frac{1}{2}} e^{-\frac{1}{2}x^2} W_{-\frac{1}{4},\frac{1}{4}}(x)$$

$$\mathrm{Erf}(x^{\frac{1}{2}} e^{i\frac{\pi}{4}}) = C(x) + S(x) + i[C(x) - S(x)]$$

$$\mathrm{Erfc}(x^{\frac{1}{2}} e^{i\frac{\pi}{4}}) = 1-C(x) - S(x) + i[S(x) - C(x)]$$

Fresnel's integrals

$$C(x) = (2\pi)^{-\frac{1}{2}} \int_0^x t^{-\frac{1}{2}} \cos t \, dt ; S(x) = (2\pi)^{-\frac{1}{2}} \int_0^x t^{-\frac{1}{2}} \sin t \, dt$$

Exponential integrals

$$-Ei(-z) = -\gamma - \log z - \sum_{n=1}^{\infty} (n \cdot n!)^{-1} (-z)^n = z^{-\frac{1}{2}} e^{-\frac{1}{2}z} W_{-\frac{1}{2},0}(z)$$

$$-Ei(-x) = \int_x^{\infty} t^{-1} e^{-t} dt, \qquad x > 0$$

$$\overline{E}i(z) = \gamma + \log z + \sum_{n=1}^{\infty} (n \cdot n!)^{-1} z^n$$

$$\overline{E}i(x) = \tfrac{1}{2}[Ei(-xe^{i\pi}) + Ei(-xe^{-i\pi})] = -P \cdot V \cdot \int_{-x}^{\infty} t^{-1} e^{-t} dt, \; x > 0$$

$$Ei(-ze^{\pm i\pi}) = \pm i\pi + \overline{E}i(z); \; \overline{E}i(ze^{\pm i\pi}) = \pm i\pi + Ei(-z)$$

$$Ei(-ze^{\pm \frac{1}{2} i\pi}) = Ci(z) \pm i[\tfrac{1}{2}\pi - Si(z)]; \overline{E}i(ze^{\pm \frac{1}{2} i\pi}) = Ci(z) \pm i[\tfrac{1}{2}\pi + Si(z)]$$

$$Ei(-xe^{\pm \frac{1}{2} i\pi}) = Ci(x) \mp i si(x); \overline{E}i(xe^{\pm \frac{1}{2} i\pi}) = Ci(x) \pm i[\pi + si(x)]$$

$$x > 0$$

Sine and cosine integral

$$Si(z) = \sum_{n=0}^{\infty} (-1)^n [(2n+1)(2n+1)!]^{-1} z^{2n+1} = \int_0^z t^{-1} \sin t \, dt$$

$$Ci(z) = \gamma + \log z + \sum_{n=1}^{\infty} (-1)^n [2n(2n)!]^{-1} z^{2n}$$

$$Ci(x) = -\int_x^{\infty} t^{-1} \cos t \, dt, \; si(x) = -\int_x^{\infty} t^{-1} \sin t \, dt = Si(x) - \frac{\pi}{2}$$

$$x > 0$$

$$\mathrm{Ci}(x) = \gamma+\log x - \int_0^x t^{-1}(1-\cos t)dt, \qquad x > 0$$

$$\mathrm{Ci}(z)=\tfrac{1}{2}[\mathrm{Ei}(-ze^{\frac{1}{2}i\pi})+\mathrm{Ei}(-ze^{-\frac{1}{2}i\pi})]=\tfrac{1}{2}[\overline{\mathrm{E}}\mathrm{i}(ze^{\frac{1}{2}i\pi})+\overline{\mathrm{E}}\mathrm{i}(ze^{-\frac{1}{2}i\pi})]$$

$$\mathrm{Si}(z) = \tfrac{1}{2}\pi+\tfrac{1}{2}i[\mathrm{Ei}(-ze^{\frac{1}{2}i\pi}) - \mathrm{Ei}(-ze^{-\frac{1}{2}i\pi})]$$

Hyperbolic sine and cosine integral

$$\mathrm{Si}_h(x) = \int_0^x t^{-1}\sinh t dt = \sum_{n=0}^{\infty} [(2n+1)(2n+1)!]^{-1}x^{2n+1}$$

$$\mathrm{Ci}_h(x) = \gamma+\log x + \int_0^x t^{-1}(\cosh t-1)dt$$

$$= \gamma+\log x + \sum_{n=1}^{\infty} 2n[(2n)!]^{-1}x^{2n}$$

Incomplete gamma funtion

$$\gamma(\nu,x) = \int_0^x t^{\nu-1}e^{-t}dt=\nu^{-1}x^{\nu}\ {}_1F_1(\nu,\nu+1;-x), \qquad \mathrm{Re}\ \nu > 0$$

$$= \nu^{-1}x^{\frac{1}{2}\nu-\frac{1}{2}}e^{-\frac{1}{2}x}\ M_{\frac{1}{2}\nu-\frac{1}{2},\frac{1}{2}\nu}(x)$$

$$\Gamma(\nu,x) = \Gamma(\nu)-\gamma(\nu,x)= \int_x^{\infty} t^{\nu-1}e^{-t}dt=x^{\frac{1}{2}\nu-\frac{1}{2}}e^{-\frac{1}{2}x}\ W_{\frac{1}{2}\nu-\frac{1}{2},\frac{1}{2}\nu}(x)$$

$$\Gamma(\tfrac{1}{2},z^2) = \pi^{\frac{1}{2}}\mathrm{Erfc}(z);\ \Gamma(0,z) = -\mathrm{Ei}(-z)$$

$$\Gamma(\tfrac{1}{2},z^2) = \pi^{\frac{1}{2}}\mathrm{Erf}(z);\ \gamma(1,z) = 1-e^{-z},\ \Gamma(1,z) = e^{-z}$$

12. Particular cases of Whittaker's functions

$$M_{-\frac{1}{4},\frac{1}{4}}(z) = \tfrac{1}{2}\pi^{\frac{1}{2}}z^{\frac{1}{4}}e^{\frac{1}{2}z}\mathrm{Erf}(z^{\frac{1}{2}})$$

$$M_{\frac{1}{4},\frac{1}{4}}(z) = -i\tfrac{1}{2}\pi^{\frac{1}{2}}z^{\frac{1}{4}}e^{-\frac{1}{2}z}\mathrm{Erf}(iz^{\frac{1}{2}})$$

$$M_{k,0}(z) = z^{\frac{1}{2}}e^{-\frac{1}{2}z}L_{k-\frac{1}{2}}(z)$$

$$M_{0,\mu}(z) = 2^{2\mu}\Gamma(1+\mu)z^{\frac{1}{2}}I_{\mu}(\tfrac{1}{2}z)$$

$$M_{\mu+\frac{1}{2},\mu}(z) = z^{\mu+\frac{1}{2}}e^{-\frac{1}{2}z}$$

$$M_{k,k+\frac{1}{2}}(z) = (2k+1)z^{-k}\, e^{\frac{1}{2}z}\, \gamma(2k+1,z)$$

$$W_{-\frac{1}{4},\frac{1}{4}}(z) = \pi^{\frac{1}{2}}z^{\frac{1}{4}}e^{\frac{1}{2}z}\ \mathrm{Erfc}(z^{\frac{1}{2}})$$

$$W_{k,\frac{1}{4}}(z) = 2^{\frac{1}{4}-k}z^{\frac{1}{4}}\ D_{2k-\frac{1}{2}}[(2z)^{\frac{1}{2}}]$$

$$W_{0,\mu}(z) = \pi^{-\frac{1}{2}}z^{\frac{1}{2}}\ K_{\mu}(\tfrac{1}{2}z)$$

$$W_{-\frac{1}{2},0}(z) = -z^{\frac{1}{2}}e^{\frac{1}{2}z}\ \mathrm{Ei}(-z)$$

$$W_{k,\frac{1}{2}+k}(z) = z^{-k}e^{\frac{1}{2}z}\ \Gamma(2k+1,z)$$

$$W_{k,k-\frac{1}{2}}(z) = z^{\mu+\frac{1}{2}}e^{-\frac{1}{2}z}$$

13. Elliptic integrals and elliptic theta functions

Complete elliptic integrals

$$K(k) = \int_0^{\frac{1}{2}\pi} (1-k^2\sin^2 x)^{-\frac{1}{2}}dx = \tfrac{1}{2}\pi\ {}_2F_1(\tfrac{1}{2},\tfrac{1}{2};1;k^2)$$

$$E(k) = \int_0^{\frac{1}{2}\pi} (1-k^2\sin^2 x)^{\frac{1}{2}}dx = \tfrac{1}{2}\pi\ {}_2F_1(-\tfrac{1}{2},\tfrac{1}{2};1;k^2)$$

Theta functions

$$\theta_1(z|t) = (\pi t)^{-\frac{1}{2}} \sum_{n=-\infty}^{\infty} (-1)^n \exp[-(z+n-\tfrac{1}{2})^2/t]$$

$$= 2\sum_{n=0}^{\infty} (-1)^n \exp[-\pi^2 t(n+\tfrac{1}{2})^2]\sin[(2n+1)\pi z]$$

$$\theta_2(z|t) = (\pi t)^{-\frac{1}{2}} \sum_{n=-\infty}^{\infty} (-1)^n \exp[-(z+n)^2/t]$$

$$= 2 \sum_{n=0}^{\infty} \exp[-\pi^2 t(n+\tfrac{1}{2})^2] \cos[(2n+1)\pi z]$$

$$\theta_3(z|t) = (\pi t)^{-\frac{1}{2}} \sum_{n=-\infty}^{\infty} \exp[-(z+n)^2/t]$$

$$= \sum_{n=0}^{\infty} \varepsilon_n \exp(-\pi^2 t n^2) \cos(2\pi n z)$$

$$\theta_4(z|t) = (\pi t)^{-\frac{1}{2}} \sum_{n=-\infty}^{\infty} \exp[-(z+n+\tfrac{1}{2})^2/t]$$

$$= \sum_{n=0}^{\infty} (-1)^n \varepsilon_n \exp(-\pi^2 t n^2) \cos(2\pi n z)$$

Modified theta functions

$$\hat{\theta}_1(z|t) = (\pi t)^{-\frac{1}{2}} \{ \sum_{n=0}^{\infty} (-1)^n \exp[-(z+n+\tfrac{1}{2})^2/t]$$

$$- \sum_{n=-1}^{-\infty} (-1)^n \exp[-(z+n+\tfrac{1}{2})^2/t] \}$$

$$\hat{\theta}_2(z|t) = (\pi t)^{-\frac{1}{2}} \{ \sum_{n=0}^{\infty} (-1)^n \exp[-(z+n)^2/t]$$

$$- \sum_{n=-1}^{-\infty} (-1)^n \exp[-(z+n)^2/t] \}$$

$$\hat{\theta}_3(z|t) = (\pi t)^{-\frac{1}{2}} \{ \sum_{n=0}^{\infty} \exp[-(z+n)^2/t]$$

$$- \sum_{n=-1}^{-\infty} \exp[-(z+n)^2/t] \}$$

$$\hat{\theta}_4(z|t) = (\pi t)^{-\frac{1}{2}}\{\sum_{n=0}^{\infty} \exp[-(z+n+\tfrac{1}{2})^2/t]$$

$$- \sum_{n=-1}^{-\infty} \exp[-(z+n+\tfrac{1}{2})^2/t]\}$$

14. <u>Generalized hypergeometric functions</u>

$${}_pF_q(a_1,a_2,\cdots a_p;b_1,b_2,\cdots b_q;z) = \sum_{n=0}^{\infty} \frac{(a_1)_n\cdots(a_p)_n}{(b_1)_n\cdots(b_q)_n}\frac{z^n}{n!}$$

$$p,q = 0,\ 1,\ 2,\ \cdots$$

$|z|<1$ if $p = q+1$, $|z|<\infty$ if $p\leqq q$; divergent otherwise

Hypergeometric functions of two variables

$$\Phi_1(a,b,c;x,z) = \sum_{m,n=0}^{\infty} \frac{(a)_{m+n}(b)_m}{(c)_{m+n}m!n!} x^m z^n$$

$$\Phi_2(a,b,c;x,z) = \sum_{m,n=0}^{\infty} \frac{(a)_m(b)_n}{(c)_{m+n}m!n!} x^m z^n$$

$$\Phi_3(b,c;x,z) = \sum_{m,n=0}^{\infty} \frac{(b)_m}{(c)_{m+n}m!n!} x^m z^n$$

$$Y_1(a,b,c,d;x,z) = \sum_{m,n=0}^{\infty} \frac{(a)_{m+n}(b)_m}{(c)_m(d)_n m!n!} x^m z^n$$

$$Y_2(a,c,d;x,z) = \sum_{m,n=0}^{\infty} \frac{(a)_{m+n}}{(c)_m(d)_n m!n!} x^m z^n$$

$$Y_3(a,b,c,d;x,z) = \sum_{m,n=0}^{\infty} \frac{(a)_m(b)_n(c)_m}{(d)_{m+n}m!n!} x^m z^n$$

$$Y_4(a,b,c;x,z) = \sum_{m,n=0}^{\infty} \frac{(a)_m(b)_m}{(c)_{m+n}m!n!} x^m z^n$$

$$F_1(a;b,c;d;x,z) = \sum_{m,n=0}^{\infty} \frac{(a)_{m+n}(b)_m(c)_n}{(d)_{m+n}m!n!} x^m z^n$$

$$F_2(a;b,c;d,e;x,z) = \sum_{m,n=0}^{\infty} \frac{(a)_{m+n}(b)_m(c)_n}{(d)_m(e)_n m!n!} x^m z^n$$

$$F_3(a,b,c,d;e;x,z) = \sum_{m,n=0}^{\infty} \frac{(a)_m(b)_n(c)_m(d)_n}{(e)_{m+n}m!n!} x^m z^n$$

$$F_4(a,b;c,d;x,z) = \sum_{m,n=0}^{\infty} \frac{(a)_{m+n}(b)_{m+n}}{(c)_m(d)_n m!n!} x^m z^n$$

15. Miscellaneous functions

Riemann's zeta function $\zeta(z) = \sum_{n=1}^{\infty} n^{-z}$, Re $z > 1$

Hurwitz's zeta function $\zeta(z,a) = \sum_{n=0}^{\infty} (n+a)^{-z}$, Re $z > 1$

Functions of the Volterra type

$$\nu(x) = \int_0^{\infty} \frac{x^s}{\Gamma(s+1)}\, ds, \qquad \nu(x,a) = \int_0^{\infty} \frac{x^{s+a}}{\Gamma(s+a+1)}\, ds$$

$$\mu(x,a) = \int_0^{\infty} \frac{x^s s^a}{\Gamma(s+1)}\, ds$$

$\delta(t)$ = Delta function

Unit step function

$H(t) = 1$, $t > 0$; $H(t) = 0$, ~~$t > 0$~~ $t < 0$

List of Functions

Symbol	Name of the Function	Listed under
$C(x)$	Fresnel's integral	11
$Ci(x)$	Cosine integral	11
Ci_h	Hyperbolic cosine integral	11
$C_n^{\nu}(x)$	Gegenbauer's polynomial	2
$\delta(x)$	Delta function	15
$D_{\nu}(z)$	Parabolic cylinder function	11
$E(k)$	Complete elliptic integral	13
$Ei(-x)$, $\overline{E}i(x)$	Exponential integrals	11
$Erf(z)$, $Erfc(z)$	Error integrals	11
$\mathbf{E}_{\nu}(z)$	Anger-Weber functions	7
${}_mF_n(z)$, F_r	Hypergeometric functions	10, 11, 12, 14
$H(x)$	Unit step function	15
$He_n(x)$	Hermite's polynomial	2
$H_{\nu}^{(1),(2)}(z)$	Hankel's functions	5
$\mathbf{H}_{\nu}(z)$	Struve's function	8
$I_{\nu}(z)$	Modified Bessel function	6
$J_{\nu}(z)$	Bessel function	5
$\mathbf{J}_{\nu}(z)$	Anger-Weber function	7

Symbol	Name of the Function	Listed under
$K(k)$	Complete elliptic integral	13
$K_\nu(z)$	Modified Hankel function	6
$L_\nu(z)$	Laguerre's function	11
$L_n^\alpha(x)$	Laguerre's polynomial	2
$\mathbf{L}_\nu(z)$	Struve's function	8
$M_{k,\mu}(z)$, $W_{k,\mu}(z)$	Whittaker's functions	11
$O_n(z)$	Neumann polynomials	9
$P_n(x)$	Legendre's polynomials	2
$P_n^{(\alpha,\beta)}(x)$	Jacobi's polynomials	2
$p_\nu^\mu(z)$, $P_\nu^\mu(x)$, $q_\nu^\mu(z)$, $Q_\nu^\mu(x)$	Legendre functions	4
$S(x)$	Fresnel's integral	11
$S_n(z)$	Schläfli polynomials	9
$si(x)$, $Si(x)$	Sine integrals	11
$Si_h(x)$	Hyperbolic sine integral	11

Symbol	Name of the Function	Listed under
$s_{\mu,\nu}(z)$, $S_{\mu,\nu}(z)$	Lommel's functions	9
$T_n(x)$, $U_n(x)$	Chebycheff's polynomials	2
$U_\nu(w,z)$, $V_\nu(w,z)$	Lommel functions of two variables	9
$W_{\mu,\nu}(z)$	Whittaker's function	11
$Y_\nu(z)$	Neumann's function	5
$B(x,y)$	Beta function	3
$\Gamma(z)$	Gamma function	3
$\Gamma(\nu,z)$, $\gamma(\nu,z)$	Incomplete gamma functions	11
$\psi(z)$	Psi function	3
$\zeta(z)$	Riemann's zeta function	15
$\zeta(z,a)$	Hurwitz's zeta function	15
$\nu(x)$, $\nu(x,a)$, $\mu(x,a)$	Volterra-type functions	15
$\Theta_1(z\|t)$, $\Theta_2(z\|t)$, $\Theta_3(z\|t)$, $\Theta_4(z\|t)$	Elliptic theta functions	13

Symbol	Name of the Function	Listed under
$\hat{\Theta}_1(z\|t)$	Modified elliptic theta functions	13
$\hat{\Theta}_2(z\|t)$		
$\hat{\Theta}(z\|t)$		
$\hat{\Theta}(z\|t)$		
Φ_1	Generalized hypergeometric functions	14
Φ_2		
Y_1		
Y_2		
Y_3		
Y_4		
F_1		
F_2		
F_3		
F_4		

8480-34
5-17

F. Oberhettinger: Tabellen zur Fourier Transformation

X, 214 Seiten. 1957. (Die Grundlehren der mathematischen Wissenschaften, Bd. 90)
Gebunden DM 48,–; US $17.80
ISBN 3-540-02149-3

W. Magnus, F. Oberhettinger, R. P. Soni: Formulas and Theorems for the Special Functions of Mathematical Physics

Third enlarged edition
VIII, 508 pages. 1966. (Die Grundlehren der mathematischen Wissenschaften, Bd. 52)
Cloth DM 66,–; US $24.50
ISBN 3-540-03518-4

From the Reviews: "Compared to the 1948 German edition the book contains several additions which present further facts on the special functions in question, and enlarge the list of formulae concerning them. Furthermore some of these functions, for instance Kummer's function, the Whittaker function, parabolic cylinder functions, etc., on which only relatively short accounts were given in the previous edition, appear here as subjects of individual chapters...
... The scope of the present book will be best seen from its table of contents. Chapter I: The gamma functions and related functions (the Riemann zeta function, Bernoulli and Euler polynomials, etc.) Chapter II: The hypergeometric function. Chapter III: Bessel functions. Chapter IV: Legendre functions (including Gegenbauer functions, toroidal and conical functions). Chapter V: Orthogonal polynomials (Jacobi, Gegenbauer, Legendre, generalized Laguerre, Hermite, Chebyshev polynomials). Chapter VI: Kummer's function. Chapter VII: Whittaker function. Chapter VIII: Parabolic cylinder functions and parabolic functions. Chapter IX: The incomplete gamma function and special cases. Chapter X: Elliptic integrals, theta functions and elliptic functions. Chapter XI: Integral transforms (including Fourier transforms of various kind, Laplace, Mellin, Hankel, Lebedev, Mehler and Gauss transforms). Chapter XII: Transformation of systems of coordinates.
There is no doubt that this excellently written book, similarly to its earlier editions, will be of great value for mathematicians and physicists."
Acta Scientiarum Mathematicarum

F. Oberhettinger: Tables of Bessel Transforms

IX, 289 pages. 1972
DM 27,60; US $10.30
ISBN 3-540-05997-0

This book contains a comprehensive collection of integrals for integral transforms which have cylindrical functions as kernels; the selection reflects the extensive experience of the author. This particular type of integral transforms is of great importance for applied mathematicians, physicists, and engineers.

Prices are subject to change without notice

Springer-Verlag Berlin Heidelberg New York

München · London · Paris
Sydney · Tokyo · Wien

Applied Mathematical Sciences

Edited by F. John, J. P. LaSalle, L. Sirovich

The Applied Mathematical Sciences series provides an outlet for material less formally presented and more anticipatory of needs than finished texts or monographs, yet of immediate interest because of the novelty of its treatment of an application or of mathematics being applied or lying close to applications.
Through rapid publication in an inexpensive format, the series makes material of current interest widely accessible to the user of mathematics, the mathematician interested in applications, and the student scientist.
Many of the books will originate out of and will stimulate the development of new undergraduate courses in the applications of mathematics.

Prices are subject to change without notice

Springer-Verlag
Berlin
Heidelberg
New York
München · London · Paris
Sydney · Tokyo · Wien

Volume 1
F. John: Partial Differential Equations
31 figs. IX, 221 pages. 1971. DM 24,–; US $8.90
ISBN 3-540-90021-7

Volume 2
L. Sirovich: Techniques of Asymptotic Analysis
23 figs. IX, 306 pages. 1971. DM 24,–; US $8.90
ISBN 3-540-90022-5

Volume 3
J. Hale: Functional Differential Equations
15 figs. IX, 238 pages. 1971. DM 24,–; US $8.90
ISBN 3-540-90023-3

Volume 4
J. Percus: Combinatorial Methods
58 figs. IX, 194 pages. 1971. DM 24,–; US $8.90
ISBN 3-540-90027-6

Volume 5
R. von Mises, K. O. Friedrichs: Fluid Dynamics
216 figs. IX, 353 pages. 1971. DM 24,–; US $8.90
ISBN 3-540-90028-4

Volume 6
W. Freiberger, U. Grenander: A Course in Computational Probability and Statistics
35 figs. XII, 155 pages. 1971. DM 24,–; US $8.90
ISBN 3-540-90029-2

Volume 7
A. C. Pipkin: Lectures on Visco-elasticity Theory
16 figs. IX, 180 pages. 1972. DM 20,80; US $7.70
ISBN 3-540-90030-6

Volume 8
G. E. O. Giacaglia: Pertubation Methods in Non-Linear Systems
IX, 369 pages. 1972. DM 27,80; US $10.30
ISBN 3-540-90054-3

Distribution rights for U.K., Commonwealth, and Traditional British Market (excluding Canada): Allen & Unwin Ltd., London